AF360843

Carbon and Nutrient Transfer via Above and Belowground Litter in Forests

Carbon and Nutrient Transfer via Above and Belowground Litter in Forests

Editors

Fuzhong Wu
Zhenfeng Xu
Wanqin Yang

MDPI • Basel • Beijing • Wuhan • Barcelona • Belgrade • Manchester • Tokyo • Cluj • Tianjin

Editors

Fuzhong Wu
Key Laboratory for Humid
Subtropical Eco-Geographical
Processes of the Ministry of
Education, School of
Geographical Sciences,
Fujian Normal University,
Fuzhou, China

Zhenfeng Xu
Institute of Ecology & Forestry,
Sichuan Agricultural University,
Chengdu, China

Wanqin Yang
School of Life Science,
Taizhou University,
Taizhou, China

Editorial Office
MDPI
St. Alban-Anlage 66
4052 Basel, Switzerland

This is a reprint of articles from the Special Issue published online in the open access journal *Forests* (ISSN 1999-4907) (available at: https://www.mdpi.com/journal/forests/special_issues/Carbon_Litter).

For citation purposes, cite each article independently as indicated on the article page online and as indicated below:

LastName, A.A.; LastName, B.B.; LastName, C.C. Article Title. *Journal Name* **Year**, *Volume Number*, Page Range.

ISBN 978-3-0365-6501-9 (Hbk)
ISBN 978-3-0365-6502-6 (PDF)

Contents

Editorial

Carbon and Nutrient Transfer via Above- and Below-Ground Litter in Forests

Fuzhong Wu [1,*], Zhenfeng Xu [2] and Wanqin Yang [3]

[1] Key Laboratory for Humid Subtropical Eco-Geographical Processes of the Ministry of Education, School of Geographical Sciences, Fujian Normal University, Fuzhou 350007, China
[2] Institute of Ecology & Forestry, Sichuan Agricultural University, Chengdu 611130, China
[3] School of Life Science, Taizhou University, Taizhou 318000, China
* Correspondence: wufzchina@fjnu.edu.cn

Citation: Wu, F.; Xu, Z.; Yang, W. Carbon and Nutrient Transfer via Above- and Below-Ground Litter in Forests. *Forests* **2022**, *13*, 2176. https://doi.org/10.3390/f13122176

Received: 11 December 2022
Accepted: 16 December 2022
Published: 18 December 2022

Publisher's Note: MDPI stays neutral with regard to jurisdictional claims in published maps and institutional affiliations.

Plants periodically shed more than 90% of their biomass production as above- and below-ground litter, including leaves, twigs, flowers, logs, roots and other tissues [1,2]. Litter production and decomposition are two of the most critical biogeochemical processes in transferring carbon and nutrients from plants back into the soil within forests. The returned litter components are the most important sources of soil organic matter and are crucial for maintaining soil fertility. Sequestering more carbon in forest soils by litter decomposition is currently a potential efficient strategy in serving "Carbon Neutrality". It is well-documented that, globally, land has been greening since the 1980s [3]. Consequently, litter production could inevitably increase with the increase in net productivity. However, attention has not been fully paid to the changes in litter quality and quantity, and the transfer processes of carbon and nutrients during litter production in forest ecosystems. Moreover, different forests could exhibit divergent responses in litter production to ongoing climate change. Up-to-date knowledge and theory are urgently needed.

To address this urgent topic, we aimed to collect different manuscripts that illustrate "Carbon and Nutrient Transfer via Above- and Below-ground Litter in Forests". A total of 15 high-quality articles were published after online peer review. The articles present topical examples of world-class research, including litterfall production, litter humification, litter quality, litter ^{13}C NMR spectroscopy components, nutrient return and resorption, the roles of soil fauna and microorganisms in litter decomposition, and any other related key points between litter and soil organic matter. We are delighted to present this book which is a compilation of these selected articles.

Here, we outline the key research activities and highlights of the publications included in this book. First of all, litter production is one of the primary nutrients returning processes in forest ecosystems. Yang et al. gave us a full picture of the production of total litterfall and its components (including leaves, twigs, reproductive organs and miscellaneous litterfall) from three cypress (*Cupressus funebris*) plantations with strip filling and ecological thinning [4]. Liu et al. investigated the seasonal and annual litterfall and calculated the returns of organic carbon, nitrogen, phosphorus, potassium, sulfur, calcium and magnesium in an evergreen broad-leaved forest [5]. Moreover, Hou et al. shed new light on the budget of plant litter and litter carbon in forest streams to check the biogeochemical linkages among mountain forests, riparian vegetation, and aquatic ecosystems [6]. Secondly, three of the articles addressed the initial litter quality and the changes in organic components during litter decomposition. Yang et al. used proximate analysis and ^{13}C nuclear magnetic resonance (NMR) with spectral editing techniques to quantify the variations in the initial C quality of newly shed foliar litter for fir, spruce, willow and rose over eight months in an alpine forest on the eastern Tibetan Plateau [7]. Liao et al. found that the content of acid hydrolysable components was 16%–21% in fresh litter across species, and only 4%–5% of these components remained in the litter after four years of decomposition when 53%–66%

of litter mass was lost in an alpine forest [8]. Xu et al. measured tree fresh tissue-leached DOM quantity and biodegradability in the fresh leaves and twigs of two broadleaf trees and two coniferous trees in subtropical plantations [9]. Thirdly, environmental changes and decomposers including soil fauna and microorganisms play essential roles in litter humification and decomposition. Tan et al. observed that the contribution rates of soil fauna to the accumulation of humic substances were 109.06%, 71.48%, 11.22% and −44.43% for the litter of fir, cypress, birch and willow after one year of incubation in cold forests, respectively [10]. Wang et al. checked humus accumulations in five foliar litters during four years of decomposition and their responses to reduced snow cover in an alpine forest [11]. Guo et al. monitored the litter decomposition associated with microbial organisms using microbial phospholipid fatty acids (PLFAs) in the forest floor, headwater stream, and intermittent stream [12]. Li et al. assessed the potential seasonal effects of forest gaps on litter mass loss and carbon release from a four-year litterbag decomposition experiment along an elevation gradient (3000, 3300, and 3600 m) [13]. Wang et al. investigated the composition and diversity of soil microbial communities across a subalpine forest successional series on the eastern Qinghai–Tibet Plateau [14]. Fang et al. elucidated the changes in the community structure of soil fungi and bacteria as well as alteration in litter and soil chemical characteristics in Japanese cedar plantations experiencing the expansion of moso bamboo [15]. Additionally, Zhang et al. studied the monthly dynamics of leaf nitrogen (N) and phosphorus (P) resorption efficiencies and C:N:P stoichiometric ratios in two selected subtropical plantations during a growing season [16]. He et al. compared soil macronutrient stocks at soil depths of 0−20 and 20−40 cm across a chronosequence of Masson pine plantations [17]. McNabb and Startsev reported that the bulk density of the trafficked soil failed to recover after 7 years to a depth of 20 cm at nine boreal forest sites in Alberta, Canada [18].

In summary, the articles included in this book highlight the ecological importance of forest litter in nutrients cycling, soil carbon sequestrating, and soil organic matter forming. The results would be of great scientific value in improving forest services in the context of global climate change.

Our sincere thanks go to the authors of each article for their excellent contributions. We are also grateful to the editors and all reviewers for their suggestions and assistance with the book's publication.

Funding: This project was supported by the National Natural Science Foundation of China (32171641, 32022056, U22A20450).

Conflicts of Interest: The authors declare no conflict of interest.

References

1. Hobbie, S.E. Plant species effects on nutrient cycling: Revisiting litter feedbacks. *Trends Ecol. Evol.* **2015**, *30*, 357–363. [CrossRef] [PubMed]
2. Gessner, M.O.; Swan, C.M.; Dang, C.K.; McKie, B.G.; Bardgett, R.D.; Wall, D.H.; Hättenschwiler, S. Diversity meets decomposition. *Trends Ecol. Evol.* **2010**, *25*, 372–380. [CrossRef] [PubMed]
3. Liu, Y.; Chai, Y.; Yue, Y.; Huang, Y.; Yang, Y.; Zhu, B.; Lou, D.; Li, Y.; Shi, D.; Ullah, W. Effects of global greening phenomenon on water sustainability. *Catena* **2022**, *208*, 105732. [CrossRef]
4. Yang, Y.; Yang, H.; Wang, Q.; Dong, Q.; Yang, J.; Wu, L.; You, C.; Hu, J.; Wu, Q. Effects of Two Management Practices on Monthly Litterfall in a Cypress Plantation. *Forests* **2022**, *13*, 1581. [CrossRef]
5. Liu, X.; Wang, Z.; Liu, X.; Lu, Z.; Li, D.; Gong, H. Dynamic Change Characteristics of Litter and Nutrient Return in Subtropical Evergreen Broad-Leaved Forest in Different Extreme Weather Disturbance Years in Ailao Mountain, Yunnan Province. *Forests* **2022**, *13*, 1660. [CrossRef]
6. Hou, J.; Li, F.; Wang, Z.; Li, X.; Yang, W. Budget of Plant Litter and Litter Carbon in the Subalpine Forest Streams. *Forests* **2021**, *12*, 1764. [CrossRef]
7. Yang, J.; Mu, J.; Zhang, Y.; Fu, C.; Dong, Q.; Yang, Y.; Wu, Q. Initial Carbon Quality of Newly Shed Foliar Litter in an Alpine Forest from Proximate Analysis and 13C NMR Spectroscopy Perspectives. *Forests* **2022**, *13*, 1886. [CrossRef]
8. Liao, S.; Yue, K.; Ni, X.; Wu, F. Acid Hydrolysable Components Released from Four Decomposing Litter in an Alpine Forest in Sichuan, China. *Forests* **2022**, *13*, 876. [CrossRef]

9.	Xu, J.-W.; Ji, J.-H.; Hu, D.-N.; Zheng, Z.; Mao, R. Tree Fresh Leaf- and Twig-Leached Dissolved Organic Matter Quantity and Biodegradability in Subtropical Plantations in China. *Forests* **2022**, *13*, 833. [CrossRef]

10.	Tan, Y.; Yang, K.; Xu, Z.; Zhang, L.; Li, H.; You, C.; Tan, B. The Contributions of Soil Fauna to the Accumulation of Humic Substances during Litter Humification in Cold Forests. *Forests* **2022**, *13*, 1235. [CrossRef]

11.	Wang, D.; Ni, X.; Guo, H.; Dai, W. Alpine Litter Humification and Its Response to Reduced Snow Cover: Can More Carbon Be Sequestered in Soils? *Forests* **2022**, *13*, 897. [CrossRef]

12.	Guo, H.; Wu, F.; Zhang, X.; Wei, W.; Zhu, L.; Wu, R.; Wang, D. Effects of Habitat Differences on Microbial Communities during Litter Decomposing in a Subtropical Forest. *Forests* **2022**, *13*, 919. [CrossRef]

13.	Li, H.; Du, T.; Chen, Y.; Zhang, Y.; Yang, Y.; Yang, J.; Dong, Q.; Zhang, L.; Wu, Q. Effects of Forest Gaps on Abies faxoniana Rehd. Leaf Litter Mass Loss and Carbon Release along an Elevation Gradient in a Subalpine Forest. *Forests* **2022**, *13*, 1201. [CrossRef]

14.	Wang, Z.; Bai, Y.; Hou, J.; Li, F.; Li, X.; Cao, R.; Deng, Y.; Wang, H.; Jiang, Y.; Yang, W. The Changes in Soil Microbial Communities across a Subalpine Forest Successional Series. *Forests* **2022**, *13*, 289. [CrossRef]

15.	Fang, H.; Liu, Y.; Bai, J.; Li, A.; Deng, W.; Bai, T.; Liu, X.; Lai, M.; Feng, Y.; Zhang, J.; et al. Impact of Moso Bamboo (Phyllostachys edulis) Expansion into Japanese Cedar Plantations on Soil Fungal and Bacterial Community Compositions. *Forests* **2022**, *13*, 1190. [CrossRef]

16.	Zhang, Y.; Yang, J.; Wei, X.; Ni, X.; Wu, F. Monthly Dynamical Patterns of Nitrogen and Phosphorus Resorption Efficiencies and C:N:P Stoichiometric Ratios in *Castanopsis carlesii* (Hemsl.) Hayata and *Cunninghamia lanceolata* (Lamb.) Hook. Plantations. *Forests* **2022**, *13*, 1458. [CrossRef]

17.	He, J.; Dai, Q.; Xu, F.; Yan, Y.; Peng, X. Variability in Soil Macronutrient Stocks across a Chronosequence of Masson Pine Plantations. *Forests* **2021**, *13*, 17. [CrossRef]

18.	McNabb, D.H.; Startsev, A. Seven-Year Changes in Bulk Density Following Forest Harvesting and Machine Trafficking in Alberta, Canada. *Forests* **2022**, *13*, 553. [CrossRef]

 *forests*

Article

Initial Carbon Quality of Newly Shed Foliar Litter in an Alpine Forest from Proximate Analysis and ^{13}C NMR Spectroscopy Perspectives

Jiaping Yang [1], Junpeng Mu [1], Yu Zhang [2], Changkun Fu [3], Qing Dong [1], Yulian Yang [1] and Qinggui Wu [1,*]

[1] Ecological Security and Protection Key Laboratory of Sichuan Province, Mianyang Normal University, Mianyang 621000, China
[2] School of Management, Lanzhou University, Lanzhou 730000, China
[3] College of Life Science, Sichuan Normal University, Chengdu 610101, China
* Correspondence: qgwu30@mtc.edu.cn

Abstract: The initial carbon (C) quality of plant litter is one of the major factors controlling the litter decomposition rate and regulating C sequestration, but a comprehensive understanding is still lacking. Here, we used proximate analysis and ^{13}C nuclear magnetic resonance (NMR) with spectral editing techniques to quantify the variations in the initial C quality for four dominant species (fir: *Abies faxoniana* Rehd. et Wils.; spruce: *Picea asperata* Mast; willow: *Salix paraplesia* Schneid; and rosa: *Rosa omeiensis* Rolfe.), including the organic compositions and C-based chemical structures of newly shed foliar litter over eight months in an alpine forest on the eastern Tibetan Plateau. The results indicated that the fractions of acid-soluble extractives (ASE) and acid-unhydrolyzable residues (AUR) were the main fractions of organic components, and aliphatic C and O-alkyl C were the main functional C groups for all plant species. Under the effects of the plant species, higher levels of ASE (37.62%) and aliphatic C (35.44%) were detected in newly shed rosa foliar litter, while higher levels of AUR (fir: 37.05%; spruce: 41.45%; and willow: 40.04%) and O-alkyl C (fir: 32.03%; spruce: 35.02%; and willow: 32.34%) were detected in newly shed fir, spruce and willow foliar litter. Moreover, the A/O-A and HB/HI ratios in rosa litter were 0.88 and 1.15, respectively, which were higher than those in fir, spruce and willow litter. The C quality of newly shed foliar litter varied seasonally due to the litter quality and environmental conditions, especially nitrogen (N), dissolved organic carbon (DOC), manganese (Mn) and monthly air temperature. We also found that C loss during 4-year litter decomposition was highly related to the aromatic C and phenolic C contents in newly shed foliar litter, suggesting that litter decomposition was strongly controlled by the initial recalcitrant C fractions. We conclude that the C quality of newly shed foliar litter in rosa might be structurally stable and more resistant to degradation than that of fir, spruce and willow, which contain abundant labile C fractions, and the initial recalcitrant C fractions are closely related to C loss during litter decomposition, which might contribute to soil C sequestration in alpine forests.

Keywords: litter C quality; soil C sequestration; dissolved organic C; acid-unhydrolyzable residues; litter decomposition

Citation: Yang, J.; Mu, J.; Zhang, Y.; Fu, C.; Dong, Q.; Yang, Y.; Wu, Q. Initial Carbon Quality of Newly Shed Foliar Litter in an Alpine Forest from Proximate Analysis and ^{13}C NMR Spectroscopy Perspectives. *Forests* **2022**, *13*, 1886. https://doi.org/10.3390/f13111886

Academic Editor: Marcin Pietrzykowski

Received: 13 September 2022
Accepted: 8 November 2022
Published: 10 November 2022

Publisher's Note: MDPI stays neutral with regard to jurisdictional claims in published maps and institutional affiliations.

1. Introduction

The initial litter quality, mainly referred to as the initial carbon (C) quality, which is commonly present in newly shed litter, is a critical factor controlling the processes of litter decomposition and soil C sequestration [1–4]. Generally, litter with abundant labile components has been widely documented to be more decomposable and to undergo rapid decomposition [5,6]. Simultaneously, as the main carrier of soil organic C, recalcitrant components mainly manage the accumulation and stabilization of soil organic matter [7,8]. The initial qualities of labile and recalcitrant components could be characterized by the organic

C fractions [9–11], thus facilitating the evaluation and prediction of litter decomposition and subsequent soil C sequestration, but there remains a lack of comprehensive knowledge.

Traditionally, proximate C fraction analysis (PA) has been used to separate water-soluble extractives (WSE), organic-soluble extractives (OSE), and acid-soluble extractives (ASE) from acid-unhydrolyzable residues (AUR), which can indicate the C fractions of organic components in plant litter and soils [12,13]. This method provides a close approximation of the lignin content in materials such as wood containing a low tannin content [12,13]. Nevertheless, the insolubility and complicated chemical structures of AUR fractions make it difficult to be characterized via this method, especially the AUR in foliar litter which generally includes substantial contributions of tannins and aliphatic C (cutin and, surface waxes) [14]. Solid-state ^{13}C cross polarization magic angle spinning magnetic resonance (CPMAS-NMR) spectroscopy has been proven to be an effective method that allows direct and nondestructive characterization of newly shed plant materials and litter and provides more specific information on the C quality, especially in terms of the AUR in foliar litter, which can be better defined [3,15,16]. Based on this method, previous studies demonstrated that O-alkyl C, which ranges from 60 to 95 ppm chemical shifts, is the most labile component originating from decomposable carbohydrates. Aliphatic C (0–45 ppm) is commonly referred to as a biologically stable structure, and it mainly contains long-chain aliphatic compounds, cutin and waxes [2,16–18]. Likewise, aromatic C (110–140 ppm) and phenolic C (140–165 ppm) have been determined as chemically recalcitrant fractions [2,18,19]. In addition, studies have found that the ratio of aliphatic C to O-alkyl C (A/O-A) or hydrophobic C to hydrophilic C (HB/HI) ((0–45 ppm + 110–165 ppm)/(45–110 ppm + 165–190 ppm)) could reflect the enhancement in chemical structures against degradation, and the higher these two ratios are, the more stable the chemical structures against degradation [20].

Seasonal climate change regulates the C quality in litter due to the plant phenology. The consumption of decomposable C fractions is accelerated by strong rainfall and transpiration, which results from the increased temperature and solar radiation as a direct consequence of the decline in the fractions of WSE, OSE fractions and carbohydrate O-alkyl C [21,22]. Similarly, the aforementioned climatic changes could lead to the accumulation of recalcitrant C fractions [23]. Moreover, as the main fractions determining the litter quality, structurally complex C fractions notably determine and vary with the various litter decomposition stages, particularly the early stage of decomposition, which is susceptible to climatic conditions, rather than the late stage of decomposition [24–26]. A large proportion of labile C input derived from newly shed litter rapidly decreases at the early stage of decomposition [7,22], whereas the limitation on litter decomposition at the late stage of decomposition is enhanced, and more recalcitrant C fractions are formed [27]. In addition, vegetation is often considered the key factor resulting in variation in the litter C quality [28]. It has been reported that ASE and AUR were the main C fractions in foliar litter, while the content of recalcitrant AUR was higher in coniferous litter than that in broadleaf litter [26,29]. The chemical structures and compositions of AUR vary with vegetation, which results in the differences in the decomposable capacity and C fraction contents [25,30]. It has been considered that a strong signal intensity is mostly found for O-alkyl C, which indicates that newly shed foliar litter comprises labile carbohydrates. Moreover, Ono et al. [19,31] found that more aliphatic C was detected in newly shed coniferous litter (cedar and cypress) than in broadleaf litter. Conversely, broadleaved litter (beech and oak) retained more recalcitrant aromatic C, which is more decomposable than aliphatic C. In addition, it was reported that phenolic C (lignin fractions) and methoxyl C in coniferous and broadleaf species were significantly related to the accumulation of AUR at the initial stage of decomposition, suggesting that the contribution of phenolic C and methoxyl C to the AUR fraction is higher and not limited by the species at this decomposition stage [26]. Therefore, we addressed the hypothesis that (1) the contents of C fractions would vary with the species, and more labile C fractions would be released from newly shed foliar litter, while more recalcitrant C fractions might accumulate during seasons with heavy rainfall and severe temperatures, while (2) the labile C fractions might be the major factor controlling C accumulation during litter decomposition.

To test the above hypotheses, we used PA analysis and ^{13}C CPMAS NMR spectroscopy to detect the initial C quality of newly shed foliar litter for four dominant tree species, namely fir, spruce, willow and rosa, which were collected on a monthly basis from August 2015 to July 2016 in an alpine forest on the eastern Tibetan Plateau. In this study, we aimed to address three scientific questions: (1) How do the plant species and litter quality affect the C quality of newly shed foliar litter? (2) How does C quality of newly shed foliar litter vary seasonally? (3) How does initial C quality regulate decomposition processes? Therefore, our objective was to evaluate the capacity of the initial C input from plant litter and the initial potential of litter-derived C for C sequestration in the decomposition process in this study.

2. Materials and Methods

2.1. Site Description

This study was conducted at the Long-Term Research Station of the Alpine Forest Ecosystem in the Miyaluo Nature Reserve (31°14′ N, 102°53′ E, 2982–3020 m above sea level (masl)) which is located on the eastern Tibetan Plateau. This region is a typical and transitional winter-cold zone between the Tibetan Plateau and the Sichuan Basin. The annual mean temperature is 2.7 °C. The maximum and minimum temperatures are 23 °C (July) and −18 °C (January), respectively. The annual mean precipitation reaches approximately 850 mm. Winter normally extends from late October until the following April with snow accumulating, and the growing season with the most concentrated litterfall extends from April until November. The forest is dominated by coniferous fir (*Abies faxoniana*) and spruce (*Picea asperata*). The main understory shrubs include willow (*Salix paraplesia*), rosa (*Rosa omeiensis*) and azalea (*Rhododendron lapponicum*). The forest soil in this area is dark brown soil and brown soil, and it is considered to be poor and barren due to the frequent natural disasters and low temperature conditions [22,32,33].

2.2. Experimental Design and Sample Collection

Based on a previous field investigation, three replicate 50 × 50 m sampling sites exhibiting similar aspects and slopes were established in a spruce-fir forest (2982–3020 m above sea level (masl)), and the three sites were at least 50 m apart from each other. Twenty litterfall funnel-shaped trap collectors were placed randomly at each site and fixed approximately 1 m above the ground at the end of July 2015.

Newly shed foliar litter was collected on a monthly basis from August 2015 to April 2016. The collected litter samples were air-dried at room temperature for one week, and foliar litter was classified as fir, spruce, willow and rosa according to the litter species. All foliar litter materials were ground and passed through a 0.25-mm sieve as subsamples for chemical analysis.

2.3. Chemical Analysis

The samples were analyzed using a sequential extraction technique according to Hilli et al. [34] and Preston et al. [14]. The samples were sequentially treated with trichloromethane to extract organic-soluble extractives (OSE), followed by boiling water to extract water-soluble extractives (WSE) and finally 72% (*v/v*) sulfuric acid to extract acid-soluble extractives (ASE), and the remaining residues were considered acid-unhydrolyzable residues (AUR). We performed 3 laboratory replicates for each sample during the experiment.

During solid-state ^{13}C NMR measurement, all samples were treated with hydrofluoric acid (HF) to remove paramagnetic materials such as iron oxide from samples as follows: 0.5 g subsamples were weighed into 50-mL falcon tubes, and the tubes were tightly capped and horizontally shaken after adding 10 mL of 46% HF. Then, 40 mL of deionized water was added to the tubes, and the tubes were centrifuged at 3000 rpm for 10 min. The supernatant was removed and discarded, and the residue was washed 3–5 times with deionized water to completely remove HF, and then dried at 60 °C in an oven. The HF-treated samples were used for solid-state ^{13}C NMR analysis [19]. We did not perform laboratory replicate analyses of the samples. Generally, the measurements were better than the chemical extractions, and the replicates at each site were adequate to ensure the reliability of the data.

Solid-state ^{13}C NMR spectra were obtained on a Bruker AVANCE III 400 spectrometer (Bruker BioSpin AG, Fällanden, Switzerland) at 100 MHz. The dried and finely powdered samples were packed in a zirconium dioxide (ZrO_2) rotor topped with a Kel-F cap, which was spun at a 5 kHz rate. The contact time for the ramp sequence of the 1H ramp was 2 ms. A total of 2048 scans were recorded with a 1 s recycle delay for each sample, and the plotted spectral regions ranged from 0–220 ppm.

The contributions of various functional C groups to the total organic C (TOC) content were determined via the integration of their ^{13}C signal intensity in the respective chemical shifts using MestreNova-9.0.1 (Mestrelab Research S.L., Santiago de Compostela, Spain). Functional C groups were assigned to aliphatic C (0–50 ppm), methoxyl C (50–60 ppm), O-alkyl C (60–95 ppm), di-O-alkyl C (95–110 ppm), aromatic C (110–140 ppm), phenolic C (140–165 ppm), and carboxyl C (165–220 ppm) [17,35]. Several hydrophobicity indices, including the ratio of aliphatic C to O-alkyl C (A/O-A) and the ratio of hydropholic-C to hydrophilic-C (HB/HI), i.e., (aliphatic C + functional C groups of the 110–165 ppm peak area)/(functional C groups of the 45–110 ppm peak area + functional C groups of the 165–190 ppm peak area), were used to reflect the stability of organic C. The higher these two ratios are, the higher the stability of organic C [20,36,37].

Total organic C, N and phosphorus (P) concentrations were determined via the dichromate oxidation, Kjeldahl determination (KDN, Top Ltd., Zhejiang, China) and phospho-molybdenum yellow spectrophotometry (TU-1901, Puxi Ltd., Beijing, China) methods, respectively [38]. Litter DOC was extracted by shaking 0.5 g of dried sample with 50 mL of deionized water for 30 min at room temperature, and the suspension was filtered through a 0.45-μm filter membrane. A total organic C analyzer (multi N/C 2100, Analytik Jena, Thüringen, Germany) was used to determine the litter DOC concentration [39]. The concentration of Mn via inductively coupled plasma-mass spectroscopy (ICP-MS, IRIS Advantage 1000, Thermo Elemental, Waltham, MA, USA) after litter materials (1.00 g) were digested in HNO_3-$HClO^4$ (5:1, v/v) at 160 °C for 5 h [40].

2.4. Statistical Analysis

A two-way ANOVA was performed to evaluate the effects of the collection time and plant species on the initial C quality and hydrophobicity indices. Multiple comparisons were used to assess the significance of the differences among the various plant species. The above analyses were performed in SPSS 22.0 (IBM SPSS Statistics Inc., Chicago, IL, USA). Linear analysis was used to determine the effects of methoxyl C, aromatic C and phenolic C in newly shed foliar litter on C loss during litter decomposition. Stepwise regression analysis was conducted to evaluate the effects of the environmental conditions and litter quality on the initial C quality. The above analyses were performed in Origin Pro 9.0 (OriginLab, Northampton, MA, USA). All graphics were generated in Origin Pro 9.0 (OriginLab, Northampton, MA, USA). C loss data for fir and willow litter in the 4-year decomposition process were collected from Ni et al. [41], which were used to detect the effects of the C quality of newly shed fir and willow foliar litter on C loss during litter decomposition.

3. Results

3.1. Proportions of Organic Components in Newly Shed Foliar Litter

The fractions of ASE and AUR were the main components among all plant species. In rosa litter, the content of ASE (37.62%) was higher than that of other C fractions, while in fir, spruce and willow litter, the content of AUR was higher than that of other C fractions (fir: 37.05%; spruce: 41.45%; willow: 40.04%) (Table 1). ASE and AUR were significantly affected by the collection time (ASE: $p = 0.015$; AUR: $p = 0.003$) and plant species (ASE: $p = 0.020$; AUR: $p < 0.001$) (Table 2). Regarding WSE, similar trends were observed between the fir and spruce litter materials, with peaks occurring in April (Figure 1). Regarding AUR, the content in fir, willow and rosa litter generally decreased from August to November (Figure 1).

Table 1. Properties of the C quality of newly shed foliar litter for each plant species. Different lowercase letters indicate significant ($p < 0.05$) differences among the various plant species for each C quality aspect (plant species: fir, spruce, willow and rosa; functional C groups: aliphatic C, methoxyl C, O-alkyl C, di-O-alkyl C, aromatic C, phenolic C and carboxyl C; organic components: WSE, water-soluble extractives, OSE, organic-soluble extractives, ASE, acid-soluble extractives, AUR, acid-unhydrolyzable residues; HB/HI, (0–45 ppm + 110–165 ppm)/(45–110 ppm + 165–190 ppm), and A/O-A, aliphatic C/O-alkyl C)).

Carbon Quality	Fir	Spruce	Willow	Rosa
Aliphatic (%)	23.66 ± 0.65 bc	22.06 ± 1.31 c	28.45 ± 3.45 b	35.44 ± 3.49 a
Methoxyl (%)	5.60 ± 0.20 b	6.67 ± 0.27 a	6.79 ± 0.49 a	5.39 ± 0.17 b
O-alkyl (%)	32.03 ± 0.35 b	35.02 ± 0.64 a	32.34 ± 1.56 b	27.60 ± 1.27 c
di-O-alkyl (%)	10.58 ± 0.12 a	10.24 ± 0.21 a	9.38 ± 0.59 a	8.10 ± 0.69 b
Aromatic (%)	13.48 ± 0.23 a	12.94 ± 0.49 a	10.84 ± 1.01 b	10.68 ± 0.89 b
Phenolic (%)	10.48 ± 0.29 a	8.74 ± 0.41 b	7.86 ± 0.79 bc	7.10 ± 1.02 c
Carboxyl (%)	4.19 ± 0.35 b	4.33 ± 0.32 b	4.34 ± 0.49 b	5.68 ± 0.90 a
A/O-A	0.49 ± 0.02 c	0.43 ± 0.03 d	0.60 ± 0.09 b	0.88 ± 0.11 a
HB/HI	0.91 ± 0.02 b	0.78 ± 0.02 d	0.90 ± 0.06 c	1.15 ± 0.08 a
OSE (%)	9.26 ± 0.97 a	6.40 ± 0.77 b	3.98 ± 0.63 b	6.48 ± 0.93 b
WSE (%)	20.98 ± 1.41 a	19.65 ± 1.91 a	20.53 ± 3.25 a	19.54 ± 1.90 a
ASE (%)	34.91 ± 0.82 a	32.50 ± 2.91 a	35.45 ± 1.02 a	37.62 ± 2.97 a
AUR (%)	37.05 ± 1.31 a	41.45 ± 4.00 a	40.04 ± 3.57 a	36.35 ± 1.67 a

Table 2. Results of two-way ANOVA testing for the effects of the collection time (from August 2015 to July 2016) and plant species on the C quality of newly shed foliar litter (plant species: fir, spruce, willow and rosa; functional C groups: aliphatic C, methoxyl C, O-alkyl C, di-O-alkyl C, aromatic C, phenolic C and carboxyl C; organic components: WSE, water-soluble extractives, OSE, organic-soluble extractives, ASE, acid-soluble extractives, AUR, acid-unhydrolyzable residues; HB/HI, (0–45 ppm + 110–165 ppm)/(45–110 ppm + 165–190 ppm), and A/O-A, aliphatic C/O-alkyl C)).

Carbon Quality	Plant Species			Collection Time		
	df	*F*	*p*	*df*	*F*	*p*
Aliphatic	3	146.85	<0.001	7	27.661	<0.001
Methoxyl	3	33.17	<0.001	7	7.419	<0.001
O-alkyl	3	236.10	<0.001	7	34.761	<0.001
Di-O-alkyl	3	66.82	<0.001	7	9.365	<0.001
Aromatic	3	54.54	<0.001	7	16.017	<0.001
Phenolic	3	54.30	<0.001	7	9.073	<0.001
Carboxyl	3	9.24	<0.001	7	3.039	0.007
HB/HI	3	352.77	<0.001	7	37.097	<0.001
A/O-A	3	415.89	<0.001	7	55.054	<0.001
OSE	3	45.938	<0.001	7	28.316	<0.001
WSE	3	0.941	0.426	7	16.048	<0.001
ASE	3	3.776	0.015	7	2.621	0.020
AUR	3	5.149	0.003	7	10.741	<0.001

3.2. Proportions of Functional C Groups in Newly Shed Foliar Litter

Aliphatic C and O-alkyl C were the main functional C groups, while methoxyl C and carboxyl C were less abundant than the other functional groups (Table 1). In rosa litter, the content of aliphatic C (35.44%) was higher than that of the other functional C groups, while fir, spruce and willow litter contained more O-alkyl C (fir: 32.03%; spruce: 35.02%; willow: 32.34%). The plant species ($p < 0.001$) and collection time ($p < 0.01$) remarkably affected the functional C groups (Table 2). Smaller proportions of O-alkyl C in the fir and spruce litter samples were observed in April (Figure 2). The aromatic C and phenolic C contents in the fir and spruce litter samples remained stable with low proportions (Figure 2). The aliphatic C content in willow litter continuously decreased from August to November, while the aliphatic C content in rosa litter increased from August to October (Figure 2).

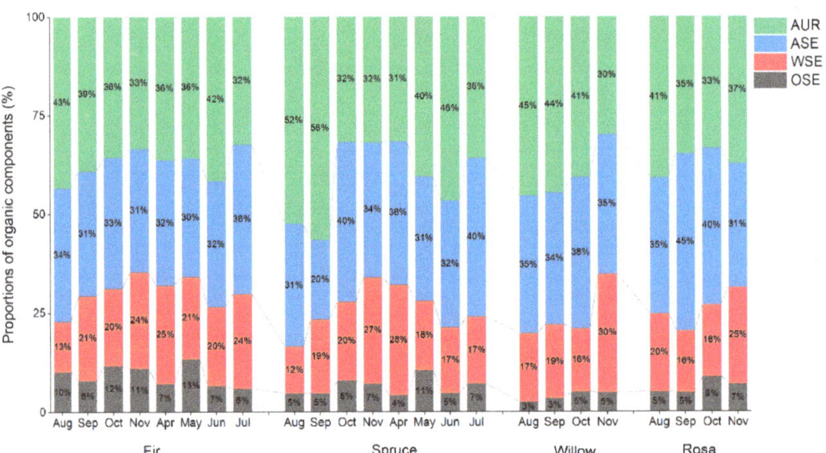

Figure 1. Monthly dynamics of the proportions of organic components (%) (WSE: water-soluble extractives; OSE: organic-soluble extractives; ASE: acid-soluble extractives; AUR: acid-unhydrolyzable residues) in newly shed foliar litter determined via proximate C fraction analysis.

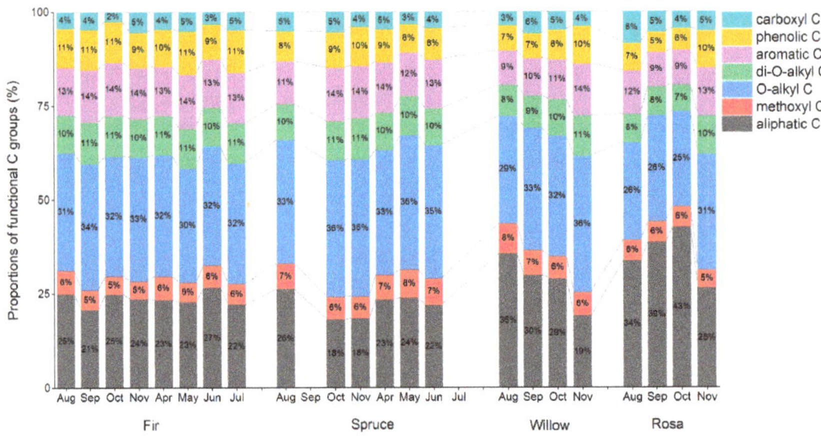

Figure 2. Monthly dynamics of the proportions of functional C fractions (%) (aliphatic C, methoxyl C, O-alkyl C, di-O-alkyl C, aromatic C, phenolic C, and carboxyl C) in newly shed foliar litter obtained from the integration of different chemical shifts determined via CP/MAS ^{13}C NMR spectroscopy.

3.3. A/O-A and HB/HI Ratios of Newly Shed Foliar Litter

The ratios of A/O-A (0.88) and HB/HI (1.15) in rosa litter were higher than those in the litter of the other species, while those in spruce litter were the lowest (A/O-A: 0.43; HB/HI: 0.78) (Table 1). The plant species ($p < 0.001$) and collection time ($p < 0.001$) significantly influenced A/O-A and HB/HI (Table 2). The ratios of A/O-A in the fir and spruce litter samples were significantly lower than those in the willow and rosa litter samples (Figure 3a). The ratios of HB/HI in the fir and spruce litter samples were more stable than those in the willow and rosa litter samples, respectively (Figure 3b).

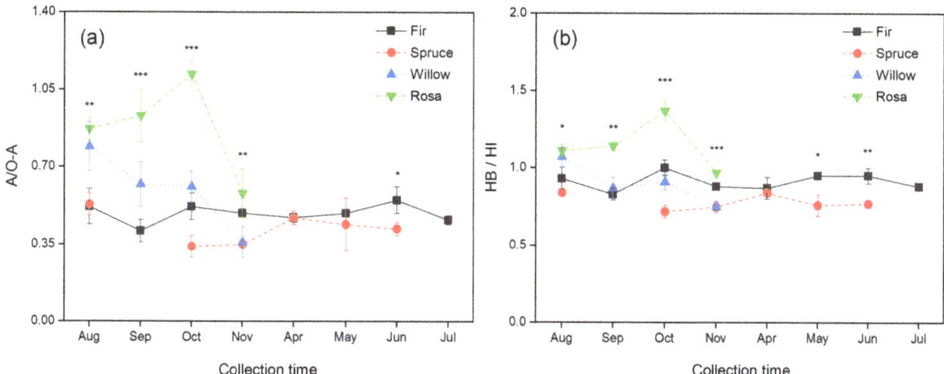

Figure 3. Monthly dynamics of the ratio of (**a**) aliphatic C to O-alkyl C (A/O-A) and (**b**) the index of hydrophobicity (HB/HI, (0–45 ppm + 110–165 ppm)/(45–110 ppm + 165–190 ppm)) in newly shed foliar litter determined via CP/MAS ^{13}C NMR spectroscopy. The * indicates significant ($p < 0.05$) differences among the various plant species. * $p < 0.05$, ** $p < 0.01$, and *** $p < 0.001$.

3.4. Relationships between the Initial Functional C Groups and C Loss during Litter Decomposition

Among all the functional C groups, methoxyl C, aromatic C and phenolic C significantly ($p < 0.05$) affected C loss during litter decomposition. Methoxyl C (coefficient: 0.49) and aromatic C (coefficient: -0.54) markedly controlled C loss in the third and first years of litter decomposition, respectively, and phenolic C significantly controlled C loss in both the second (coefficient: -0.51) and fourth (coefficient: -0.47) years of litter decomposition (Figure 4).

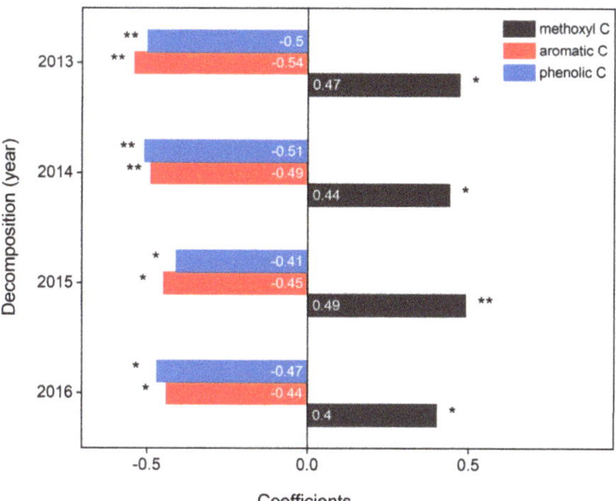

Figure 4. Results of linear regression analysis between functional C groups and C loss in litter decomposition processes. The * indicates the significance ($p < 0.05$) between functional C groups and C loss at different decomposition stages in the regression model. * $p < 0.05$, ** $p < 0.01$, and Pearson's r values from linear regression are shown in each panel.

4. Discussion

In this study, ASE and AUR were the main organic C fractions for all species, which is consistent with the study results of Yang et al. [29]. Consistent with our hypothesis, a large amount of labile O-alkyl C was detected in the newly shed foliar litter, so that it could be sufficiently consumed by a large microbial population in the subsequent decomposition process, which has been commonly reported [3,11]. Moreover, the functional C group contents in newly shed foliar litter varied with the plant species, and labile O-alkyl C was mostly detected in the fir, spruce and willow litter materials, whereas rosa litter contained abundant recalcitrant aliphatic C. Therefore, this result suggested that the initial C quality might be species-dependent, which is consistent with our hypothesis. This might occur because rosa litter could highly protect the leaf surface from leaching as a result of the abundant recalcitrant aliphatic C and higher hydrophobicity indices (Table 1) [11,18], while the initial C fractions in fir, spruce and willow litter might potentially contribute to C turnover in alpine forests. Moreover, in accordance with the results of other studies on litter decomposition, the initial C quality in foliar litter was controlled by DOC, N and Mn (Figure 5) [1,4,6], which illustrated the key role of the litter quality in the formation and preservation of litter-derived C in soils. Methoxyl C, aromatic C and phenolic C in newly shed foliar litter significantly regulated C loss during litter decomposition, which indicated that the variation in the C quality of newly shed litter was considered a critical factor influencing litter decomposition and greatly contributed to soil C sequestration in alpine forests.

The plant species significantly regulated the initial C quality of newly shed foliar litter. In our study, the initial C quality of newly shed fir, spruce and willow foliar litter was dominated by labile carbohydrates because strong signals were detected in the O-alkyl C regions [35,42]. Furthermore, previous studies found that the rapid decrease in O-alkyl C attributed to polysaccharide materials at the initial decomposition stage generally occurred due to the preferential degradation of labile cellulosic compounds [2,43]. Hence, the result whereby O-alkyl C release in rosa litter was higher than that in fir, spruce and willow litter in this study might mostly occur because carbohydrates in O-alkyl C in rosa litter might be easier to degrade than those in fir, spruce and willow litter. Moreover, xylose is considered the dominant hemicellulose monosaccharide in broad-leaved tissues, and its degradation occurs at least one year after broad-leaved litter decomposition. Nevertheless, arabinose and galactose, the main dominant hemicellulose monosaccharides in coniferous tissues, immediately degrade upon litter formation [27,44]. Therefore, the ASE content in rosa and willow litter was higher than that in fir and spruce litter. Our study also found that the aliphatic C content and A/O-A and HB/HI ratios in rosa litter were higher than those in the litter of the other species, suggesting that the C structure in rosa litter was mainly assigned to aliphatic C and was more complex than that in the litter of the other species [20,45]. The results are consistent with the studies of Hishinuma et al. [2] and De Marco et al. [26] but inconsistent with the studies of Ono et al. [31] and Ono et al. [25], which could be explained by the variations in biosynthetic lipid processes among plant species. In addition, the less complex chemical structure determined by the indices of A/O-A and HB/HI in coniferous species versus broadleaved species might also be attributable to the plant species [45,46]. Thus, rosa litter might be more resistant to decomposition. In addition, the abundance of guaiacyl lignins and stable tannins in coniferous fir and spruce and broadleaf willow and rosa might lead to differences in the proportions of aromatic C and phenolic C among the various litter types [25,45]. In general, the contents of ASE and aliphatic C might be the key C quality factors controlling the chemical structure of newly shed foliar litter, particularly that of broad-leaved rosa in this study.

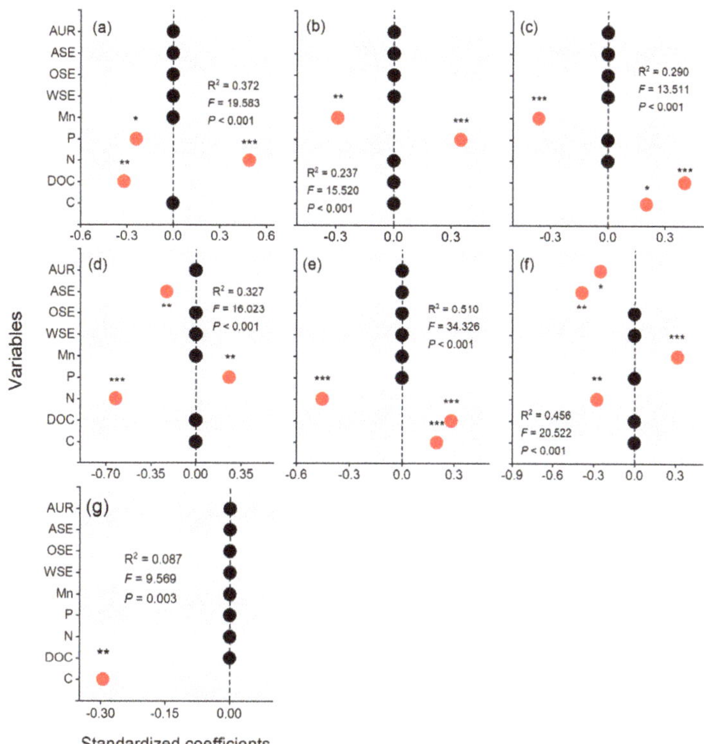

Figure 5. Results of stepwise regression analysis of the functional C groups ((**a**): aliphatic C, (**b**): methoxyl C, (**c**): O-alkyl C, (**d**): di-O-alkyl C, (**e**): aromatic C, (**f**): phenolic C, and (**g**): carboxyl (C)) using the litter quality (C: carbon; DOC: dissolved organic carbon; N: nitrogen; P: phosphorus; Mn: manganese; WSE: water-soluble extractives; OSE: organic-soluble extractives; ASE: acid-soluble extractives; AUR: acid-unhydrolyzable residues). The dominant variables are emphasized in red in the regression model. The values represent standardized coefficients, and the * indicates significant effects of the variables on the functional C fractions in a given regression model * $p < 0.05$, ** $p < 0.01$, and *** $p < 0.001$. For example, aliphatic C was significantly impacted by P ($p < 0.05$), DOC ($p < 0.01$) and N ($p < 0.001$) in the stepwise regression testing the effects of litter chemical quality on aliphatic C (**a**).

The litter quality also dramatically determined the C quality of newly shed foliar litter, and this might lead to seasonal variations in the C quality together with environmental conditions. Our analysis results showed that the N content also dominated the recalcitrant C quality of newly shed foliar litter (Figure 5), which is consistent with the decomposition process. This might be due to foliar litter releasing a large amount of nutrients, such as N, at the early stage of decomposition under the notable influences of the litter quality and environmental conditions and then further stimulating microbial activity that might easily accelerate the degradation of recalcitrant substances [47–49]. This might also be the reason that stable aromatic C and phenolic C and low ratios of A/O-A and HB/HI in willow and rosa litter were detected in November. The insulating effect of snow cover at our study site during this period might promote microbial activity, thus resulting in fast decomposition [41,50]. Similarly, soluble nutrients, such as DOC, were rapidly released at the early stage of decomposition. In our study, DOC was the most significant factor influencing WSE and O-alkyl C and strongly responded to environmental conditions (Figure 5); therefore, a large amount of easily available compounds may be formed as a result of the thick litter layer and could be easily leached from foliar litter by light snowfall in October and November of this region, thus leading to more labile WSE and O-alkyl C released from foliar litter [51]. Moreover, our analysis results showed that DOC significantly affected the recalcitrant AUR (Figure S1), aliphatic C and aromatic C, which might have resulted from DOC derived from the oxidative degradation of lignin. Water-soluble components primarily controlled the formation and persistence of recalcitrant substances in forest soils by significantly affecting microbial activity [52,53]. This might be the main reason why the temporal characteristics of the aliphatic C and AUR fraction were opposite to those of WSE and O-alkyl C during the periods of October and November. In addition, the initial Mn generally functioned as a cofactor of lignin-degrading enzymes to regulate litter decomposition and humus stability, revealing a high contribution to the control of C sequestration in the organic layer [4,53]. The significant correlations between Mn and C quality, especially O-alkyl C and O-alkyl C, in newly shed foliar litter in our study might also indicate that the initial Mn might potentially contribute to C sequestration in soils.

In accordance with our hypothesis (2), the initial C quality notably controlled litter decomposition processes and could further affect C sequestration in soils. In combination with the litter decomposition experiment of Ni et al. [41], we found that the C loss in the 4-year decomposition process was highly related to the initial aromatic C and phenolic C, which are commonly referred to as structurally complex guaiacyl- and syringyl-lignin, respectively. This result notably illustrated that the initial recalcitrant C quality mainly determined C sequestration in soils in the decomposition process, which is contrary to the results of recent studies whereby labile components were the main factor influencing the decomposition process [18,20]. Nevertheless, the initial methoxyl C, which is also commonly reported as the aromatic rings of guaiacyl and syringyl units in lignin and could be structurally stable [11,36], was positively related to the C loss of decomposing litter in our study. This might be due to the region of initial methoxyl C also arising from other sources including proteins and amino acids [14].

5. Conclusions

Our results indicated that the initial C quality of newly shed foliar litter significantly varied with the plant species over time. Newly shed foliar litter of fir, spruce and willow comprising more labile C might contribute to the magnitude of accumulation of the soil C pool in alpine forests. Rosa litter, obtaining more structurally complex recalcitrant C fractions, might qualitatively contribute to the stability and sequestration of soil organic C. Moreover, other litter quality-related factors, such as N, DOC, and Mn, strongly controlled the initial C quality of newly shed foliar litter. In addition, the initial methoxyl C, aromatic C and phenolic C were the best predictors of C loss during litter decomposition in the alpine forests. However, our study did not consider the C quality in soils, and we could speculate that labile and recalcitrant C fractions could exhibit different characteristics in soils. Therefore, future efforts should be devoted to thoroughly studying the C quality based on the continuous process of "newly shed litter-decomposing litter-soils" to obtain the contributions of litter-derived C to soil C sequestration in alpine forests.

Supplementary Materials: The following supporting information can be downloaded at: https://www.mdpi.com/article/10.3390/f13111886/s1, Figure S1: Results of stepwise regression analyses of initial organic components ((a): WSE, water-soluble extractives; (b): OSE, organic-soluble extractives; (c): ASE, acid-soluble extractives; (d): AUR, acid-unhydrolyzable residues) using litter quality and environmental conditions (C: carbon; DOC: dissolved organic carbon; N: nitrogen; P: phosphorus; Mn: manganese; MAT: monthly average temperature; MAP: monthly average precipitation). The dominant variables are emphasized by red color in the regression model, Values represent the standardized coefficients, The * indicates significant effects of variables on C fractions in a given regression model * $p < 0.05$, ** $p < 0.01$, and *** $p < 0.001$. For example, OSE was significantly impacted by MAT ($p < 0.05$), Mn ($p < 0.001$), P ($p < 0.001$), and DOC ($p < 0.001$) in the stepwise regression testing the effects of litter chemical quality and environmental conditions on OSE (a).

Author Contributions: Conceptualization, methodology, validation, resources, supervision and writing—review and editing, J.M. and Q.W.; Project administration and funding acquisition, Q.W.; Conceptualization, data curation, investigation, methodology and writing—original draft, J.Y.; Investigation, Q.D., Y.Y., Y.Z. and C.F. All authors have read and agreed to the published version of the manuscript.

Funding: This work was supported by the National Natural Science Foundation of China (31800521, 32022056, 31800373, 31922052 and 32071747), the Fok Ying-Tong Education Foundation for Young Teachers (161101), the National Natural Science Foundation of Sichuan Province (2022NSFSC0087), the Key R&D Program of Sichuan (18ZDYF0307), the Research Fund of Mianyang Normal University (QD2020A18), and the Open Fund of Ecological Security and Protection Key Laboratory of Sichuan Province, Mianyang Normal University (ESP1807).

Institutional Review Board Statement: Not applicable.

Acknowledgments: We wish to thank Zhuang Wang, Fan Yang, Xinyu Wei, Ling Mou, Qun Liu, Fujia Wu, Ji Yuan, and Long Jiang for the sample collecting.

Conflicts of Interest: The authors declare no conflict of interest.

References

1. Li, Q.; Zhang, M.; Geng, Q.; Jin, C.; Zhu, J.; Ruan, H.; Xu, X. The roles of initial litter traits in regulating litter decomposition: A "common plot" experiment in a subtropical evergreen broadleaf forest. *Plant Soil* **2020**, *452*, 207–216. [CrossRef]
2. Hishinuma, T.; Osono, T.; Fukasawa, Y.; Azuma, J.-i.; Takeda, H. Application of ^{13}C NMR spectroscopy to characterize organic chemical components of decomposing coarse woody debris from different climatic regions. *Ann. For. Res.* **2015**, *58*, 3–13. [CrossRef]
3. Sarker, T.C.; Maisto, G.; De Marco, A.; Esposito, F.; Panico, S.C.; Alam, M.F.; Mazzoleni, S.; Bonanomi, G. Explaining trajectories of chemical changes during decomposition of tropical litter by ^{13}C-CPMAS NMR, proximate and nutrients analysis. *Plant Soil* **2019**, *436*, 13–28. [CrossRef]
4. De Marco, A.; Fioretto, A.; Giordano, M.; Innangi, M.; Menta, C.; Papa, S.; Virzo De Santo, A. C stocks in forest floor and mineral soil of two mediterranean beech forests. *Forests* **2016**, *7*, 181. [CrossRef]
5. Mastný, J.; Kaštovská, E.; Bárta, J.; Chroňáková, A.; Borovec, J.; Šantrůčková, H.; Urbanová, Z.; Edwards, K.R.; Picek, T. Quality of DOC produced during litter decomposition of peatland plant dominants. *Soil Biol. Biochem.* **2018**, *121*, 221–230. [CrossRef]
6. Ni, X.Y.; Yang, W.Q.; Liao, S.; Li, H.; Tan, B.; Yue, K.; Xu, Z.F.; Zhang, L.; Wu, F.Z. Rapid release of labile components limits the accumulation of humic substances in decomposing litter in an alpine forest. *Ecosphere* **2018**, *9*, e02434. [CrossRef]
7. Kögel-Knabner, I. The macromolecular organic composition of plant and microbial residues as inputs to soil organic matter: Fourteen years on. *Soil Biol. Biochem.* **2017**, *105*, A3–A8. [CrossRef]
8. Pisani, O.; Lin, L.H.; Lun, O.O.Y.; Lajtha, K.; Nadelhoffer, K.J.; Simpson, A.J.; Simpson, M.J. Long-term doubling of litter inputs accelerates soil organic matter degradation and reduces soil carbon stocks. *Biogeochemistry* **2016**, *127*, 1–14. [CrossRef]
9. Grossman, J.J.; Cavender-Bares, J.; Hobbie, S.E. Functional diversity of leaf litter mixtures slows decomposition of labile but not recalcitrant carbon over two years. *Ecol. Monogr.* **2020**, *90*, e01407. [CrossRef]
10. Wang, Y.; Zheng, J.; Boyd, S.E.; Xu, Z.; Zhou, Q. Effects of litter quality and quantity on chemical changes during eucalyptus litter decomposition in subtropical Australia. *Plant Soil* **2019**, *442*, 65–78. [CrossRef]
11. De Marco, A.; Spaccini, R.; Virzo De Santo, A. Differences in nutrients, organic components and decomposition pattern of *Phillyrea angustifolia* leaf litter across a low maquis. *Plant Soil* **2021**, *464*, 559–578. [CrossRef]
12. Preston, C.M.; Nault, J.R.; Trofymow, J.A.; Smyth, C. Chemical changes during 6 years of decomposition of 11 litters in some Canadian forest sites. Part 1. elemental composition, tannins, phenolics, and proximate fractions. *Ecosystems* **2009**, *12*, 1053–1077. [CrossRef]
13. Hilli, S.; Stark, S.; Willför, S.; Smeds, A.; Reunanen, M.; Hautajärvi, R. What is the composition of AIR? Pyrolysis-GC–MS characterization of acid-insoluble residue from fresh litter and organic horizons under boreal forests in southern Finland. *Geoderma* **2012**, *179–180*, 63–72. [CrossRef]
14. Preston, C.M.; Nault, J.R.; Trofymow, J.A. Chemical changes during 6 years of decomposition of 11 litters in some Canadian forest sites. Part 2. ^{13}C abundance, solid-state ^{13}C NMR spectroscopy and the meaning of "lignin". *Ecosystems* **2009**, *12*, 1078–1102. [CrossRef]
15. Assuncao, S.A.; Pereira, M.G.; Rosset, J.S.; Berbara, R.L.L.; Garcia, A.C. Carbon input and the structural quality of soil organic matter as a function of agricultural management in a tropical climate region of Brazil. *Sci. Total Environ.* **2019**, *658*, 901–911. [CrossRef]
16. Yeasmin, S.; Singh, B.; Smernik, R.J.; Johnston, C.T. Effect of land use on organic matter composition in density fractions of contrasting soils: A comparative study using ^{13}C NMR and DRIFT spectroscopy. *Sci. Total Environ.* **2020**, *726*, 138395. [CrossRef]
17. Mathers, N.J.; Jalota, R.K.; Dalal, R.C.; Boyd, S.E. ^{13}C-NMR analysis of decomposing litter and fine roots in the semi-arid Mulga Lands of southern Queensland. *Soil Biol. Biochem.* **2007**, *39*, 993–1006. [CrossRef]

18. Lorenz, K.; Lal, R.; Preston, C.M.; Nierop, K.G.J. Strengthening the soil organic carbon pool by increasing contributions from recalcitrant aliphatic bio(macro)molecules. *Geoderma* **2007**, *142*, 1–10. [CrossRef]

19. Ono, K.; Hirai, K.; Morita, S.; Ohse, K.; Hiradate, S. Organic carbon accumulation processes on a forest floor during an early humification stage in a temperate deciduous forest in Japan: Evaluations of chemical compositional changes by ^{13}C NMR and their decomposition rates from litterbag experiment. *Geoderma* **2009**, *151*, 351–356. [CrossRef]

20. Ji, H.; Han, J.; Xue, J.; Hatten, J.A.; Wang, M.; Guo, Y.; Li, P. Soil organic carbon pool and chemical composition under different types of land use in wetland: Implication for carbon sequestration in wetlands. *Sci. Total Environ.* **2020**, *716*, 136996. [CrossRef]

21. de Wit, H.A.; Ledesma, J.L.J.; Futter, M.N. Aquatic DOC export from subarctic Atlantic blanket bog in Norway is controlled by seasalt deposition, temperature and precipitation. *Biogeochemistry* **2016**, *127*, 305–321. [CrossRef]

22. Zhang, Y.; Yang, J.P.; Yang, W.Q.; Tan, B.; Fu, C.K.; Wu, F.Z. Climate, plant organs and species control dissolved nitrogen and phosphorus in fresh litter in a subalpine forest on the eastern Tibetan Plateau. *Ann. Sci.* **2018**, *75*, 51. [CrossRef]

23. Ni, X.Y.; Yang, W.Q.; Tan, B.; He, J.; Xu, L.Y.; Li, H.; Wu, F.Z. Accelerated foliar litter humification in forest gaps: Dual feedbacks of carbon sequestration during winter and the growing season in an alpine forest. *Geoderma* **2015**, *241–242*, 136–144. [CrossRef]

24. Prescott, C.E.; Maynard, D.G.; Laiho, R. Humus in northern forests: Friend or foe? *Ecol. Manag.* **2000**, *133*, 23–36. [CrossRef]

25. Ono, K.; Hiradate, S.; Morita, S.; Hirai, K. Fate of organic carbon during decomposition of different litter types in Japan. *Biogeochemistry* **2013**, *112*, 7–21. [CrossRef]

26. De Marco, A.; Spaccini, R.; Vittozzi, P.; Esposito, F.; Berg, B.; Virzo De Santo, A. Decomposition of black locust and black pine leaf litter in two coeval forest stands on Mount Vesuvius and dynamics of organic components assessed through proximate analysis and NMR spectroscopy. *Soil Biol. Biochem.* **2012**, *51*, 1–15. [CrossRef]

27. Berg, B. Decomposition patterns for foliar litter—A theory for influencing factors. *Soil Biol. Biochem.* **2014**, *78*, 222–232. [CrossRef]

28. Chapin, F.S. Effects of plant traits on ecosystem and regional processes: A conceptual framework for predicting the consequences of global change. *Ann. Bot.* **2003**, *91*, 455–463. [CrossRef]

29. Yang, J.; Zhang, Y.; Fu, C.; Liang, Z.; Yue, K.; Xu, Z.; Ni, X.; Wu, F. Seasonal dynamics of organic components in fresh foliar litters at different gap positions in an alpine forest on the eastern Tibetan Plateau. *J. Soils Sediments* **2021**, *21*, 810–820. [CrossRef]

30. Khan, M.U.; Ahring, B.K. Lignin degradation under anaerobic digestion: Influence of lignin modifications—A review. *Biomass Bioenergy* **2019**, *128*, 105325. [CrossRef]

31. Ono, K.; Hiradate, S.; Morita, S.; Ohse, K.; Hirai, K. Humification processes of needle litters on forest floors in Japanese cedar (*Cryptomeria japonica*) and Hinoki cypress (*Chamaecyparis obtusa*) plantations in Japan. *Plant Soil* **2010**, *338*, 171–181. [CrossRef]

32. Wu, F.Z.; Yang, W.Q.; Zhang, J.; Deng, R.J. Litter decomposition in two subalpine forests during the freeze–thaw season. *Acta Oecol.* **2010**, *36*, 135–140. [CrossRef]

33. IUSS Working Group WRB. World Reference Base for Soil Resources 2014, update 2015. In *International Soil Classification System for Naming Soils and Creating Legends for Soil Maps*; Schad, P., van Huyssteen, C., Micheli, E., Eds.; World Soil Resources Reports No. 106; FAO: Rome, Italy, 2015; 192p.

34. Hilli, S.; Stark, S.; Derome, J. Carbon quality and stocks in organic horizons in boreal forest soils. *Ecosystems* **2008**, *11*, 270–282. [CrossRef]

35. Carrasco, B.; Cabaneiro, A.; Fernandez, I. Exploring potential pine litter biodegradability as a natural tool for low-carbon forestry. *Ecol. Manag.* **2017**, *401*, 166–176. [CrossRef]

36. Bonanomi, G.; Incerti, G.; Giannino, F.; Mingo, A.; Lanzotti, V.; Mazzoleni, S. Litter quality assessed by solid state ^{13}C NMR spectroscopy predicts decay rate better than C/N and Lignin/N ratios. *Soil Biol. Biochem.* **2013**, *56*, 40–48. [CrossRef]

37. He, Y.; Chen, C.; Xu, Z.; Williams, D.; Xu, J. Assessing management impacts on soil organic matter quality in subtropical Australian forests using physical and chemical fractionation as well as ^{13}C NMR spectroscopy. *Soil Biol. Biochem.* **2009**, *41*, 640–650. [CrossRef]

38. Zhu, J.X.; He, X.H.; Wu, F.Z.; Yang, W.Q.; Tan, B. Decomposition of abies faxoniana litter varies with freeze–thaw stages and altitudes in subalpine/alpine forests of southwest China. *Scand. J. For. Res.* **2012**, *27*, 586–596. [CrossRef]

39. Wang, D.; Abdullah, K.M.; Xu, Z.; Wang, W. Water extractable organic C and total N: The most sensitive indicator of soil labile C and N pools in response to the prescribed burning in a suburban natural forest of subtropical Australia. *Geoderma* **2020**, *377*, 114586. [CrossRef]

40. Peng, Y.; Yang, W.Q.; Yue, K.; Tan, B.; Huang, C.P.; Xu, Z.F.; Ni, X.Y.; Zhang, L.; Wu, F.Z. Temporal dynamics of phosphorus during aquatic and terrestrial litter decomposition in an alpine forest. *Sci. Total Environ.* **2018**, *642*, 832–841. [CrossRef]

41. Ni, X.Y.; Berg, B.; Yang, W.Q.; Li, H.; Liao, S.; Tan, B.; Yue, K.; Xu, Z.F.; Zhang, L.; Wu, F.Z. Formation of forest gaps accelerates C, N and P release from foliar litter during 4 years of decomposition in an alpine forest. *Biogeochemistry* **2018**, *139*, 321–335. [CrossRef]

42. Baldock, J.A.; Masiello, C.A.; Gélinas, Y.; Hedges, J.I. Cycling and composition of organic matter in terrestrial and marine ecosystems. *Mar. Chem.* **2004**, *92*, 39–64. [CrossRef]

43. McKee, G.A.; Soong, J.L.; Caldéron, F.; Borch, T.; Cotrufo, M.F. An integrated spectroscopic and wet chemical approach to investigate grass litter decomposition chemistry. *Biogeochemistry* **2016**, *128*, 107–123. [CrossRef]

44. Schädel, C.; Blöchl, A.; Richter, A.; Hoch, G. Quantification and monosaccharide composition of hemicelluloses from different plant functional types. *Plant Physiol. Biochem.* **2010**, *48*, 1–8. [CrossRef] [PubMed]

45. Wang, H.; Liu, S.; Wang, J.; Shi, Z.; Lu, L.; Guo, W.; Jia, H.; Cai, D. Dynamics and speciation of organic carbon during decomposition of leaf litter and fine roots in four subtropical plantations of China. *Ecol. Manag.* **2013**, *300*, 43–52. [CrossRef]
46. Ussiri, D.A.N.; Johnson, C.E. Characterization of organic matter in a northern hardwood forest soil by ^{13}C NMR spectroscopy and chemical methods. *Geoderma* **2003**, *111*, 123–149. [CrossRef]
47. Yue, K.; Peng, C.; Yang, W.; Peng, Y.; Zhang, C.; Huang, C.; Wu, F. Degradation of lignin and cellulose during foliar litter decomposition in an alpine forest river. *Ecosphere* **2016**, *7*, e01523. [CrossRef]
48. Maisto, G.; De Marco, A.; Meola, A.; Sessa, L.; Virzo De Santo, A. Nutrient dynamics in litter mixtures of four Mediterranean maquis species decomposing in situ. *Soil Biol. Biochem.* **2011**, *43*, 520–530. [CrossRef]
49. Hopkins, D.W.; Sparrow, A.D.; Elberling, B.; Gregorich, E.G.; Novis, P.M.; Greenfield, L.G.; Tilston, E.L. Carbon, nitrogen and temperature controls on microbial activity in soils from an Antarctic dry valley. *Soil Biol. Biochem.* **2006**, *38*, 3130–3140. [CrossRef]
50. Saccone, P.; Morin, S.; Baptist, F.; Bonneville, J.-M.; Colace, M.-P.; Domine, F.; Faure, M.; Geremia, R.; Lochet, J.; Poly, F.; et al. The effects of snowpack properties and plant strategies on litter decomposition during winter in subalpine meadows. *Plant Soil* **2013**, *363*, 215–229. [CrossRef]
51. Yuan, Z.Y.; Chen, H.Y.H. Global trends in senesced-leaf nitrogen and phosphorus. *Glob. Ecol. Biogeogr.* **2009**, *18*, 532–542. [CrossRef]
52. Klotzbücher, T.; Kaiser, K.; Guggenberger, G.; Gatzek, C.; Kalbitz, K. A new conceptual model for the fate of lignin in decomposing plant litter. *Ecology* **2011**, *92*, 1052–1062. [CrossRef] [PubMed]
53. Trum, F.; Titeux, H.; Ponette, Q.; Berg, B. Influence of manganese on decomposition of common beech (*Fagus sylvatica* L.) leaf litter during field incubation. *Biogeochemistry* **2015**, *125*, 349–358. [CrossRef]

Article

Dynamic Change Characteristics of Litter and Nutrient Return in Subtropical Evergreen Broad-Leaved Forest in Different Extreme Weather Disturbance Years in Ailao Mountain, Yunnan Province

Xingyue Liu [1], Ziyuan Wang [1], Xi Liu [1], Zhiyun Lu [2], Dawen Li [2] and Hede Gong [1,*]

1   School of Geography and Ecotourism, Southwest Forestry University, Kunming 650224, China
2   Ailaoshan Station of Subtropical Forest Ecosystem Studies, Xishuangbanna Tropical Botanical Garden, Chinese Academy of Sciences, Jingdong, Kunming 676209, China
*   Correspondence: gonghede3@163.com; Tel.: +86-15-887-221-978

Abstract: By studying the dynamic change characteristics of litter production, composition, nutrient content, and return amount of different components in different extreme weather interference years of Ailao Mountain evergreen broad-leaved forest, the paper provides theoretical support for the post-disaster nutrient cycle, ecological recovery, and sustainable development of the subtropical mid-mountain humid evergreen broad-leaved forest. Square litter collectors were randomly set up to collect litter. After drying to a constant mass, we calculated the seasonal and annual litter volume and the contents of organic carbon (C), total nitrogen (N), total phosphorus (P), total potassium (k), total sulfur (S), total calcium (Ca), and total magnesium (Mg). Finally, the nutrient return amount is comprehensively calculated according to the litter amount and element content. We tracked dynamic changes in litter quantity, nutrient composition, and nutrient components across different years. The results showed that the amount of litter from 2005 to 2015 was 7704–8818 kg·hm^{-2}, and the order of magnitude was: 2005 (normal year) > 2015 (extreme snow and ice weather interference) > 2010 (extreme drought weather interference); the composition mainly included branches, leaves, fruit (flowers), and other components (bark, moss, lichen, etc.), of which the proportion of leaves was the largest, accounting for 41.70%–61.52%; The monthly changes and total amounts in different years exhibited single or double peak changes, and the monthly litter components in different years showed significant seasonality. In this study, the nutrient content of litter was higher than that of litter branches each year. The total amount of litter and the nutrient concentration of each component are C, Ca, N, K, Mg, S, and P, from large to small. The order of nutrient return in different years was the same as that of litter, and the returns of nutrients in litter leaves were greater than that of litter branches. The ratio of nutrient returns of litter and litter branches from 2005 to 2010 was 2.03, 1.23, and 3.69, respectively. The research shows that the litter decreased correspondingly under the extreme weather disturbance, and the impact of the extreme dry weather disturbance was greater than that of the extreme ice and snow weather disturbance. However, the evergreen broad-leaved forest in the study area recovers well after being disturbed. The annual litter amount and nutrient return amount is similar to that of evergreen broad-leaved forests in the same latitude and normal years in other subtropical regions. The decomposition rate and seasonal dynamics of litter nutrients are not greatly affected by extreme weather.

Keywords: litter quantity; nutrient content; return of nutrients; subtropical evergreen broad-leaved forest; extreme weather disturbance

Citation: Liu, X.; Wang, Z.; Liu, X.; Lu, Z.; Li, D.; Gong, H. Dynamic Change Characteristics of Litter and Nutrient Return in Subtropical Evergreen Broad-Leaved Forest in Different Extreme Weather Disturbance Years in Ailao Mountain, Yunnan Province. *Forests* **2022**, *13*, 1660. https://doi.org/10.3390/f13101660

Academic Editors: Fuzhong Wu, Zhenfeng Xu and Wanqin Yang

Received: 25 August 2022
Accepted: 4 October 2022
Published: 10 October 2022

Publisher's Note: MDPI stays neutral with regard to jurisdictional claims in published maps and institutional affiliations.

1. Introduction

Forest litter refers to the general term for all organic matter in the forest ecosystem that is produced by aboveground plants and other biological components and returned to the forest surface as a source of material and energy for decomposers to maintain ecosystem functions, including litter leaves, litter branches, flower, and fruit reproductive organs and debris [1,2]. Leaf litter is considered to be an important survival strategy for plants to cope with adverse growth conditions, such as soil drying due to temperature reduction or drought [3,4]. Forest litter is an important structural and functional unit of material circulation and energy flow in the forest ecosystem. Its litter, accumulation, and decomposition are basic ecosystem processes. Important functions have irreplaceable ecological roles, and a large amount of organic matter and mineral elements are transported from the canopy of plants to the soil surface through the litter. Therefore, the collection and measurement of forest litter are important means of studying the structure and function of forest ecosystems [5–8]. To a certain extent, litter yield and nutrient return are suitable indicators of the overall function of forest ecosystems and play a key role in forest ecosystem dynamics, nutrient cycling, and forest productivity [9–12].

The forest types in subtropical regions are important forest ecosystems unique to the same latitude in the world. They have high primary productivity and are biodiversity hotspots, which play an important role in carbon storage in global terrestrial ecosystems [13]. The montane moist evergreen broad-leaved forest in the Ailao Mountain Nature Reserve in Yunnan is currently the largest and most well-preserved subtropical evergreen broad-leaved forest in my country. It is one of the valuable zonal vegetation. It is of great significance to study the dynamic change law of yield, nutrient cycle law and composition characteristics of the litter of the evergreen broad-leaved forest in Ailao Mountain to understand the nutrient availability and productivity of the forest system [14–17].

Forest litter has a distinct seasonal pattern, mainly in a unimodal, bimodal or irregular pattern [18–21]. The temporal variability of forest litter is an important source of uncertainty in the forest carbon cycle, and biological factors such as forest type, origin, forest age, tree species richness, phenological rhythm and genetic characteristics are important factors affecting the seasonal pattern of litter [22–25]. Different forest communities in different climatic zones around the world are not the same. According to research, the annual litter of the main forest types in each climatic zone can be specifically expressed as rainforest > evergreen broad-leaved forest > mixed coniferous and broad-leaved forest > deciduous broad-leaved forest > coniferous forest [26–28]. Although in the past few decades, litter yield, structure and composition, decomposition rate, and its influencing factors have been extensively studied around the world, the existing dynamic changes of litter are mostly related to litter yield, nutrient The results of a single study on the return of elements or nutrients for one year are mostly planted forests, and there are relatively few studies on primitive natural forests [29–31]. In addition, the research on the evergreen broad-leaved forest in this region mainly includes the vegetation type and diversity research, and the multi-year comparative comprehensive research combining the dynamic law of litter yield and nutrient return is rarely reported [7,10,32–36].

Forest litter is an important carbon pool for forest ecosystems. Its nutrient cycle and nutrient balance are strongly affected by climate, and play a complex source-sink effect in the process of global change. It is one of the basic parameters of carbon exchange with the atmosphere and is closely related to global change and the circulation of materials in the global ecosystem [37]. Climatic factors (including temperature, precipitation, long-term extreme weather, etc.) are important factors affecting forest productivity, in addition, trees take a long time to regenerate and increase biomass, so forest ecosystems are sensitive to extreme weather disturbances. In recent years, forest litter has begun to be studied in the context of the global environment, focusing on its important role in carbon and nutrient cycling [38,39]. Climate change will not only directly affect the community structure and vegetation composition of forest ecosystems by changing climatic factors such as temperature and precipitation, but also indirectly affect the primary productivity level of forest ecosystems and the nutrients of forest ecosystems by changing the area and intensity of natural disturbances. The distribution and material circulation process have an impact, which in turn affects the amount, composition and dynamic changes of litter in forest ecosystems [40–42]. Specifically, with global warming since the 20th century, the scope, frequency, and intensity of extreme weather events around the world have increased significantly. Compared with conventional weather change, extreme weather will have a greater impact on forest ecosystems, increasing the The survival pressure of forest communities and the potential risk of local extinction have caused a series of crises to the security of forest ecosystems [43]. In particular, extreme weather disasters such as drought, ice and snow have become one of the most important climate disasters due to their high frequency and large scope, and they have always been widely concerned [44,45].

From a geographical point of view, Yunnan, as one of the areas with the most frequent drought disasters in Southwest China, has attracted extensive attention of many scholars. However, most of the current related researches focus on the analysis of the spatial and temporal distribution characteristics and causes of drought and flood disasters in Yunnan. The research results on the combination of litter dynamics and drought in typical subtropical evergreen broad-leaved forests have not yet been reported [46,47]. In 2010, parts of Yunnan suffered continuous severe drought, which was the most severe drought event ever recorded in Southwest China, which caused serious damage to the forest ecosystem in the region [48,49]. Subtropical evergreen broad-leaved forests have become the main victims of ice and snow disasters due to their evergreen and relatively wide canopies [50], and mechanical damage is particularly serious. Therefore, the disturbance of extreme freezing rain and snow weather disturbances to subtropical forest ecosystems is also very serious [51,52]. As far as ice and snow disasters are concerned, because they mainly occur in Europe and eastern North America, the research on the impact of snow and ice disasters on forest vegetation is also mainly concentrated in this region, resulting in the response model of subtropical forests to snow and ice disasters is still not completely clear [53–55]. However, in January 2015, Ailao Mountain encountered a catastrophic ice and snow weather disturbance, resulting in a certain degree of fragmentation of the evergreen broad-leaved forest in the area, and overall changes in the woodland habitat and forest structure. The forest ecosystem and diversity have caused serious damage [56]. Once the dynamics of litter changes, the material cycle of the forest ecosystem will be affected, and the productivity and service functions of the forest will change [42,57]. Therefore, it is very important to study the dynamics of litter in the study area. In addition, human beings and forest ecosystems are closely related, and these changes will inevitably have a direct impact on human beings. How to promote the flow of material circulation and energy in post-disaster ecosystems, strengthen the protection of biodiversity in damaged forest ecosystems, and restore ecological functions is an urgent problem to be solved. However, the solution to these problems is inseparable from the research on the changes of litter and its nutrient cycle after disasters [37,58].

However, the current research on whether extreme drought will change, how it affects, and the degree of impact on the litter dynamics of subtropical forest ecosystems and the development process of forest ecosystems is relatively scarce, and the comparison of the degree of damage to the same forest ecosystem caused by different extreme weather disturbances not clear [37,42]. To this end, this study selects three special years with equal intervals of growth in the virgin forest of Ailao Mountain National Nature Reserve, including 2005 (a normal year not disturbed by extreme weather) and 2010 (a year disturbed by extreme arid weather), 2015 (the year disturbed by extreme ice and snow weather), the amount of litter in different parts and the nutrient element content and return amount of each component were measured to explore the components and total amount of litter in normal years and before and after being disturbed by different extreme weathers. The interannual variation, monthly dynamic variation, seasonal dynamic variation characteristics of litter volume and its significant differences with each component and nutrient element content and return amount. On the basis of a comprehensive analysis of the impact of extreme weather change on litter dynamics, this paper attempts to summarize the variation laws and nutrient cycling laws of litter under different extreme weather disturbances, in order to enrich the ecological structure, ecological structure, and nutrient cycle of the montane humid evergreen broad-leaved virgin forest. The research on the function provides a theoretical basis for the material cycle, nutrient balance, and ecological restoration of the disturbed forest ecosystem. This paper discusses how to better apply the ecological function of litter to forest ecosystem management, so as to improve the management level of forest ecosystem and give full play to the self-sustaining mechanism of forest ecosystem.

According to previous studies, we know that extreme weather interference can form a large number of abnormal litters, which refers to the fresh residues and litters of individual plants or plant organs caused by external forces under extreme weather, fire, or geological disasters, such as a large number of fallen trees, twigs and litter leaves on the forest land due to the impact of natural disasters such as low temperature, snow, freezing, or typhoon, to affect the amount of litter and its nutrient content [44–47,59]. Based on this, we propose the following main research questions: (1) Can extreme drought and ice and snow weather disturbance affect litter yield and nutrient return? (2) Will extreme arid weather disturbance hinder or destroy the normal succession process, seasonal dynamic changes, and decomposition rate of forest ecosystems? (3) If there is an impact, is the impact of different extreme weather disturbances on the components of litter consistent?

2. Materials and Methods

2.1. Overview of the Study Area

The study area (Figure 1) is located in the Xujiaba area (24°32′ N, 101°01′ E) in the core area of Ailao Mountain National Nature Reserve, with an altitude of 2400–2600 m, and the soil is fertile acid yellow-brown soil [14]. According to the long-term monitoring data of the Ailao Mountain Forest Ecosystem Research Station, the average annual temperature in this area is 11.3 °C, the average annual precipitation is 1931 mm, and the average annual evaporation is 1192 mm. The climate belongs to the southwest monsoon region, with distinct dry and rainy seasons. The precipitation in the rainy season (May to October) accounts for about 85% of the annual precipitation, and the average annual relative humidity is 85% [15,16]. The zonal vegetation is the mid-mountain humid evergreen broad-leaved forest, the canopy is highly closed, the community tree, shrub, and grass layers are clearly layered, the interlayer plants are common with woody vines, and epiphytes are abundant, mainly mosses and ferns. The dominant species are *Castanopsis rufescens* Hook. f. et Thoms., *Schima noronhae* Reinw. ex Bl. Bijdr., *Lithocarpus xylocarpus* (Kurz) Markgraf, *Camellia forrestii* (Diels) Coh. St., *Machilus bombycina* King ex J. D. Hooker and other large trees, as well as shrubs such as *Sinarundinaria nitida* Franch. widely distributed in the forest and Plagiogyria communis and other herbs.

2.2. Sample Plot Setting and Litter Collection

The evergreen broad-leaved forest community in this area was selected, and a square fixed plot with a plot area of 1 hm^2 (100 m × 100 m) was established, and the established fixed plot was divided into 100 small plots of 10 m × 10 m. A total of 25 of them were randomly selected for long-term litter observation. A box-type litter collector with an area of 1 m^2 was placed horizontally on the forest floor or at a certain distance from the ground. The bottom of the collector was 0.5 m from the ground, and nylon mesh screens were used, and the surrounding areas were fixed with PVC pipes. Then, all the fresh litter in the collector is collected at the end of every five years at equal intervals, that is, at the end of each month in 2005, 2010, and 2015. Stones and other sundries are numbered and placed in sterile plastic bags, and all litter is used to record the amount of litter recovered [2,60,61]. We separated the collected litter by branches, leaves, flowers, and fruits (reproductive organs), other components (bark, moss, lichen, debris), and cleaned up the attached impurities. Then the litter were dried to constant weight in the oven at 65 °C and weighed and recorded for calculating the litter amount of each component of the litter [62,63]. The litter amount of each litter component was converted from the average value of the corresponding components in 25 collection frames. The monthly litter amount was the sum of the litter components of the month, and the annual litter amount is the sum of the litter amount of 12 months [25,33,64–66].

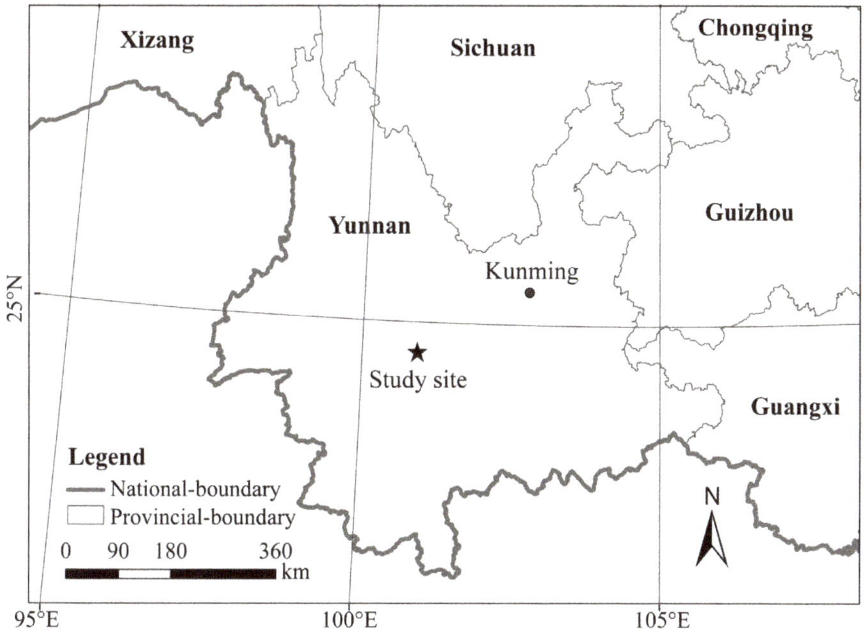

Figure 1. Study Area Map.

2.3. Determination of Litter Nutrients

The dried and weighed litter analysis samples were ground with a plant crusher and sieved through a 60 mesh to determine the nutrient contents. Carbon (C) and nitrogen (N) were measured with a carbon element analyzer (EA3000, EuroVector, Milan, Italy) [33–35]. For the determination of other elements, the sample was digested by the H_2O_2–H_2SO_4 digestion method and then prepared into the solution to be measured. The content of total phosphorus (P) is determined by the Mo-Sb colorimetry method, and the content of total potassium (K), total sulfur (S), total calcium (Ca), and total magnesium (Mg) were determined by the flame photometer and spectrophotometer [6,36]. The amount of nutrients returned by litter is closely related to the amount of litter and nutrient content in the litter.

2.4. Calculation of Nutrient Return of Litter

The amount of nutrient return of litter is closely related to the amount of litter and the nutrient content in the litter. The amount of nutrient return is equal to the product of the amount of litter and the nutrient content in the litter. The specific calculation is that the monthly nutrient return amount of each component of the litter is the product of the nutrient content of the component of the litter in the current month and the litter amount of the component in the current month. The total annual nutrient return amount of the litter is the total return amount of each component of the litter in 12 months [25,33,37,38].

2.5. Data Analysis

After using the Shapiro–Wilk test to test the normality of the data, one-way ANOVA and LSD were used to compare the difference in the amount of litter in different parts of different years and its components, the content of nutrient elements, and the amount of return. For seasonal dynamic changes, the coefficient of variation was used to characterize the annual variation of litter volume, which was obtained by dividing the standard error of litter volume in different years by its mean. The annual variation of litter was obtained by dividing the difference between the maximum monthly litter volume and the minimum monthly litter volume by the monthly average litter volume during the observation period. All data statistical analysis and chart production were completed on Excel 2010 and SPSS 25.

3. Results

3.1. Annual Changes in Litter and Composition in Different Years

3.1.1. Annual Variation in Litter Components and Total Amounts in Different Years

It can be seen from Table 1 that the total amount of litter from 2005 to 2015 was between 7704.15 and 8817.50 $kg \cdot hm^{-2} \cdot a^{-1}$ and the order of magnitude was 2005 > 2015 > 2010. The litter amount fluctuated from year to year, and the coefficient of variation was 7.03%. The composition of litter mainly includes litter branches, litter leaves, litter fruits (flowers), and other litter components (bark, moss lichens, debris, etc.), among which litter branches and other components are significantly different from litter leaves and litter fruits (flowers) ($p < 0.05$), but the difference between litter branches and other components is not significant ($p > 0.05$). Each component of litter and its proportion to the total forest litter in different years are as follows: litter leaves > litter branches> other components> litter fruits (flowers), and the total average percentage is: 50.89% > 25.28% > 15.86% > 7.97%. Litter leaves are the absolute dominant component of the litter in the evergreen broad-leaved forest community in Ailao Mountain, accounting for 41.70%–61.52% of the total litter in different years, the highest in 2010 and the lowest in 2015. Litter branches, litter fruits (flowers), and other components accounted for 17.26%–32.05%, 6.29%–9.37%, and 14.57%–16.88% of the total forest litter from 2005 to 2015, respectively. It accounted for the highest percentage of total forest litter, followed by 2005 and 2010.

Table 1. Annual variation of litter components in different years (kg/hm^2).

Year	Litterfall				
	Litter Branch	**Litter Leaf**	**Litter Fruit (Flower)**	**Other Litter**	**Litter Total**
2005	2276.36 ± 260.55 [A] (25.82%)	4404.97 ± 170.26 [A] (49.96%)	719.90 ± 27.88 [A] (8.16%)	1416.27 ± 60.31 (16.06%)	8817.50 ± 319.67 [A] (100%)
2010	1357.74 ± 83.05 [B] (17.26%)	4739.35 ± 268.05 [A] (61.52%)	484.72 ± 27.16 [B] (6.29%)	1122.34 ± 67.98 [C] (14.57%)	7704.15 ± 400.3 [B] (100%)
2015	2567.25 ± 465.77 [A] (32.05%)	3340.59 ± 142.72 [B] (41.70%)	751.02 ± 37.73 [A] (9.37%)	1352.02 ± 88.71 [B] (16.88%)	8010.87 ± 643.65 [B] (100%)
Total average	2067.12 ± 631.32 [Ac] (25.28%)	4161.64 ± 730.44 [Ab] (50.89%)	651.88 ± 145.60 [Ae] (7.97%)	1296.88 ± 154.53 [Bd] (15.86%)	8177.51 ± 575.08 [Ba] (100%)

Note: the percentage of litter amount of each component in the total litter amount of the whole year is shown in brackets. Different capital letters in the same column indicate significant differences between different years, and different lowercase letters in the same row indicate significant differences between different litter components in the same year ($p < 0.05$).

3.1.2. Monthly Dynamics of Litter Components in Different Years

It can be seen from Figure 2 that in 2005, the total amount of litter and each component exhibited an irregular pattern, with multiple peaks, in which the total amount of litter showed a four-peak pattern with peaks in February, April, August, and November (727.71, 943.44, 1647.86, and 764.36 kg/hm^2, respectively). The total amount of litter in August was significantly higher than in the other months. Leaves showed a bimodal pattern, with the first peak in April at 693.59 kg/hm^2 and the second peak in November at 499.74 kg/hm^2. The number of litter branches also significantly increased in February (310.23 kg/hm^2) and August (982.01 kg/hm^2), whereas the monthly variations in fruits (flowers) and other components were relatively flat. In 2010, the total amount of litter and the amount of litter leaf showed a double peak curve, one in April and the other in November. The difference between the two peaks was obvious, but they both appeared in the dry season, among which the peak of the litter was in April. The span of the twigs was larger, and the amount of litter was higher, which was significantly higher than that in November, and the litter branches and other components also showed obvious unimodal monthly variation characteristics, which were all significantly increased in April. In 2015, the highest value of the total amount of litter and the amount of each component was concentrated in January, and then the number dropped sharply and then slightly rebounded in March and September. Overall, the litter in 2015 showed a single peak pattern throughout the year. The amount of litter in January was higher than the annual average, and the peaks were mainly concentrated in the dry season rather than the rainy season. In 2005, 2010, and 2015, the annual changes in the total amount of litter in the study area were 161.50%, 209.12%, and 359.37%, respectively, of which 2015 had the largest change, indicating that the intermonthly variation of the litter amount was more severe, while in 2005 The smallest variation range indicated that the monthly variation of litter in this year was relatively gentle.

3.1.3. Seasonal Dynamics of Litter Fractions in Different Years

Seasons in this experiment were divided as follows: March–May for spring, June–August for summer, September–November for autumn, and December–February for winter. The monthly litter of different components in different years exhibited significantly different seasonal variation (Figure 3). The total amount of litter in different years was significantly higher than that of other components. In 2005, the components with the highest amount of litter in the other three seasons were leaves, except in summer, which was dominated by branches. The distribution of branch components across seasons peaked in summer. Leaves represented the highest proportion in spring, which then fluctuated and decreased with the passage of summer, autumn, and winter; fruit drop (flower) slowly rose from spring to winter. Other components rose from spring to summer, and then decreased gradually after reaching a high point in summer, with small fluctuations. In 2010, litter was composed mostly of leaves and least by fruit (flower). The seasonal variation in fruit (flower) was lowest in spring. After reaching a peak in summer, fruit (flower) remained stable until autumn, decreased from autumn to winter, and reached the lowest value in winter. The seasonal variation in other components was similar, showing a peak in spring, followed by a sharp decline. In 2015, the maximum value of litter component in winter was branches, and the component with the highest amount of litter in the other three seasons was leaves. The seasonal variation in each component was highest in spring, then decreased reaching the lowest values in summer, and rose thereafter.

3.2. Dynamic Characteristics of Litter Nutrient Contents in Different Years

3.2.1. Annual Average Nutrient Contents of Litter Components in Different Years

The nutrient content of all years was higher in litter than in branches, while the concentration of different nutrients was C > Ca > N > K > Mg > S > P (Table 2). The C content was not significantly different ($p > 0.05$) between branches and leaves in all years except for 2010. The overall distribution of C in the two organs was relatively average, and the distribution of C content in the leaves was only slightly higher than that in the branches. Branches and leaves showed significant differences ($p < 0.05$) between 2005, 2010, and 2015 in N and K contents, whereas P, S, Ca, and Mg contents were not significantly different ($p > 0.05$) between the years. C, Ca, and Mg contents in branches were the same in different years, with the highest in 2005 and the lowest in 2015. The trends in N and S content were also 2010 > 2005 > 2010, and the P and K contents showed the highest levels in 2015 (0.48 and 0.87 g/kg). The P content and K content were the lowest in 2005 and 2010, respectively. In leaves, the annual average contents of N, P, and S increased with the year, while the content of C, K, Ca, and Mg is 2005 > 2010 > 2015.

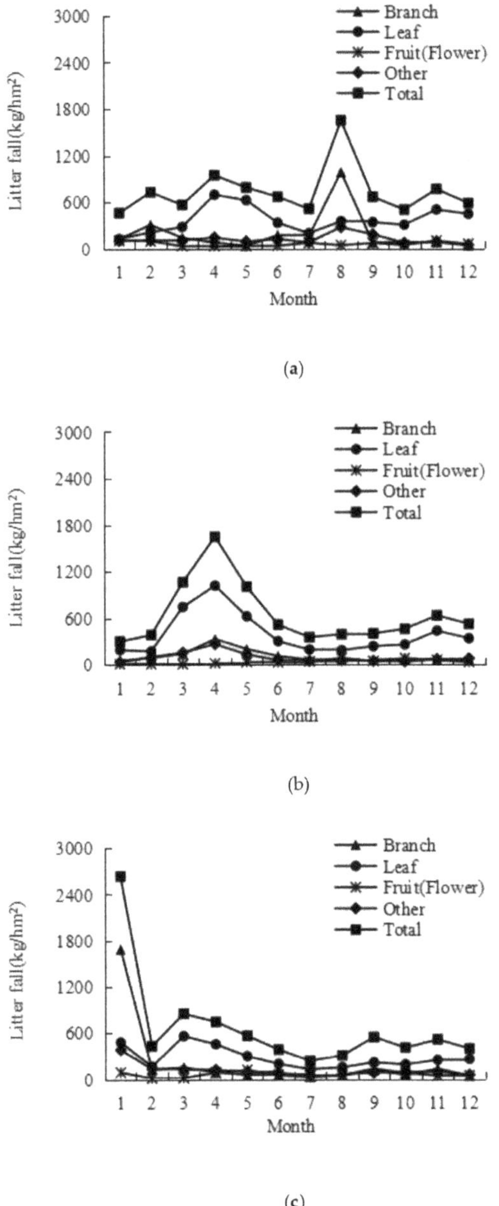

(a)

(b)

(c)

Figure 2. Monthly dynamics of litter components and totals in 2005, 2010, and 2015. (**a**) Monthly dynamics of litter components and totals in 2005; (**b**) monthly dynamics of litter components and totals in 2010; (**c**) monthly dynamics of litter components and totals in 2015.

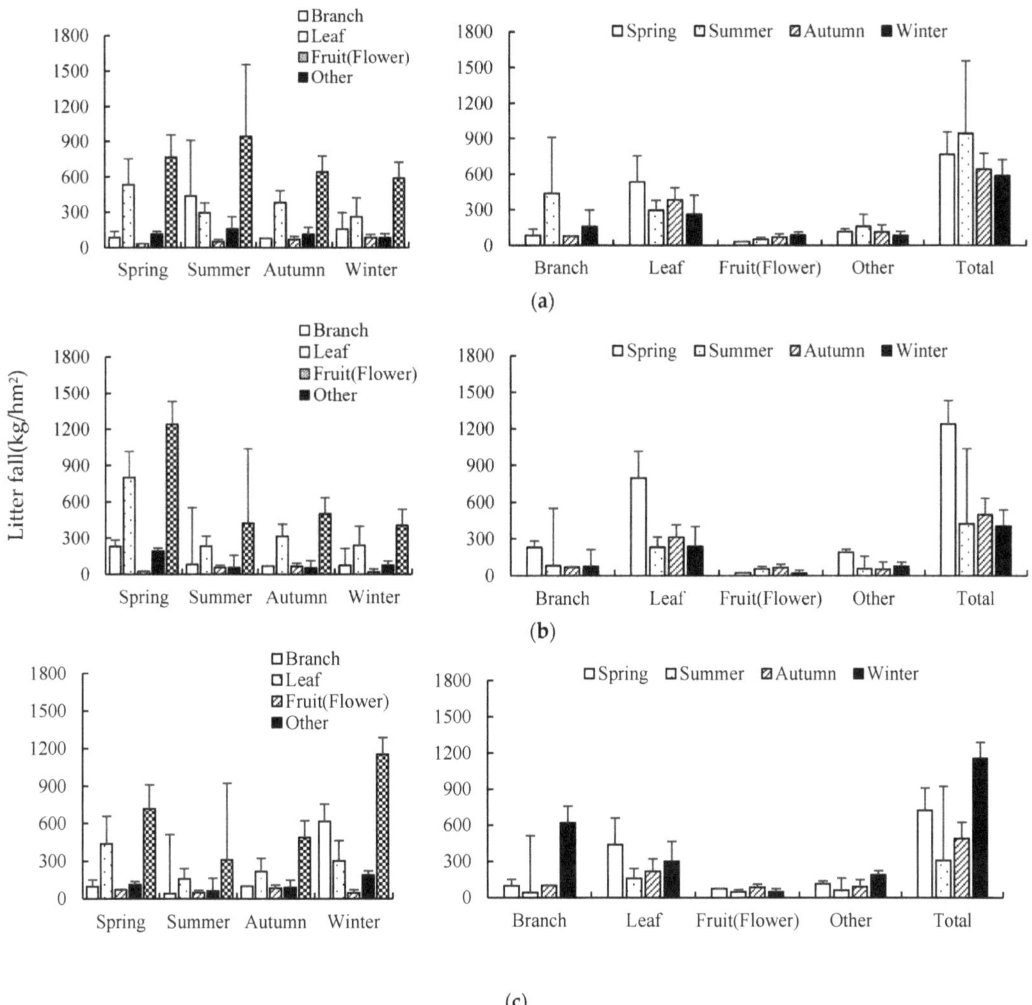

Figure 3. Seasonal dynamics of litter in 2005, 2010, and 2015. (**a**) Seasonal dynamics of litter in 2005; (**b**) seasonal dynamics of litter in 2005; (**c**) seasonal dynamics of litter in 2005.

The values of C/N, C/P and N/P of litter in 2005, 2010, and 2015 were 57.66, 658.05, 19.02; 47.88, 998.59, 20.86; 55.65, 869.55, 15.63, respectively. The ratios of C/P and N/P were 42.27, 772.86, 18.29; 36.73, 685.47, 18.66; 33.78, 573.80, 16.99, respectively. The mean ratios of C/N, C/P and N/P of litter were 48.65, 904.69, 18.59; 41.45, 803.68, 18.66; 41.89, 684.83, 16.35, respectively.

Table 2. Annual average nutrient content of litter components in different years (g/kg).

Nutrient	2005			2010			2015		
	Litterfall			Litterfall			Litterfall		
	Litter Branches	Litter Leaves	Average	Litter Branches	Litter Leaves	Average	Litter Branches	Litter Leaves	Average
C	526.44 ± 22.42 aA	541.10 ± 21.06 aA	533.77 ± 10.37 aA	489.31 ± 10.80 aB	507.25 ± 10.92 aA	498.28 ± 12.69 aB	486.95 ± 9.62 aB	499.21 ± 11.80 aB	493.08 ± 8.67 aB
N	9.13 ± 1.06 bB	12.80 ± 2.39 bA	10.97 ± 2.60 bA	10.22 ± 0.87 bB	13.81 ± 2.55 bA	12.02 ± 2.54 bA	8.75 ± 1.92 bB	14.78 ± 2.07 bA	11.77 ± 4.26 bA
P	0.48 ± 0.06 dA	0.70 ± 0.17 dA	0.59 ± 0.16 dA	0.49 ± 0.02 dA	0.74 ± 0.15 dA	0.62 ± 0.18 dA	0.56 ± 0.19 dA	0.87 ± 0.17 dA	0.72 ± 0.22 dA
K	1.99 ± 0.60 cB	4.79 ± 1.70 cA	3.39 ± 1.98 cA	1.80 ± 0.05 cB	4.55 ± 0.73 cA	3.18 ± 1.94 cA	2.82 ± 1.79 cB	5.46 ± 0.59 cA	4.14 ± 1.87 cA
S	0.87 ± 0.13 dA	1.20 ± 0.19 dA	1.04 ± 0.23 dA	0.92 ± 0.05 dA	1.25 ± 0.09 dA	1.09 ± 0.23 dA	0.76 ± 0.16 dA	1.26 ± 0.15 dA	1.01 ± 0.35 dA
Ca	15.19 ± 1.87 bA	13.25 ± 0.99 bA	14.22 ± 1.37 bA	14.38 ± 0.95 bA	12.18 ± 0.61 bA	13.28 ± 1.56 bA	10.77 ± 2.78 bA	10.30 ± 2.33 bA	10.54 ± 0.33 bB
Mg	1.55 ± 0.20 cA	2.48 ± 0.43 cA	2.02 ± 0.66 cA	1.54 ± 0.06 cA	2.47 ± 0.25 cA	2.01 ± 0.66 cA	1.37 ± 0.24 cA	2.41 ± 0.25 cA	1.89 ± 0.74 cA

Note: the percentage of litter amount of each component in the total litter amount of the whole year is shown in brackets. Different capital letters in the same column indicate significant differences between different years, and different lowercase letters in the same row indicate significant differences between different litter components in the same year ($p < 0.05$).

3.2.2. Seasonal Dynamics of the Nutrient Contents of Each Component of Litter in Different Years

It can be seen from Table 3 that the order of nutrient content of each component in each season is C > Ca > N > K > Mg > S > P. Therefore, C, Ca, N, K and Mg are the major elements of the evergreen broad-leaved forest, and S and P are the trace elements. The seasonal dynamic change characteristics of nutrient content of each component of litter in different years are quite different, and the seasonal change regularity of the content of each element is uncertain. In 2005, the seasonal variation of C nutrient content was spring > winter > summer > autumn, while the seasonal sequence of N content was completely opposite; P. The contents of K, Mg, S, and Ca are higher in winter and lower in summer. In 2010, the seasonal variation of C and Mg nutrient content was the same, which was summer > spring > autumn > winter; N. The nutrient contents of P, K, and S were the highest in summer and the lowest in spring. In 2015, C and Mg nutrient contents were the highest in spring and the lowest in summer; N. On the contrary, the contents of P, Ca, and other macroelements were the highest in summer and the lowest in spring; K content sequence is consistent with that in 2005; S content is similar to 2010.

Table 3. Seasonal dynamics of nutrient content of branches and leaves in different years (g/kg).

Year	Components	Season	C	N	P	K	S	Ca	Mg
2005	Litter branches	Spring	547.44	9.30	0.49	2.53	0.87	15.34	1.58
		Summer	517.44	8.48	0.42	1.46	0.75	13.31	1.25
		Autumn	506.06	10.28	0.55	2.03	1.04	16.80	1.69
		Winter	534.83	8.48	0.46	1.93	0.82	15.32	1.67
	Litter leaves	Spring	557.11	11.29	0.60	5.98	1.14	13.74	2.75
		Summer	521.50	13.96	0.74	3.50	1.20	13.29	2.04
		Autumn	531.17	12.74	0.65	4.35	1.17	13.24	2.49
		Winter	554.06	12.79	0.78	5.58	1.25	12.82	2.65
	Average	Spring	552.28	10.30	0.54	4.26	1.01	14.54	2.16
		Summer	519.47	11.22	0.58	2.48	0.98	13.30	1.64
		Autumn	518.61	11.51	0.60	3.19	1.11	15.02	2.09
		Winter	544.44	10.64	0.62	3.76	1.03	14.07	2.16
2010	Litter branches	Spring	488.75	6.63	0.36	1.68	0.60	10.20	1.38
		Summer	502.50	11.60	0.69	4.36	0.97	12.00	1.84
		Autumn	484.25	7.95	0.46	3.40	0.71	9.70	1.19
		Winter	484.48	9.23	0.68	3.19	0.82	9.68	1.27
	Litter leaves	Spring	506.00	12.51	0.62	5.05	1.06	9.60	2.46
		Summer	520.00	17.92	0.97	4.86	1.42	9.19	2.57
		Autumn	496.50	15.46	0.89	5.62	1.38	11.69	2.62
		Winter	490.50	13.80	0.85	5.55	1.21	11.43	2.36
	Average	Spring	497.38	9.57	0.49	3.36	0.83	9.90	1.92
		Summer	511.25	14.76	0.83	4.61	1.20	10.60	2.20
		Autumn	490.38	11.71	0.68	4.51	1.05	10.70	1.91
		Winter	487.49	11.51	0.77	4.37	1.01	10.55	1.82
2015	Litter branches	Spring	488.75	9.05	0.51	1.81	0.91	13.17	1.55
		Summer	474.25	10.33	0.50	1.73	0.86	14.32	1.48
		Autumn	497.75	11.16	0.50	1.85	0.97	14.58	1.63
		Winter	496.50	10.34	0.47	1.82	0.96	15.47	1.51
	Litter leaves	Spring	520.75	11.42	0.60	5.36	1.27	12.02	2.58
		Summer	494.00	17.22	0.95	4.51	1.37	12.81	2.23
		Autumn	507.25	14.24	0.72	3.59	1.15	12.51	2.30
		Winter	507.00	12.38	0.70	4.73	1.22	11.41	2.78
	Average	Spring	504.75	10.24	0.56	3.58	1.09	12.59	2.06
		Summer	484.13	13.77	0.72	3.12	1.12	13.57	1.85
		Autumn	502.50	12.70	0.61	2.72	1.06	13.54	1.96
		Winter	501.75	11.36	0.59	3.28	1.09	13.44	2.14

3.3. Characteristics of Nutrient Return of Litter in Different Years

3.3.1. Annual Nutrient Return from Litter

The nutrient return in other years generally showed C > Ca > N > K > Mg > S > P (Table 4), and litter was higher than branches, which is consistent with the order of annual average nutrient concentrations of each component. Among branches, the returns of C and Ca were highest in 2005 (1191 kg/hm^2), and then decreased year by year. The return of N, P, K, S, and Mg was highest in 2010, followed by 2005, and lowest in 2015. Among leaves, C, N, P, K, S, and Mg restitution was highest in 2015, and only Ca restitution was different in order of size, with the highest in 2005 and the lowest in 2010.

3.3.2. Seasonal Dynamics of Nutrient Return from Litter in Different Years

As shown in Table 5, In 2005, the maximum value of the nutrient return of each element in the litter branches was concentrated in summer and winter, the minimum value was concentrated in spring and autumn, and the seasonal change order of nutrient return of each element in the litter was opposite, the maximum value was in spring and autumn, and the minimum value was concentrated in summer and winter. In 2010, the maximum value of the nutrient return of each element in the litter branches was spring, followed by summer, and the minimum value fluctuated in winter and autumn. The order of nutrient return of most elements in the litter leaves was consistent with that in 2005. The nutrient return amount of most elements in the litter branches and leaves in 2015 was basically consistent with the seasonal fluctuation law of the litter leaves in 2010.

Table 4. Dynamic characteristics of litter nutrient return in different years (kg/hm^2).

Nutrient	2005			2010			2015		
	Litterfall			Litterfall			Litterfall		
	Branches	Leaves	Total	Branches	Leaves	Total	Branches	Leaves	Total
C	1191.09 ± 261.69 [aB]	2390.65 ± 209.59 [aA]	3581.74 ± 848.22	1127.58 ± 257.77 [aA]	1385.35 ± 615.05 [aA]	2512.93 ± 182.27	663.66 ± 113.75 [aB]	2426.98 ± 430.36 [aA]	3090.64 ± 1246.86
N	19.93 ± 4.18 [cB]	54.92 ± 3.42 [bA]	74.85 ± 24.74	23.17 ± 6.42 [bA]	29.59 ± 11.84 [bA]	52.76 ± 4.54	13.31 ± 1.93 [bB]	61.48 ± 8.20 [bA]	74.79 ± 34.06
P	1.03 ± 0.20 [dA]	2.96 ± 0.16 [dA]	3.99 ± 1.37	1.43 ± 0.38 [dA]	1.81 ± 0.74 [dA]	3.24 ± 0.27	0.68 ± 0.12 [dB]	3.27 ± 0.42 [dA]	3.95 ± 1.83
K	3.95 ± 0.65 [dB]	21.98 ± 2.82 [cA]	25.93 ± 12.75	8.46 ± 2.45 [cA]	10.91 ± 4.25 [cA]	19.37 ± 1.73	2.45 ± 0.42 [cB]	22.70 ± 4.77 [cA]	25.15 ± 14.32
S	1.84 ± 0.36 [dA]	5.19 ± 0.38 [dA]	7.04 ± 2.37	1.98 ± 0.53 [dA]	2.51 ± 1.03 [dA]	4.49 ± 0.38	1.24 ± 0.21 [dA]	5.92 ± 1.03 [dA]	7.17 ± 3.31
Ca	32.59 ± 6.43 [bB]	58.83 ± 5.28 [bA]	91.42 ± 18.56	25.23 ± 6.40 [bA]	31.62 ± 13.32 [bA]	56.85 ± 4.52	18.96 ± 2.89 [bB]	57.54 ± 9.71 [bA]	76.50 ± 27.28
Mg	3.22 ± 0.58 [dB]	11.12 ± 1.16 [cA]	14.34 ± 5.58	3.65 ± 1.01 [cA]	4.66 ± 1.86 [dA]	8.31 ± 0.72	2.09 ± 0.36 [cB]	11.85 ± 2.15 [cA]	13.94 ± 6.90
Total	1253.66 ± 446.40	2545.66 ± 894.11	3799.32	1191.49 ± 422.27	1466.45 ± 518.65	2657.94	702.39 ± 248.50	2589.75 ± 907.37	3292.13

Note: the percentage of litter amount of each component in the total litter amount of the whole year is shown in brackets. Different capital letters in the same column indicate significant differences between different years, and different lowercase letters in the same row indicate significant differences between different litter components in the same year ($p < 0.05$).

Table 5. Seasonal dynamics of litter nutrient return in different years (kg/hm^2).

Year	Components	Season	C	N	P	K	S	Ca	Mg
2005	Litter branches	Spring	139.79	2.37	0.12	0.65	0.22	3.92	0.40
		Summer	680.30	11.14	0.56	1.92	0.99	17.50	1.64
		Autumn	118.58	2.41	0.13	0.48	0.24	3.94	0.40
		Winter	252.42	4.00	0.22	0.91	0.39	7.23	0.79
	Litter leaves	Spring	890.75	18.05	0.96	9.56	1.82	21.96	4.39
		Summer	459.91	12.31	0.65	3.08	1.06	11.72	1.80
		Autumn	606.15	14.54	0.74	4.96	1.34	15.11	2.85
		Winter	433.84	10.01	0.61	4.37	0.98	10.04	2.08
2010	Litter branches	Spring	336.38	6.23	0.35	1.24	0.62	9.06	1.06
		Summer	114.76	2.50	0.12	0.42	0.21	3.47	0.36
		Autumn	102.50	2.30	0.10	0.38	0.20	3.00	0.34
		Winter	110.02	2.29	0.10	0.40	0.21	3.43	0.34
	Litter leaves	Spring	1246.19	27.33	1.44	12.83	3.03	28.77	6.16
		Summer	342.38	11.93	0.66	3.13	0.95	8.88	1.55
		Autumn	476.20	13.37	0.67	3.37	1.08	11.74	2.16
		Winter	362.20	8.84	0.50	3.38	0.87	8.15	1.98
2015	Litter branches	Spring	124.81	1.69	0.09	0.43	0.15	2.61	0.35
		Summer	660.65	15.25	0.91	5.73	1.28	15.78	2.42
		Autumn	113.47	1.86	0.11	0.80	0.17	2.27	0.28
		Winter	228.65	4.36	0.32	1.51	0.39	4.57	0.60
	Litter leaves	Spring	809.03	20.01	0.99	8.07	1.69	15.35	3.93
		Summer	458.59	15.81	0.86	4.29	1.25	8.11	2.26
		Autumn	566.59	17.64	1.02	6.41	1.58	13.34	2.99
		Winter	384.07	10.80	0.67	4.34	0.94	8.95	1.85

4. Discussion

4.1. Litter Dynamics in Different Years

Litter volume is a component of the forest ecosystem biomass, which reflects the primary productivity level of the forest ecosystem and reflects the functions of the forest ecosystem [26,67–70]. The global annual change range of forest litter is 0.13 (Alaska's Picea crassifolia forest in the United States)—15.3 (Congo's tropical rain forest) t·hm^{-2} [28,42], and it can be divided into tropical, subtropical, temperate, and cold temperate zones according to climatic conditions. The forest litter yield in different climatic zones shows a downward trend with the increase in latitude. On average, Eurasian forests are tropical (8.40 t·hm^{-2}) > subtropical (5.25 t·hm^{-2}) > temperate (3.81 t·hm^{-2}) > cold temperate (t·hm^{-2}) [21,37,43,71]. The annual litter amount of different forest types in different climate zones in the world also varies greatly. The annual litter amount of main forest types in different climate zones can be shown as the maximum average annual litter amount of tropical rain forest and seasonal rain forest, which is 9.98 t·hm^{-2}. It is followed by evergreen broad-leaved forest, with an average annual litter of 6.96t·hm^{-2}. Then, 5.79 t·hm^{-2} in coniferous and broad-leaved mixed forest; deciduous broad-leaved mixed forest 5.1 t·hm^{-2}; The coniferous forest with the smallest litter is 4.77 t·hm^{-2} [72]. The subtropical evergreen broad-leaved forest in the study area is located in the north-south transition zone between the middle subtropical climate and the south subtropical climate. In this study, the average annual total litter in 2005, 2010, and 2015 is 8.18 t·hm^{-2}, which is within the range of subtropical forest changes (1.01–13.00 t·hm^{-2}). Compared with the evergreen broad-leaved forest in different regions of the subtropical zone, its annual litter is less than that of the Dinghushan monsoon evergreen broad-leaved forest in the south subtropical zone (8.45 t·hm^{-2}) [17], greater than that of the Tiantong evergreen broad-leaved forest in the north subtropical zone (5.55 t·hm^{-2}) [73–75], but similar to that of the Xiaokeng subtropical evergreen broad-leaved forest in the middle subtropical zone (7.99–8.450 t·hm^{-2}), It shows that the litter of wet evergreen broad-leaved forest in Ailao Mountain is similar to the geographical location and stand composition structure of the middle subtropical forest [70].

In different years, each component of litter and its proportion in the total forest litter are the largest, and the sum of the two reaches 76.17% of the total, which is consistent with the general situation of plant growth and litter, and basically consistent with the proportion sequence of each component of litter in other research results [71,76–79]. The total amount of litter in the study area in 2005 (normal year without extreme weather interference), 2010 (affected by extreme arid weather), and 2015 (affected by extreme ice and snow weather) are 8.82, 7.70, and 8.01 t·hm^{-2}, respectively, of which the proportion of litter is 49.96%, 61.52%, and 41.30%, respectively, and the proportion of litter branches is 25.82%, 17.26%, and 32.05%. Compared with other subtropical forests of the same type, the annual average percentage of leaves (50.89%) is lower, while the three-year average percentage of branches (25.28%) is higher. In 2005, the proportion of litter branches in the study area was less than 30%, which was still in the middle and early stages of succession. The proportion of litter branches was small. With the succession, the proportion of litter branches should have gradually increased, but the proportion of litter branches in the region from 2005 to 2010 was gradually decreasing, reached the valley in 2010, and then gradually increased from 2010 to 2015. This is because of the rare serious drought in Yunnan in 2010. After the forest was disturbed by the extreme arid weather for a long time, the characteristics of the plant community degenerated toward the type of low-grade community, and secondary succession occurred within the community. At the same time, because the soil was lack of water for plant growth in this dry season, some leaves fell, so the proportion of branches and litter was less than that when the forest was not disturbed, The proportion of leaf litter also increased correspondingly [42,43], indicating that extreme drought weather interference seriously hindered the normal succession process of forests. In 2015, the proportion of litter leaves was less than that of undisturbed years, but the proportion of litter branches increased accordingly, which is consistent with the short duration of this extreme ice and snow weather disturbance. The wood litter in the forest, such as branches, is quite random. The litter collected every month is the litter branches that died on the trees before. Therefore, the litter of branches is usually not directly related to phenology but is greatly affected by climate factors. It also shows that although the evergreen broad-leaved forest in Ailao Mountain is affected by extreme ice and snow weather, However, due to the rich species and complex community structure in Ailao Mountain, the forest ecosystem has the ability of self-regulation and high early post-disaster recovery, which makes the evergreen broad-leaved forest vegetation in Ailao Mountain recover slowly. Thus, the litter amount of evergreen broad-leaved forest in Ailao Mountain after being disturbed by extreme weather is close to that of evergreen broad-leaved forest in the same latitude in normal years [37,71]. To sum up, the impact of extreme drought weather on litter yield and nutrients in the study area is greater than that of extreme ice and snow weather interference, and it mainly has a greater impact on litter branches, litter leaves, and the total amount of litter, while the impact on other litter components is relatively small [56].

Due to the influence of external environmental factors, the dynamic changes of litter in different years show some regularity but also show some differences [70–72]. The cause of the occurrence of the litter rhythm is mainly determined by the climate change factors and the biological characteristics factors of the tree species that make up the community [1,3,23,47]. The monthly dynamic performance patterns of litter in 2005, 2010, and 2015 are different, including single peak, two peaks, and four peaks, with different peak sizes, and all of them play a leading role in litter amount. In the normal year 2005, the maximum peak value of withering occurred in the rainy season, while in the two special years, 2010 and 2015, the peak value of withering mainly occurred in the dry season.

In 2005, the total amount and each component of litter reached the maximum in August, which was affected by the external temperature and accelerated the rate of metabolism, prompting a large number of plants to litter. The temperature in August is suitable for the vigorous growth of plants. The rising temperature of evergreen trees during the germination period will promote the sprouting of new leaves and accelerate the aging and apoptosis of old leaves. When the nutrients required for the growth of new leaves exceed the nutrients absorbed from the soil, the plants will preferentially transfer the nutrients stored in the old leaves for the growth of new branches and leaves, and the transfer and reabsorption of nutrients will accelerate the apoptosis of old leaves [17,39,71]. The other peak months in 2005 are basically the same as those in 2010. Although the plants in the evergreen broad-leaved forest gradually change leaves throughout the year, the temperature rises at the beginning of the rainy season (April and May), and a large number of new leaves germinate and grow vigorously. In these growing seasons, new leaves compete with old leaves for limited mineral elements. Because the strong vitality of new leaves makes the elements that can be transported from old leaves to new leaves, and finally causes the senescent leaves to fall off one after another, During this period, the rainfall increased, so the first short and concentrated peak of leaf litter appeared in April, basically in line with the physiological characteristics of evergreen broad-leaved forest. Another secondary peak of litter in the study area is at the end of the rainy season in November. The occurrence of the second peak of litter is related to the defoliation period of the dominant tree species of the evergreen broad-leaved forest in this area, Kaempferous, and because of the sudden drop in humidity and temperature in autumn and winter, the leaves of some evergreen broad-leaved forests lose their vitality in dry air and wither [17,70,71].

In January 2015, the study area had an extremely strong ice and snow weather. Because a large number of leaves on the crown of evergreen tree species condensed with ice and snow when the ice and snow disaster occurred, and some tree species were easy to break, strong wind, heavy rain and heavy snow can blow and break the litter branches of the previous period to the ground [5,14,37,42,70], After the interference of extreme ice and snow weather, the leaves were violently shaken by strong external forces, resulting in non physiological shedding, which led to a sharp increase in the amount of litter in January 2015 and a peak. This randomness can cause great changes in the number of litter branches at different times, which also indicates that the weather conditions in 2015 were worse than those in 2005 and 2010. In addition, after the interference of extreme ice and snow weather, a large number of trees fell down, branches were crushed and dropped by heavy snow, forest canopy was seriously damaged, forest canopy density was reduced, and insect food in habitats with severely damaged vegetation was reduced, so the amount of litter brought by insects was reduced successively, thus significantly reducing the amount of litter after the peak in January. However, the rose fluctuation amplitude of the litter after the forest was disturbed by the extreme ice and snow weather decreased significantly and tended to be stable, indicating that the extreme ice and snow weather caused great damage to the physiological structure of the tree species, leading to changes in the subsequent litter patterns. In a word, the litters in the three years have their own distinct littering rhythms, which may be because the species composition of evergreen broad-leaved forests in the study area, the climatic conditions and the degree of interference vary greatly in different years, and the litters are very vulnerable to the biological characteristics of forest species, climatic conditions, and other environmental factors, making the seasonal dynamics of each year different [20].

4.2. Nutrient Concentration Dynamics and Nutrient Return of Litter

The return of nutrients from the litter to the soil maintains soil fertility and promotes nutrient cycling in forest ecosystems, which is important for improving the habitat conditions of forest trees [2,26–30]. Analyzing the dynamics of nutrient concentrations and return in the litter is necessary for understanding the function of forest ecosystems [5,8,9,31,43,74,80].

The nutrient content of litter can reflect the nutrient utilization efficiency of plants. In this study, the total amount of litter and the different nutrient concentrations of each component have the same order, with the highest content of C, followed by the content of Ca and N, and the content of K, Mg, S, P, and other elements is relatively low [14,39,48]. The formation of this sequence is because the fresh litters in the study area are in the early stage of decomposition, and the initial content of N element is high, while the elements with low content of P and K in the litters are reusable elements, which can be transferred in large quantities before the leaves fall, and then the nutrient reuse is realized through the plant transfer mechanism, which indicates that the evergreen broad-leaved forest plants in this area have high nutrient utilization efficiency. However, the strong seasonal dynamics of nutrient concentration of each component also show that the transfer amount of different elements is different in different seasons, and the decomposition rate of litter is faster due to the rapid leaching and degradation of P and K [70,71,81]. The content of nutrient elements is greatly affected by their physiological functions. The subtropical climate is hot and rainy, so the leaching effect of litter is very strong. The rainy season in this region is long, mainly concentrated in summer and autumn, and the physiological functions of each element are different [66]. Among them, K and P are easy to be lost due to rainfall, but P is a limiting factor for plant growth and development in tropical and subtropical regions. For this reason, most organisms in the subtropical ecosystem have a mechanism to maintain that P is not leached out under high temperature and humidity conditions, which makes the K and P contents in the litter lower in spring and summer as a whole, but P content is far lower than K content [3,4,26]. Ca, N, and other elements are relatively stable and not easy to be washed away by rain, which makes the seasonal variation of various nutrient elements in different years and different components inconsistent, and there is no obvious correlation. These nutrient elements will eventually return to the soil, which is of great significance for maintaining the long-term productivity of the forest land [82].

The chemical properties of litter are the internal factors affecting litter decomposition. C/N and C/P in the litter are common indicators of litter decomposition. N/P also has a certain characterization ability for litter decomposition. Since N can affect the growth and turnover of soil microbial communities, the C/N ratio can best reflect the rate of litter decomposition [42]. N and P control the important process of litter decomposition through coupling with C. The C/N ratio of the litter in the study area in 2005 (the normal year without extreme weather interference) is 48.65, which is greater than 25, indicating that N is the main element limiting litter decomposition in the region [51]. From 2005 to 2015, the average C/N ratio was the highest in 2005, and the content of N was the lowest, indicating that the decomposition rate of litter was the slowest in that year, and the extreme weather interference did not weaken the decomposition rate of litter. The C/N ratio of litter branches and litter leaves is not the same each year, so it can be concluded that the response of nutrient content among various organs of forest litter to extreme weather disturbance is also different [82].

The annual total amount of litter in each year is in the order of 2005 (normal year not affected by extreme weather) > 2015 (affected by extreme ice and snow weather) > 2010 (affected by extreme drought weather). From the perspective of the annual amount of nutrient return of litter, the annual amount of nutrient return is roughly proportional to the amount of litter. In addition, leaf litter is the main component of the litter, accounting for more than 50% of the total litter. Compared with other components, the nutrient concentration is generally high and easy to decompose, so leaf litter is the main body of nutrient return of the litter.

5. Conclusions

Due to the longer duration and wider coverage of the extreme arid weather disturbance, the degree of damage to the forest ecosystem is greater than that of the extreme ice and snow weather disturbance. Due to the extreme arid weather disturbance, the number of litter branches of the subtropical evergreen broad-leaved forest in Ailao Mountain decreases, regresses to the lower community succession, and then slowly recovers. However, in general, the forest ecosystem of Ailao Mountain has the ability of self-regulation and high early recovery after disasters, which makes the amount of litter of the evergreen broad-leaved forest in Ailao Mountain after being disturbed by extreme weather close to the amount of litter of the evergreen broad-leaved forest in the same latitude in normal years. The nutrient element content, decomposition rate, and seasonal dynamics of the litter are not affected by extreme weather interference and have a high resistance to extreme weather interference events in the short term. It can strongly maintain soil fertility in the region. Therefore, studying the mechanism of litter change of evergreen broad-leaved forests in the same region in different extreme weather interference years can fully compare the damage caused by different extreme weather interference to the normal forest growth process, and provide a reference for the research of evergreen broad-leaved forests in other subtropical regions, At the same time, it has far-reaching significance for the recovery and evolution of subtropical forest ecosystem disturbed by extreme weather, sustainable management and improvement in ecological function to study the change of litter volume before and after extreme drought and ice and snow weather disturbance.

Author Contributions: Conceptualization, X.L. (Xingyue Liu); methodology, X.L. (Xingyue Liu); software, X.L. (Xingyue Liu) and Z.W.; validation, X.L. (Xingyue Liu), Z.W. and X.L.(Xi Liu); formal analysis, X.L. (Xingyue Liu) and H.G.; investigation, X.L. (Xingyue Liu), Z.L. and D.L.; resources, Z.L., D.L. and H.G.; data curation, X.L. (Xingyue Liu) and H.G.; writing—original draft preparation, X.L. (Xingyue Liu); writing—review and editing, X.L. (Xingyue Liu) and H.G.; visualization, X.L. (Xingyue Liu), Z.W. and X.L. (Xi Liu); supervision, X.L. (Xingyue Liu) and H.G.; project administration, H.G.; funding acquisition, H.G. All authors have read and agreed to the published version of the manuscript.

Funding: This research was funded by the Yunnan Natural Science Foundation (2019FB064).

Institutional Review Board Statement: Not applicable.

Informed Consent Statement: Not applicable.

Data Availability Statement: The data presented in this study are available on request from the corresponding author. The data are not publicly available due to privacy restrictions.

Conflicts of Interest: The authors declare no conflict of interest.

References

1. Zeng, Z.; Li, M.; Song, Y.; Yang, R.; Xia, J.; Kuang, D.; Yang, Q.; Zhang, C. Comparison of litterfall and nutrients return properties of primary and secondary forest ecosystems, the karst region of Northwest Guangxi. *Ecol. Environ. Sci.* **2010**, *19*, 146–151.
2. Ebermer, E. *Die Geamte Lechr der Waldtreu mit Rucksicht auf die Chemische Statikdes Waldbaues*; Springer: Berlin/Heidelberg, Germany, 1876.
3. Xu, W.M.; Yan, W.D.; Li, J.B.; Zhao, J.; Wang, G.J. Amount and dynamic characteristics of litterfall in four forest types in subtropical China. *Acta Ecol. Sin.* **2013**, *33*, 7570–7575.
4. Vitousek, P.M. Litterfall, nutrient cycling, and nutrient limitation in tropical forests. *Ecology* **1984**, *64*, 285–298. [CrossRef]
5. Ranger, J.; Gerard, F.; Lindemann, M.; Gelhaye, D.; Gelhaye, L. Dynamics of litterfall in a chronosequence of Douglas-fir (*Pseudotsuga menziesii Franco*) stands in the Beaujolais Mounts (France). *Annu. For. Sci.* **2003**, *60*, 475–488. [CrossRef]
6. Barlow, J.; Gardner, T.A.; Ferreira, L.V.; Peres, C.A. Litter fall and decomposition in primary, secondary and plantation forests in the Brazilian Amazon. *For. Ecol. Manag.* **2007**, *247*, 91–97. [CrossRef]
7. Guo, J.; Yu, L.H.; Fang, X.; Xiang, W.H.; Deng, X.W.; Lu, X. Litter production and turnover in four types of subtropical forests in China. *Acta Ecol. Sin.* **2015**, *35*, 4668–4677.
8. Aerts, R.; Van Bodegom, P.M.; Cornelissen, J.H.C. Litter stoichiometric traits of plant species of high-latitude ecosystems show high responsiveness to global change without causing strong variation in litter decomposition. *New Phytol.* **2012**, *196*, 181–188. [CrossRef] [PubMed]
9. Wang, Z.F. *Responses of Litterfall Production and Nutrient to Simulated Nitrogen Deposition in Subtropical Evergreen Broadleaved Forest across an Age Sequence*; Anhui Agriculture University: Hefei, China, 2018.
10. Ge, J.L.; Xie, Z.Q. Leaf litter carbon, nitrogen, and phosphorus stoichiometric patterns as related to climatic factors and leaf habits across Chinese broad-leaved tree species. *Plant Ecol.* **2017**, *218*, 1063–1076. [CrossRef]
11. Peng, S.L. *Community Dynamics in Lower Subtropical Forest*; Science Press: Beijing, China, 1996.
12. Liu, L.; Zhao, M.-C.; Xu, W.-T.; Shen, G.-Z.; Xie, Z.-Q. Litter nutrient characteristics of mixed evergreen and deciduous broadleaved forests in Shennongjia, China. *Acta Ecol. Sin.* **2019**, *39*, 7611–7620.
13. Li, J.B. Dynamics of Litter and Nutrient in 4 Types of Subtropical. Ph.D. Thesis, Central South University of Forestry and Technology, Changsha, China, December 2011.
14. Wu, C.; Sha, L.; Zhang, Y. Effect of Litter on Soil Respiration and Its Temperature Sensitivity in a Montane Evergreen Broad-Leaved Forest in Ailao Mountains, Yunnan. *J. Northeast. For. Univ.* **2012**, *40*, 37–40.

15. Yang, G.; Gong, H.; Zheng, Z.H.; Zhang, Y.; Liu, Y.; Lu, Z. Caloric values and ash content of six dominant tree species in an evergreen broadleaf forest of Ailaoshan, Yunnan Province. *J. Zhejiang For. Coll.* **2010**, *27*, 251–258.
16. Wu, B.X. Study on thedyastudy on theedynamicsand rhythmsofmidmon-tanewetevergreen broad-leaved forest atxujiaba, Ailao mountains, Yunnan Province. *Acta Bot. Sin.* **1995**, 969–977.
17. Guan, L.L.; Zhou, G.Y.; Zhang, D.Q.; Liu, J.X.; Zhang, Q.M. 20 years of litterfall dynamics in subtropic evergreen broad-leaved forest at the Dinghushan forest ecosystem research station. *Chin. J. Plant Ecol.* **2004**, *28*, 449–456.
18. Witkamp, M. Microbial populations of leaf litter in relation to environmental conditions and decomposition. *Ecology* **1963**, *44*, 370–377. [CrossRef]
19. Maguire, D.A. Branch mortality and potential litterfall from Douglas-fir trees in stands of varying density. *For. Ecol. Manag.* **1994**, *70*, 41–53. [CrossRef]
20. Xing, J.M.; Wang, K.Q.; Song, Y.L.; Zhang, Y.J.; Zhang, Z.M.; Pan, T.S. Characteristics of litter return and nutrient dynamic change in four typical forests in the subalpine of central Yunnan province. *J. Cent. South Univ. For. Technol.* **2021**, *41*, 134–144.
21. Wang, F.Y. Review on the study of forest litterfall. *Ecol. Prog.* **1989**, *6*, 82–89.
22. Bary, J.R.; Gorha, M.E. Litter production in forests of the world. *Adv. Ecol. Res.* **1964**, *2*, 101–157.
23. Descheemaeker, K.; Muys, B.; Nyssen, J.; Poesend, J.; Raesa, D.; Hailec, M.; Dechers, J. Litter production and organic matter accumulation in exclosures of the Tigary highlands, Ethiopia. *For. Ecol. Manag.* **2006**, *233*, 21–35. [CrossRef]
24. Schessl, M.; Silva, W.L.D.; Gottsberger, G. Effects of fragmentation on forest structure and litter dynamics in Atlantic rainforest in Pernambuco, Brazil. *Flora* **2008**, *203*, 215–228. [CrossRef]
25. Wang, Q.K.; Wang, S.L.; Huang, Y. Comparisions of litterfall, litter decomposition and nutrientteturn in a monoculture Cunninghamia lanceolata and a mixed stand in southern China. *For. Ecol. Manag.* **2007**, *255*, 1–9.
26. Sundarapandian, S.M.; Swamy, P.S. Litter production and leaf-litter decomposition of selected tree species in tropical forests at Kodayar in the western ghats, India. *For. Ecol. Manag.* **1999**, *123*, 231–244. [CrossRef]
27. Liu, C.-j.; Ilvesniemi, H.; Berg, B.; Kutsch, W.; Yang, Y.-s.; Ma, X.-q.; Westman, C.J. Aboveground litterfall in Eurasian forests. *J. For. Res.* **2003**, *14*, 27–34.
28. Zhang, L.; Wang, X.H.; Mi, X.C.; Chen, J.H.; Yu, M.J. Temporal dynamics of and effects of ice storm on litter production in an evergreen broad-leaved forest in Gutianshan National Nature Reserve. *Biodivers. Sci.* **2011**, *19*, 206–214.
29. Miller, D.J.; Cooper, J.M.; Miller, H.G. Amount and nutrient weights in litter-fall, and their annual cycles, from a series of fertilizer experiments on pole-stage Sitka spruce. *Forestry* **1996**, *69*, 89–302. [CrossRef]
30. An, J.Y.; Han, S.H.; Youn, W.B.; Lee, S.I.; Rahman, A.; Dao, H.T.T.; Seo, J.M.; Aung, A.; Choi, H.S.; Hyun, H.J.; et al. Comparison of litterfall production in three forest types in Jeju Island, South Korea. *J. For. Res.* **2019**, *31*, 945–952. [CrossRef]
31. Gresham, C.A. Litterfall pattems in mature loblolly and longleaf pine stands in coastal Southem Carolina. *For. Sci.* **1982**, *28*, 223–231.
32. John, T.; Marcia, J.L. Litterfall and forest floor dynamics in Eucalyptus pilularis forests. *Austral Ecol.* **2002**, *27*, 192–199.
33. Scherer-Lorenzen, M.; Bonilla, J.L.; Potvin, C. Tree species richness affects litter production and decomposition rates in a tropical biodiversity experiment. *Oikos* **2007**, *116*, 2108–2124. [CrossRef]
34. Deborah, L. Regional-scale variation in litter production and seasonality in tropical dry forests of southern Mexico. *Biotropica* **2005**, *37*, 561–570.
35. Olena, P.; Nedret, B. Impact of deciduous tree species on litterfall quality, ecomposition rates and nutrient circulation in pine stands. *For. Ecol. Manag.* **2007**, *253*, 11–18.
36. Luo, Z.S.; Xiang, C.H.; Mu, C.L. The litterfall of majir forests in Gunasi River Watershed in Mianyang City, Sichuan Province. *Acta Ecol. Sin.* **2007**, 1772–1781.
37. Zhang, L. *Temporal Dynamics of and Effects of Ice Storm on Litter Production in an Evergreen Broad-Leaved Forest in Gutianshan National Nature Reserve*; Zhejiang University: Hangzhou, China, 2010.
38. Huang, Y.H.; Guang, L.L.; Zhou, G.Y.; Luo, Y.; Tang, J.W.; Liu, Y.H. Gross caloric values of dominant species and litter layer in mid-montane moist evergreen broad-leaved forest in Ailao mountain and in tropical season rain forest in Xishuangbanna, Yunnan, Chian. *Chin. J. Plant Ecol.* **2007**, *31*, 457–463.
39. Yang, W.Q.; Deng, R.J.; Zhang, J. Forest litter decomposition and its responses to global climate change. *Chin. J. Appl. Ecol.* **2007**, *18*, 2889–2895.
40. Hua, C.; Harmon, M.E.; Hanqin, T. Effects of global change on litter decomposion in terrestrial ecosystems. *Acta Ecol. Sin.* **2001**, *21*, 1549–1563.
41. He, J.; Zhao, X.M.; Zhao, Z.J.; Fan, M.; Mao, S.Y.; Zhou, H. Effects of the ice and snow damage to the evergreen broad-leaved forest of Jiulianshan Mountain in Jiangxi Province. *Guihaia* **2013**, *33*, 780–785+851.
42. Mao, S.Y. *The Input and Decomposition Dynamics of Litterfall of a Broad-Leaved Evergreen Forest in Jiulianshan after the Frozen Rain and Snow Disater*; Beijing Forestry University: Beijing, China, 2011.
43. Xu, Y.W.; Zhu, L.R.; Wu, K.K.; Zhou, Z.P.; Peng, S.L. Litter dynamics in different forest types suffered anextreme ice storm in the subtropical region, southern China. *Ecol. Environ. Sci.* **2011**, *20*, 1443–1448.
44. He, P.; Li, J.X.; Xu, G.Q.; Li, H.B. Analysis of precipitation Distribution and Drought and Flood Disasters in Chuxiong City on the Yunnan Plateau. *Earth Environ.* **2014**, *42*, 162–167.

45. Cheng, Q.P.; Wang, P. Drought and Flood Change Characteristics Based on RDI Index from 1960 to 2013 in Yunnan Province. *Resour. Environ. Yangtze Basin* **2018**, *27*, 185–196.

46. Duan, C.C.; Zhu, Y.; You, W.H. Characteristic and Formation Cause of Drought and Flood in Yunnan Province Rainy Season. *Plateau Meteorol.* **2007**, 402–408.

47. Wang, L.M.; Liu, T.T.; Ma, S.R.; Niu, J.; Zhang, W.X. Temporal Variation Characteristics of Drought and Flood in Yunnan over the Past 620 Years. *J. Yunnan Norm. Univ. Nat. Sci. Ed.* **2019**, *39*, 67–72.

48. Zhen, J.M.; Zhang, W.C.; Chen, Y.; Ma, T. Analysis on climatologic characteristics and causes of severe drought during 2009–2010 in Yunnan province. *J. Meteorol. Sci.* **2015**, *35*, 488–496.

49. Peng, G.F.; Liu, Y.X. A Dynamic Risk Analysis of 2009–2010 Yunnan Extra-Severe Drought. *Adv. Meteorol. Sci. Technol.* **2012**, *2*, 50–52.

50. Wang, J.; Wen, X.F.; Wang, H.M.; Wang, J.Y. The effects of ice storms on net primary productivity in a subtropical coniferous planation. *Acta Ecol. Sin.* **2014**, *34*, 5030–5039.

51. Wang, X.; Huang, S.N.; Zhou, G.Y.; Li, J.X.; Qiu, Z.J.; Zhao, X.; Zhou, B. Effects of the Frozen Rain and Snow Disaster on the Dominant Species of Castanopsis Forests in Yangdongshan Shierdushui Provincial Nature Reserve of Guangdong. *Sci. Silvae Sin.* **2009**, *45*, 41–47.

52. Han, X.Y.; Zhao, F.X.; Li, W.Y. A review of researches on forest litterfall. *For. Sci. Technol. Inf.* **2007**, *03*, 1213–1216.

53. Nykänen, M.L.; Peltola, H.; Quine, C.; Kellomäki, S.; Broadgate, M. Factors affecting snow damage of trees with particular reference to European conditions. *Silva Fenn.* **1997**, *31*, 193–213. [CrossRef]

54. Bargg, D.C.; Shelton, M.G.; Zeide, B. Impacts and management implications of ice storms on forests in the southern United States. *For. Ecol. Manag.* **2003**, *186*, 99–123. [CrossRef]

55. Beach, R.H.; Sills, E.O.; Liu, T.M.; Pattanayak, S. The influence of forest management on vulnerability of forests to severe weather. In *Advances in Threat Assessment and Their Application to Forest and Rangeland Management*; Pye, J.M., Rauscher, H.M., Sands, Y., Lee, D.C., Beatty, J.S., Eds.; General Technical Report PNW-GTR-802; Forest Service U.S. Department of Agriculture: Washington, DC, USA, 2010; pp. 185–206.

56. Tang, X.H.; Zhang, Y.P.; Wu, C.S.; Luo, G.; Liang, N.S. Responses of soil temperature and soil water content to extreme snow event in a subtropical evergreen broad-leaved forest in Ailao Mountains, Yunnan, Southwest China. *Chin. J. Ecol.* **2018**, *37*, 1833–1840.

57. Wen, Y.G.; Wen, B.G.; Li, J.J. A study on the litter production and dynamics of subtropical forest. *Sci. Silvae Sin.* **1989**, *25*, 542–548.

58. Bi, Z.; Zhi-An, L.; Yong-Zhen, D.; Wan-Neng, T. Litterfall of common plantations in south subtropical China. *Acta Ecol. Sin.* **2006**, *26*, 715–721.

59. Wu, Z.M.; Li, Y.D.; Zhou, G.Y.; Chen, B.F. Abnormal litterfall and its ecological significance. *Sci. Silvae Sin.* **2008**, *44*, 28–31.

60. Vitousek, P.M.; Turner, D.R.; Parton, W.J. 1994. Litter decomposition on the Mauna Loa environmental matrix, Hawaii: Pattern, mechanisms, and models. *Ecology* **2007**, *75*, 418–429. [CrossRef]

61. Zheng, J.X.; Xiong, D.C.; Huang, J.X.; Yang, Z.J.; Chen, G.S. Litter production and nutrient return in 2 plantations of young and old Cunninghamia lanceolata. *J. Fujian Coll. For.* **2013**, *33*, 18–24.

62. Kathryn, B.; Piatek, H.; Lee, A. Site preparation effects on foliar N and Puse, retranslocation and transfer to litter in 15-years old Plillrs tnerir. *For. Ecol. Manag.* **2000**, *129*, 143–152.

63. Shi, J.; Xu, H.; Lin, M.X.; Li, Y. Dynamics of litterfall production in the tropical mountain rainforest of Jianfengling, Hainan island, China. *Plant Sci. J.* **2019**, *37*, 593–601.

64. Laskowski, R.; Niklinska, M.; Maryanski, M. The dynamics of chemical elements in forest litter. *Ecology* **1995**, *76*, 1393–1406. [CrossRef]

65. Kappelle, M.; Leal, M.E. Changes in leaf Morphology and foliar nutrient status along a successional gradient in Costa Rican upper montane Quercus forest. *Biotropica* **1996**, *28*, 331–344. [CrossRef]

66. Fan, C.N.; Guo, Z.L.; Zheng, J.P.; Li, B.; Yang, B.G.; Yue, L.; Yu, H.B. The amount and dynamics of litterfall in the natural secondary forest in Mopan mountain. *Acta Ecol. Sin.* **2014**, *34*, 633–641.

67. Liski, J.; Nissinen, A.; Erhard, M.; Taskine, O. Climate effects on litter decomposition from arctic tundra to tropical rainforest. *Glob. Chang. Biol.* **2003**, *9*, 575–584. [CrossRef]

68. Alley, J.C.; Fitzgerald, B.M.; Berben, P.H.; Haslett, S.J. Annual and seasonal patterns of litter-fall of hard beech (*Nothofagus truncata*) and silver beech (*Nothofagus menziesil*) in relation to reproduction. *N. Z. J. Bot.* **1998**, *36*, 453–464. [CrossRef]

69. Pan, K.W.; He, J.; Wu, N. Effect of forest litter on microenvironment conditions of forestland. *Chin. J. Appl. Ecol.* **2004**, *15*, 153–158.

70. Lan, C.C. *Litterfall Production and Related Nutrient Contents of Subtropical Forests in Xiaokeng*; Anhui Agricultural University: Hefei, China, 2008.

71. Hou, L.L.; Mao, Z.J.; Sun, T.; Song, Y. Dynamic of litterfall in ten typical community types of Xiaoxing'an Mountain, China. *Acta Ecol. Sin.* **2013**, *33*, 1994–2002.

72. Liao, J.; Wang, X.G. An overview of the research on forest litter. *South China For. Sci.* **2000**, *1*, 31–34.

73. Ardk, A. *The Association between Litterfall and Variation of Climate Factors and Population in an Evergreen Broad-Leaved Forest in Tiantong, Zhejiang Province*; East China Normal University: Shanghai, China, 2017.

74. Berg, B. Litter decomposition and organic matter turnover in northern forest soils. *For. Ecol. Manag.* **2000**, *133*, 13–22. [CrossRef]

75. Paudel, E.; Dossa, G.G.; Xu, J.; Harrison, R.D. Litterfall and nutrient return along a disturbance gradient in a tropical montane forest. *For. Ecol. Manag.* **2015**, *353*, 97–106. [CrossRef]

76. Lado-Monserrat, L.; Lidón, A.; Bautista, I. Litterfall, litter decomposition and associated nutrient fluxes in Pinus halepensis: Influence of tree removal intensity in a Mediterranean forest. *Eur. J. For. Res.* **2016**, *135*, 203–214. [CrossRef]
77. Estiarte, M.; Peuelas, J. Alteration of the phenology of leaf senescence and fall in winter deciduous species by climate change: Effects on nutrient proficiency. *Glob. Chang. Biol.* **2015**, *21*, 1005–1017. [CrossRef] [PubMed]
78. Peng, S.L.; Liu, Q. The Dynamics of Forest Litter and Its Responses to Global Warming. *J. Acta Ecol. Sin.* **2002**, *09*, 1534–1544.
79. Xie, K.X.; Weng, S.R.; He, G.X. Seasonal changes of nutrient of litter of Phoebe bournei plantation. *J. Cent. South Univ. For. Technol.* **2013**, *33*, 13–116.
80. Jiao, B.R.; Zhou, G.S.; Liu, Y.Z.; Jia, Y.L. Spatial pattern and environmental controls of annual litterfall production in natural forest ecosystems in China. *J. Sci. Sin. (Vitae)* **2016**, *46*, 1304–1311.
81. Liu, Y.; Han, S.J.; Lin, L. Dynamic characteristics of litterfalls in four forest types of hangbai mountains, China. *Chin. J. Ecol.* **2009**, *28*, 7–11.
82. Zhao, Y.; Wu, Z.; Fan, W.; Gao, X. Comparison of nutrient return and litter decomposition between coniferous and broad-leaved forests in hilly region of Taihang mountains. *J. Nat. Resour.* **2009**, *24*, 1616–1624.

Article

Effects of Two Management Practices on Monthly Litterfall in a Cypress Plantation

Yulian Yang [1], Honglin Yang [1], Qiang Wang [1], Qing Dong [1], Jiaping Yang [1], Lijun Wu [1], Chengming You [1,2], Jinyao Hu [1] and Qinggui Wu [1,*]

[1] Ecological Security and Protection Key Laboratory of Sichuan Province, Mianyang Normal University, Mianyang 621000, China
[2] Forestry Ecological Engineering in the Upper Reaches of the Yangtze River Key Laboratory of Sichuan Province & National Forestry and Grassland Administration Key Laboratory of Forest Resources Conservation and Ecological Safety on the Upper Reaches of the Yangtze River & Rainy Area of West China Plantation Ecosystem Permanent Scientific Research Base, Institute of Ecology & Forestry, Sichuan Agricultural University, Chengdu 611130, China
* Correspondence: qgwu30@mtc.edu.cn; Tel.: +86-816-2579941

Abstract: Optimizing stand structure can enhance plantation forest ecosystem service functions by regulating litterfall patterns; however, the effects of close-to-nature management on litterfall production remain unclear. Here, we selected three cypress (*Cupressus funebris*) plantations, including one using the practice of strip filling (SF), one using the practice of ecological thinning (ET), and one pure cypress plantation without any artificial interference. The production of total litterfall and its components (leaf, twig, reproductive organ and miscellaneous litterfall) were investigated monthly over one year from September 2019 to August 2020. Compared with that of the pure plantation, the total annual litterfall production of the SF and ET plantations decreased significantly by 10.8% and 36.44%, respectively. The annual production of leaf and reproductive organ litter was similar to that of total litterfall, but that of twig and miscellaneous litter was higher in the SF and ET plantations than in the pure plantation. Moreover, total, leaf and reproductive organ litterfall production displayed unimodal dynamics regardless of plantation, although the peaks of reproductive organ litter production occurred in different months. In contrast, the production of twig litter showed bimodal dynamics in the pure plantation, while unimodal and irregular dynamics were observed in the plantations with ET and SF, respectively. Additionally, insignificant differences in the isometric growth index of leaf litter and total litterfall were observed. The allometric indices of twig litterfall versus total litterfall, reproductive organ litterfall versus total litterfall, and leaf litterfall versus twig litterfall were higher in the plantations with SF and ET than in the pure plantation. Redundancy analysis (RDA) revealed that diameter at breast height and air temperature were the most important factors shaping the annual and monthly production of litterfall, respectively. These results provide efficient data to support the rectification of the material circulation of cypress plantations and their future management.

Keywords: litterfall; allometric relationships; forest management practice; cypress plantation

Citation: Yang, Y.; Yang, H.; Wang, Q.; Dong, Q.; Yang, J.; Wu, L.; You, C.; Hu, J.; Wu, Q. Effects of Two Management Practices on Monthly Litterfall in a Cypress Plantation. *Forests* **2022**, *13*, 1581. https://doi.org/10.3390/f13101581

Academic Editor: Benjamin L. Turner

Received: 30 August 2022
Accepted: 23 September 2022
Published: 27 September 2022

1. Introduction

The global area of plantation forests exceeds 2.84 hundred million hm^2, with plantation forests in China showing the largest area and fastest development worldwide [1]. Many ecological problems occur with the development of plantations [2,3]; thus, close-to-nature management (actively or passively developing planted forest communities into more diverse vegetation communities dominated by native species) has recently been used to maintain and improve ecosystem ecological function and stability [4–7]. Indeed, many previous studies have demonstrated that plantation management practices (e.g., thinning, strip-cutting, introducing tree species) have increased the diversity of forest species and improved soil properties [8–11]. However, the mechanisms driving the improvement of these ecological functions involved in management practices remain unclear.

Litterfall is one of the major pathways by which organic matter and nutrients return to soils from vegetation [12,13], and the production and patterns of litterfall significantly contribute to the maintenance of soil fertility and nutrient balance in forest ecosystems [14,15]. Its production and its temporal dynamics are strongly influenced by tree density, vegetation composition and climate factors (e.g., temperature, precipitation, wind etc.) [16–18], which are significantly regulated by management practices [19,20]. For instance, thinning or strip-cutting can primarily reduce stand density or canopy density. The decrease in stand density or canopy density not only directly reduces total litter production but also indirectly alters the composition of litterfall by affecting the formation and regeneration of light-loving vegetation [8,21]. In contrast, replanting or reconstruction in mixed forests could directly change the composition of litterfall as a result of adding new species. The practices mentioned above can all make the monthly production of litterfall vary. Since the production of litterfall is higher in mixed stands or species-rich forests than it is in monocultures [22–24], we hypothesized that forest management practices that increase species richness might be more conducive to litter production. However, although some studies have reported the characteristics of litterfall production at different levels under certain management practices [25,26], it is difficult to accurately predict litterfall among these practices, which is not conducive to comprehensively understanding the mechanism of litterfall among management practices.

In general, litterfall composed of materials from different organs exhibits allometric relationships [27,28]. The allometric relationship of litterfall production provides a new approach for estimating forest productivity because litterfall is an important contributor to the net primary productivity of forest ecosystems [29]. However, the biomass allocation in litterfall is still under debate. For instance, Ma et al. [27] found that the allometric relationships between total litterfall and litterfall components were approximately the same in evergreen forests and deciduous forests in China. However, Fu et al. [28] found that the allometric relationships of forest litterfall varied with litterfall components, functional types and vertical structures in alpine forests, e.g., evergreen versus deciduous and trees versus shrubs. Thus, additional studies on the allometric relationships of forest litterfall in different climatic zones, at different regional scales and in various forest types are needed. The management practice of plantations, as a form of human disturbance, inevitably create heterogeneous habitats and are expected to affect the allocation of litterfall from different organs.

As one of the major tree species in the Yangtze River shelter forests, cypress (*Cupressus funebris*) plays a critical role in regulating the regional climate and conserving water and soil. Cypress plantations have also led to a weak capacity for water storage, low biodiversity, and soil degradation, resulting in a large decrease in ecological services, although they have addressed the demand for high forest coverage. Therefore, a series of experiments implementing forms of management practices, such as strip filling (first strip-cutting and then replanting) and ecological thinning, were performed to adjust the structure of cypress plantation monocultures and improve their ecological functions. Although the water-holding capacity, plant diversity, soil microbial diversity and physicochemical properties after implementation of several management practices have been widely reported [30–33], little attention has been given to litterfall production. Here, a one-year field study, implementing the litter trap method, was carried out on three cypress plantations, including two management practices (strip filling and ecological thinning) and an unmodified pure plantation for comparison. Litter production was investigated monthly to assess the changes in annual total litterfall production and seasonal dynamics, as well as the allometric growth relationships among litter organs under different management practices. Specifically, we hypothesized that (1) the total litterfall production and its components would be higher in the plantation with management practices than in the pure cypress plantation and (2) management practices would alter the monthly dynamics of litterfall and change the allometric relationships in cypress plantations.

2. Materials and Methods

2.1. Study Area

This study was conducted in Linshan Township, Yanting County (105°45′ E, 31°27′ N), located in the hilly area of the central Sichuan Basin, China. This region has typical low hill landforms with elevations of 400 m to 650 m. The climate is a subtropical monsoon climate, and the mean annual temperature and precipitation during 2008–2019 were 17.3 °C and 826 mm, respectively. This region is rich in purple shale that is easily weathered, and the soils are primarily purple soils. The aboveground vegetation is dominated by cypress plantations which were planted at a density of 4000–6000 plants/hm^2 in the 1970s. However, due to the high initial planting density and lack of necessary management, the ecological functioning of the plantation was in a degraded state, and the stand structure urgently needed adjustment.

2.2. Experimental Design

From 2005 to 2008, pure cypress plantations with similar aspects, slopes, and altitudes were selected for management experiments, including strip filling and ecological thinning practices. In the strip filling practice, after forming a series of 4–10 m bandwidths through artificial logging, native broadleaved tree species, such as *Alnus cremastogyne* and *Camptotheca acuminata*, were replanted for striping mixing. In comparison, the ecological thinning practice involved randomly cutting down part of the cypress trees with a removal of 10%–25% of the initial basal area. Meanwhile, the logging residues under both practices were removed from the experimental areas. Previous studies have found that there is a better ecological effect (e.g., high plant diversity and soil nutrient content) in the pattern of strip filling with a bandwidth of 6 m and in the pattern of ecological thinning with a thinning intensity of 20%–25% [33,34]. Accordingly, in the present study, we intended to further study the characteristics of litterfall under these two patterns.

In August 2019, five replicate plots (20 × 20 m in size and 50 m apart) were established in plantations with strip filling, ecological thinning and no artificial interference (pure forest), respectively. The stand characteristics of the three sampling plantations in 2019 are listed in Table S1. The meteorological data (wind speed, precipitation, temperature and relative humidity) from the Yanting Agro-ecological Station of Purple Soil, Chinese Academy of Sciences, from 2019 to 2020, are shown in Figure S1.

2.3. Litter Collection

Six litter traps (1 × 1 m in size and 50 cm above the ground) made of nylon materials with a mesh size of 1 mm were distributed randomly in each plot for the three plantations. Specifically, in the strip filling practice, six litter traps were placed equally within the strip cutting area and an adjacent uncut area. Litterfall was collected approximately monthly from September 2019 to August 2020. These monthly subsamples were pooled to create composite three-month samples in approximately the early dry season (January–March), late dry season (April–June), early wet season (July–September) and late wet season (October–December), according to the tropical forest division [15] and previous climatic data for the region. All litter materials were classified as leaf litter (including broad leaf and needle leaf litter), twig litter (including bark), reproductive organ litter (including flower, fruit and seed litter) and miscellaneous litter (unidentified components, such as indistinguishable detritus, dead animals and insect excrement). All litter materials were oven-dried separately at 65 °C, and the total litterfall was calculated as the sum of the dry litter mass for all litter components.

2.4. Statistical Analyses

Repeated-measures analysis of variance (ANOVA) was employed to test for effects of sampling time and plantation on litterfall production. One-way ANOVA with Tukey's honestly significant difference (HSD) test was used to test for the effect of plantation or season on litterfall production and the proportion of each litter component relative to total litterfall. Meanwhile, to calculate the relative influence of each factor, we employed redundancy analysis (RDA), using the "vegan" package, and hierarchical partitioning analysis using the "rdacca.hp" package in R [35]. Considering the consistency of the climate in each plantation and little within-year variation in vegetation factors, we analyzed the relative effects of vegetation factors on annual litterfall production and climate factors on monthly litterfall production. All statistical analyses were performed using R 4.0.5 (R Core Team 2021).

The allometric approach was used to analyze annual litterfall production, which was expressed as $Y = \beta X^{\alpha}$ or $\log Y = \alpha \log X + \log \beta$, where X and Y represent the total litterfall production and its different components (leaf, twig and reproductive organ litterfall) of each planation, respectively; α is the regression slope (scaling exponent); and β is the regression intercept. When the 95% confidence interval contained 1, α indicated that the scaling relationship between Y and X was isometric; otherwise, the relationship was interpreted as allometric [36].

After we log10-transformed the litterfall production data, a model II regression method (i.e., reduced major axis regression, RMA) was applied to estimate the regression parameters and test whether the slopes were significantly different from 1, as well as different among the plantations [37]. When the slopes for the different plantations did not differ significantly, a common slope was first provided, and then the Wald test was performed to test whether there were intercept or coaxial drifts between plantations. The model II regression analysis was performed in SMATR Version 2.0.

3. Results

3.1. Annual Litterfall Production

The annual total litter production and its components showed significant differences among the plantations ($p < 0.05$). The total annual litterfall production in plantations with strip filling (3672.46 kg·ha^{-2}, SF) and ecological thinning (2617.67 kg·ha^{-2}, ET) was significantly decreased by 10.8% and 36.44%, respectively, compared with that in the pure cypress plantation (4118.01 kg·ha^{-2}, PC) (Figure 1a). Similarly, both reconstruction treatments decreased the annual production of leaf and reproductive organ litter, and the greatest declines were observed in the plantation with ET (Figure 1b,d). In contrast, both reconstruction treatments increased the annual production of twig and miscellaneous litter, and a significant difference was observed only between the plantation with SF and the pure plantation (Figure 1c,e).

The proportion of each litter component relative to the annual total litterfall (except the reproductive organ litter) showed significant differences among the three plantations ($p < 0.05$). In general, leaves accounted for more than 80% of the annual total litterfall in the three plantations (Figure 2). Compared with the pure plantation, both reconstruction treatments significantly decreased the proportion of leaf litter relative to total litterfall and increased the proportion of twig and miscellaneous litter relative to total litterfall (Figure 2). Moreover, there were no significant differences in the proportions of leaf, twig and miscellaneous litter between the plantations with SF and ET ($p > 0.05$).

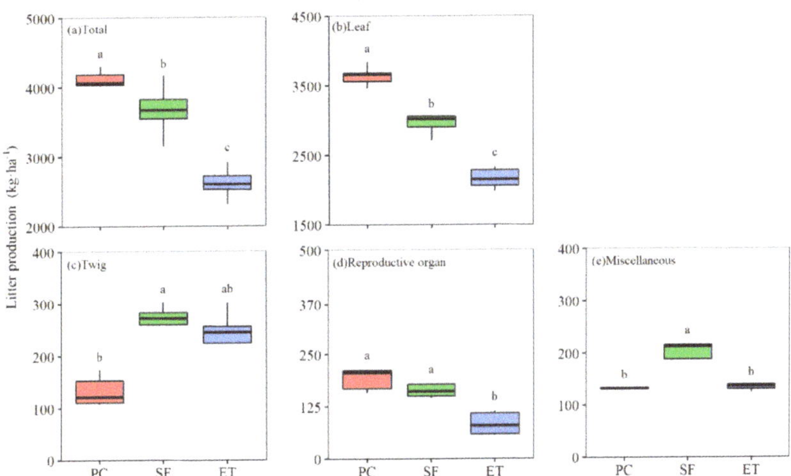

Figure 1. Annual production of total litterfall (**a**) and its components ((**b**) leaf litter; (**c**) twig litter; (**d**) reproductive organ litter; (**e**) miscellaneous litter) in different plantations. PC: pure cypress; SF: strip filling; ET: ecological thinning. Different letters denote significant differences in values at the $p < 0.05$ level.

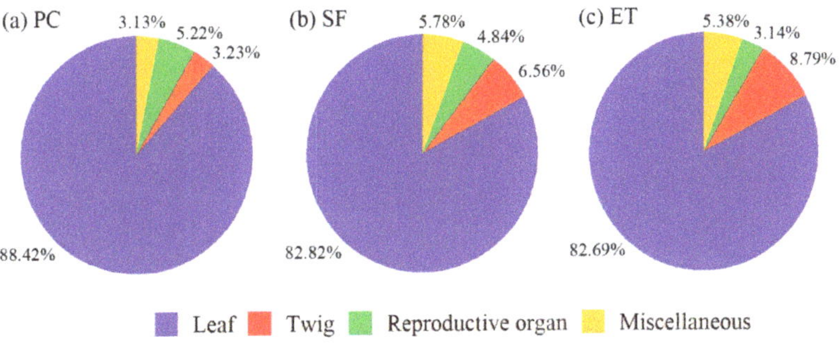

Figure 2. The proportions of each litter component in the total litterfall in different plantations. (**a**) PC: pure cypress; (**b**) SF: strip filling; (**c**) ET: ecological thinning.

3.2. Monthly Dynamics of Litterfall

Litterfall varied monthly, and this variation showed different patterns among plantations and litter components (Table S2, Figure 3). Regardless of the plantation, the production of total litter and leaf litter displayed unimodal dynamics, with the peak occurring in May (Figure 3a,b). Meanwhile, more than 70% of the total litterfall and leaf litterfall occurred during the late dry season (from April to June) and early wet season (from July to September) in each plantation, taking the rank order of pure cypress plantation > plantation with SF > plantation with ET (Figure 4a,b). However, the production of total litter and leaf litter in plantations with SF was obviously higher than that in the other two plantations during the late wet season (from October to December) and early dry season (from January to March).

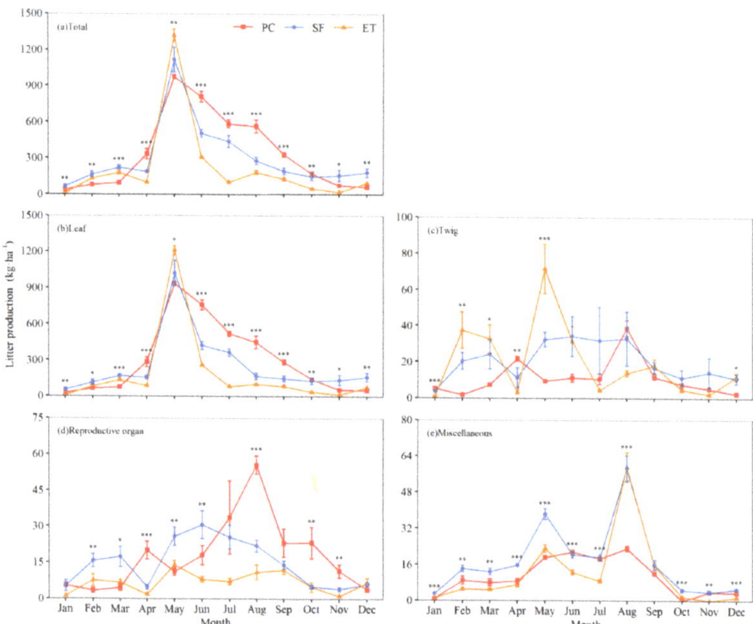

Figure 3. Monthly production of total litterfall (**a**) and its components ((**b**) leaf litter; (**c**) twig litter; (**d**) reproductive organ litter; (**e**) miscellaneous litter) in different plantations. PC: pure cypress; SF: strip filling; ET: ecological thinning. Asterisks denote significant differences among the three plantations within the same month: * indicates $p < 0.05$; ** indicates $p < 0.01$; and *** indicates $p < 0.001$.

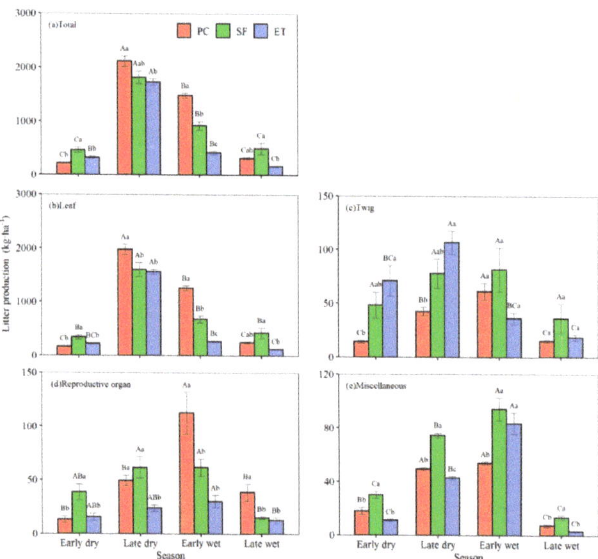

Figure 4. Seasonal production of total litterfall (**a**) and its components ((**b**) leaf litter; (**c**) twig litter; (**d**) reproductive organ litter; (**e**) miscellaneous litter) in different plantations. PC: pure cypress; SF: strip filling; ET: ecological thinning. Capital letters indicate differences among seasons within the same plantation, while lowercase letters indicate differences among the three plantations within the same seasons at the $p < 0.05$ level.

The production of twig litter showed a major peak in August and a smaller peak in April in the pure plantation. In contrast, it only peaked in May in the plantation with EF and had no clear pattern in the plantation with SF (Figure 3c). Meanwhile, more than 45% of the twig litter in the pure plantation occurred during the late dry season, while litter in the plantation with EF occurred during the early wet season. However, the production of twig litter in the plantation with ET was significantly higher than that in the pure cypress plantation (Figure 4c).

Moreover, the production of reproductive organ litter in the three plantations was unimodal, with the peaks in the pure, SF and ET plantations occurring in August, June and May, respectively (Figure 3d). In addition, it was highest in the plantation with SF and in the pure plantation during the dry and wet seasons, respectively (Figure 4d).

The production of miscellaneous litter was unimodal in the pure plantation, with the peak occurring in August, while it showed a major peak in August and a smaller peak in May, regardless of whether plantations used SF or ET (Figure 3e). For all seasons, the production of miscellaneous litter in plantations with SF was obviously higher than that in the other two plantations (Figure 4e).

3.3. Allometric Scaling Relationships of Annual Litterfall

Regardless of the plantation, significant isometric relationships were observed between leaf litterfall and total litterfall, while allometric relationships were observed for annual twig litterfall versus total litterfall and reproductive organ litterfall versus total litterfall (Table 1, Figure 5). The scaling slopes of leaf litterfall versus total litterfall in the different plantations did not differ significantly, with a common slope of 1.020. According to the Wald test, however, the intercept and coaxial drifts showed significant differences among the three plantations (Figure 5). Moreover, the scaling slopes of twig litterfall versus total litterfall in the plantation with ET and reproductive organ litterfall versus total litterfall in the plantations with SF and ET were significantly higher than those in the pure plantation.

Table 1. Allometric scaling relationships between different litter components in different plantations (L_{To}: total litterfall; L_L: leaf litterfall; L_T: twig litterfall; L_R: reproductive organ litterfall).

Parameter (y-x)	Pattern	n	Scaling Exponent			Test of Isometry	
			R^2	Slope	95%CI	*F* Value	Type
L_{To}-L_L	PC	30	0.936 ***	1.080 [a]	0.980~1.191	2.633	I
	SF	30	0.978 ***	1.020 [a]	0.963~1.081	0.496	I
	ET	30	0.962 ***	0.984 [a]	0.912~1.062	0.184	I
L_{To}-L_T	PC	30	0.498 ***	0.314 [b]	0.240~0.412	114.872 ***	A
	SF	30	0.546 ***	0.343 [b]	0.265~0.444	101.876 ***	A
	ET	30	0.527 ***	0.539 [a]	0.414~0.701	25.694 ***	A
L_{To}-L_R	PC	30	0.23 **	0.237 [b]	0.170~0.331	143.960 ***	A
	SF	30	0.316 **	0.547 [a]	0.399~0.75	16.799 ***	A
	ET	30	0.435 ***	0.443 [a]	0.333~0.591	40.725 ***	A
L_L-L_T	PC	30	0.431 ***	0.291 [b]	0.218~0.388	121.981 ***	A
	SF	30	0.444 ***	0.336 [b]	0.253~0.447	87.453 ***	A
	ET	30	0.362 ***	0.547 [a]	0.404~0.742	17.958 ***	A
L_L-L_R	PC	30	0.068	0.220	0.152~0.316	/	/
	SF	30	0.229 **	0.536 [a]	0.384~0.749	16.017 ***	A
	ET	30	0.321 ***	0.450 [a]	0.329~0.616	32.312 ***	A
L_T-L_R	PC	30	0.078	0.755	0.525~1.086	/	/
	SF	30	0.159 *	1.594 [a]	1.126~2.257	7.780 **	A
	ET	30	0.332 ***	0.823 [b]	0.603~1.123	1.620	I

Notes: PC: pure cypress; SF: strip filling; ET: ecological thinning. A indicates an allometric growth relationship, and I indicates an isometric growth relationship. Different lowercase letters indicate significant differences between the plantations ($p < 0.05$). * indicates $p < 0.05$; ** indicates $p < 0.01$; and *** indicates $p < 0.001$.

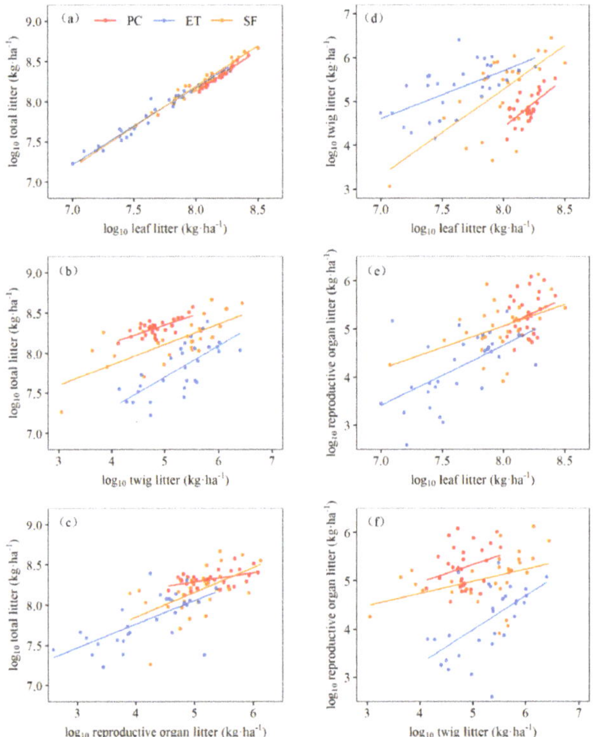

Figure 5. Allometric relationships between each litter component and total litterfall (**a–c**) and among litter components (**d–f**) in different plantations. PC: pure cypress; SF: strip filling; ET: ecological thinning.

Significant allometric relationships were observed for leaf litterfall versus twig litterfall in different plantations, and the scaling slope in the plantation with ET was significantly higher than that in the pure plantation (Table 1, Figure 5). Meanwhile, significant allometric relationships were observed for leaf litterfall versus reproductive organ litterfall in the plantations with SF and ET. Moreover, for twig litterfall and reproductive organ litterfall, a significant positive allometric relationship was observed in the plantation with SF, while an isometric relationship was observed in the plantation with ET. However, the relationships of leaf litterfall versus reproductive organ litterfall and twig litterfall versus reproductive organ litterfall in the pure plantation could not be expressed by allometric equations.

3.4. Factors Driving the Changes in Litterfall Production

RDA revealed that vegetation factors explained 62.40% of the variation in annual litterfall production (Figure 6a). Annual total litter, leaf litter and reproductive organ litter all showed high production under conditions of lower diameter at breast height and tree height. Annual twig litter and miscellaneous litter showed high production under conditions of higher species richness and lower stand density and canopy density. Meanwhile, hierarchical partitioning analysis showed that diameter at breast height and tree height were the most important factors ($p < 0.05$), explaining 17.85% and 16.14% of the variation, respectively (Table 2).

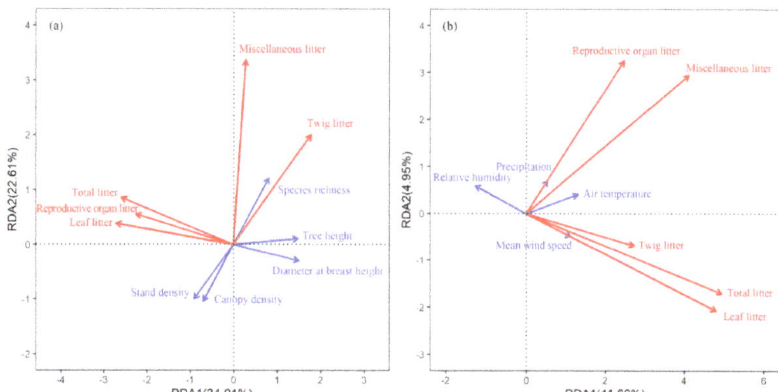

Figure 6. Relationships of vegetation factors with annual litterfall production (**a**) and of climate factors with monthly litterfall production (**b**) based on redundancy analysis (RDA). The data used in this analysis were the litterfall from the three plantations.

Table 2. The relative influences (RI%) of vegetation factors on annual litterfall and of climate factors on monthly litterfall and their *p* values based on redundancy analysis and hierarchical partitioning analysis.

Factors	RI (%)	*p* Value
Annual litterfall		
Stand density	9.66	0.104
Canopy density	7.11	0.130
Species richness	11.64	0.063
Tree height	**16.14**	**0.026**
Diameter at breast height	**17.85**	**0.024**
Monthly litterfall		
Mean wind speed	**9.09**	**0.001**
Precipitation	**5.41**	**0.001**
Air temperature	**16.38**	**0.001**
Relative humidity	**15.95**	**0.001**

Bold *p* values are significant ($p < 0.05$).

The climate factors explained 46.84% of the variation in monthly litterfall production (Figure 6b). Monthly total litter, leaf litter and twig litter all showed high production under conditions of higher mean wind speed and lower relative humidity. Monthly reproductive organ litter and miscellaneous litter showed high production under conditions of higher precipitation and air temperature. Meanwhile, in terms of the degree of influence on monthly litterfall production ($p < 0.05$), the climate factors ranked as follows: air temperature (16.38%) > relative humidity (15.95%) > mean wind speed (9.09%) > precipitation (5.41%) (Table 2).

4. Discussion

4.1. Annual Litter Production

Forest litterfall is strongly influenced by stand density, plant composition and tree age for a similar climate scenario [17,38,39]. In the present study, we observed that annual total litterfall production ranged from 2617.67 to 4118.01 kg·hm^{-2} in different plantations (Figure 1a), which was within the range of tropical and subtropical needle-leaved evergreen forests (531–10523 kg·hm^{-2}) [40]. Meanwhile, we observed that the annual production of total litterfall across the plantations exhibited the following order: pure plantation > plantation with SF > plantation with ET (Figure 1a). In general, for stand level, litterfall production increased with increasing stand density [18,39]. Accordingly, the decreased stand density in plantations with SF and ET may directly lead to decrease in litterfall. However, the variation pattern in total litter production among the investigated plantations was not consistent with their changes in tree density (Figure 6a). Bray and Gorham [41] first showed that litter production appears to be little affected by differences in plant density within closed-canopy forests, and some other studies exhibiting plantation management practices with the main purpose of reducing stand density also concluded that there was no change or significant increase in the production of litterfall [42,43]. Based on these findings the variation pattern may be related to differences in species richness caused by different management practices and the time lapse after management. In addition, the plantation with SF (lower stand density but higher species richness) showed higher total litterfall than the plantation with ET (higher stand density but lower species richness) in our study, which was consistent with previous studies suggesting that mixed plantations had higher litterfall production than pure plantations [24]. Although the trend of leaf litterfall was consistent with that of total litterfall, the proportion of broadleaf to total leaf litterfall in the plantation with SF (19.12%) was higher than that in the plantation with ET (4.57%) (Figure S2), which also suggested that the leaf litter produced by artificial replanting of broad-leaved species was more abundant than that produced by natural succession within the same time period after management. This result further indicated that the plantation with SF was more conducive to increasing the species diversity of litter, thus forming different carbon and nutrient cycling patterns. Similarly, the production of twig litterfall in the plantations with SF and ET was higher than that in the pure plantation. A likely explanation is that it was positively correlated with species richness (Figure 6a).

Consistent with many previous studies [17,20,28], we found that the production of litterfall in the three plantations was dominated by leaf litter, which emphasized the key role of leaf litter in material circulation and energy flow in cypress plantations. However, we also found that both management practices significantly decreased the proportion of leaf litter relative to total litterfall and increased the proportion of twig and miscellaneous litter relative to total litterfall (Figure 2). Our findings therefore suggest that the responses of production to forest management practices varied with the litter components. Additionally, although our study lasted one year, the reduction in leaf litterfall proportion provided a new approach to explore the material-related mechanisms of management practices, as the stand structure of cypress plantations is relatively stable after 11 years of management.

4.2. Litterfall Monthly Dynamics

The monthly dynamics of litterfall were unimodal and bimodal, which mainly depended on the climatic conditions and ecological characteristics of the tree species [40,44,45]. Therefore, the differences in canopy density and species composition under different forest management practices may result in different monthly variation characteristics of litterfall. However, many previous studies have suggested that these management practices (thinning, prescribed burning) did not alter the monthly dynamics of leaf litterfall [42,43]. Consistent with previous findings, we observed that the production of leaf litterfall showed unimodal dynamics, with the peak occurring in May in the three plantations, and reasonable explanations for this result might be as follows: (1) cypress was dominant in the three plantations, and the shedding time of old leaves was relatively concentrated. (2) The climate factors significantly influenced the intra-annual litterfall dynamics [16,46]. The obvious positive correlation between leaf litterfall and wind speed in Figure 6b could also confirm this; therefore, the higher wind speed in May might have led to the high production of leaf litterfall. Meanwhile, the production of total litterfall was controlled by leaf litter; therefore, the monthly dynamics of total litter and leaf litter were similar.

Unlike the leaf litter, we found different monthly dynamic patterns of twig litter, reproductive organ litter and miscellaneous litter in the three plantations. Previous studies showed that the monthly collected twig litterfall that came from dead branches was less directly affected by phenology but was sensitive to climatic factors, such as wind speed and rainfall [45,46]. Therefore, plantations with different management practices can vary in their response to such climate factors and, thus, result in distinct monthly dynamic patterns of twig litter. In general, the shedding time of reproductive organs varies among species, but temperature, wind speed and rainfall could also alter the characteristics of litterfall [46,47]. Therefore, although the production of reproductive organ litterfall showed unimodal dynamics in the three plantations, its peak of production could occur in different months. Nevertheless, there was no uniform pattern in the monthly dynamic changes in miscellaneous litter among different plantations resulting from the relatively complex components, although it was positively correlated with rainfall and air temperature (Figure 6b). Interestingly, we found that the production of leaf litter and reproductive organ litter was significantly higher in the early dry season and lower in the wet dry season than in the pure plantation (Figure 4). Since different plant species have different litterfall production rhythms (e.g., conifers and broadleaved trees) [15], plantations with the highest species richness are more likely to show opposite seasonal trends. Overall, based on the one-year study, we highlight the importance of using monthly or seasonal variations in litter input and subsequent decomposition to explore the effects of plantation management practices on material cycling and energy flows in ecosystems.

4.3. Litter Components and Allometric Scaling Relationships of Litterfall

Variations in species, environmental gradients, human disturbances and other factors would lead to the self-regulation of resource allocation in plants, which would be reflected by different survival strategies [48–50]. Consistent with our expectations, the effects of reconstruction management on the relationships between each litter component and the total litterfall, as well as between litter components, were related to the reconstruction treatments (Table 1, Figure 5). Leaf litterfall and total litterfall showed an isometric relationship in the three plantations, which was in agreement with the results of Ma et al. [27] and Fu et al. [28], and this might be because leaf litter was the major component (more than 80% of the total litterfall production) (Figure 2). The Wald test revealed that the differences between the intercept and coaxial drifts were significant, although there were no significant differences in scaling exponents among the three plantations, and this result indicated that the isometric growth trajectories of leaf litterfall and total litterfall were not completely consistent in the three plantations. Furthermore, the allometric scaling exponents of twig litterfall and total litterfall in the plantation with ET were significantly higher than those in the plantation with ST and the pure plantation, which suggested that within a certain range of total litterfall, the plantation with ET allocated more production to twig litter. Similarly, the plantations with reconstruction allocated more production to reproductive organ litter than the pure plantation. More reproductive organ litterfall is preferred for breeding populations [51]; therefore, this result could explain the high vegetation diversity in plantations with reconstruction treatments.

Leaves are important photosynthetic organs, and the presence of more leaves accelerates the transport of nutrients from aboveground to belowground, thus promoting plant growth [49]. In this study, the allometric scaling exponent between leaf litterfall and twig litterfall in the plantation with ET was significantly higher than that in the plantation with ST and the pure plantation (Figure 5, Table 1), which suggested that within a certain range of twig litterfall, the plantation with ET allocated less production to leaf litterfall. This result further indicated that the plantation with SF might be superior to that with ET only from the perspective of plant growth. In addition, consistent with previous studies [28], we observed that some relationships (leaf litterfall versus reproductive organ litterfall and twig litterfall versus reproductive organ litterfall) could not be expressed by allometric equations in the pure plantation (Table 1), which may be related to reproductive organ litterfall being controlled by complex micrometeorological factors [51]. Unlike in the pure plantation, significant differences in relative growth slopes between the plantations with management practices were observed. These disparate responses further emphasized that cypress plantations could modify resource allocation to organs through physiological integration after intense disturbance and that those allocations were similar to disturbance patterns.

5. Conclusions

In this study, we explored the effects of strip filling and ecological thinning on the litterfall production and monthly dynamics of cypress plantations. In contrast with our hypothesis, both management practices significantly reduced the annual production of total litter, leaf litter and reproductive organ litter, and only strip filling significantly increased twig litterfall and miscellaneous litterfall. Nevertheless, management practices obviously changed the monthly dynamic patterns of twig litterfall, reproductive organ litterfall and miscellaneous litterfall but had few effects on total litterfall and leaf litterfall. Furthermore, the dominant factors affecting annual and monthly litterfall production were diameter at breast height and wind speed, respectively. These results confirmed the significant responses of litterfall production to management practices in cypress plantations, but the investigation remains limited in terms of management time and study continuity. There is a high need to further integrate carbon and nutrient inputs, address their cycling processes, and accurately evaluate the effects of strip filling and ecological thinning via long-term continuous research. Even so, the results presented here provide a primary reference for exploring the material mechanism by which management practices affect cypress plantations.

Supplementary Materials: The following supporting information can be downloaded at: https://www.mdpi.com/article/10.3390/f13101581/s1, Figure S1: Monthly average wind speed, air temperature, relative humidity and total precipitation during the study period; Figure S2: The proportion of broad leaf and needle leaf to total leaf litterfall in different plantations; Table S1: General descriptions of different plantations in the study site; Table S2: Repeated-measures ANOVA results for the responses of litterfall production to sampling time and plantations.

Author Contributions: Conceptualization, Q.W. (Qinggui Wu), Y.Y., C.Y. and J.H.; investigation, L.W., H.Y. and Q.W. (Qiang Wang); formal analysis, Y.Y., H.Y. and C.Y.; visualization, H.Y., Y.Y. and C.Y.; writing—original draft preparation, L.W., H.Y. and Y.Y.; writing—review and editing, Y.Y., J.Y., Q.D. and Q.W. (Qinggui Wu); funding acquisition, Q.W. (Qinggui Wu), C.Y., Y.Y. and Q.D. All authors have read and agreed to the published version of the manuscript.

Funding: This research was supported by the National Natural Science Foundation of China (32071747), the National Natural Science Foundation of Sichuan Province (2022NSFSC0087, 2022NS-FSC1173), the Open Fund of Ecological Security and Protection Key Laboratory of Sichuan Province, Mianyang Normal University (ESP2203) and the Research Fund of Mianyang Normal University (QD2020A18).

Institutional Review Board Statement: Not applicable.

Informed Consent Statement: Not applicable.

Data Availability Statement: Not applicable.

Acknowledgments: We are grateful to the Ecological Security and Protection Key Laboratory of Sichuan Province.

Conflicts of Interest: The authors declare no conflict of interest. The funders had no role in the design of the study; in the collection, analyses, or interpretation of data; in the writing of the manuscript, or in the decision to publish the results.

References

1. FAO. *Global Forest Resources Assessment 2020: Main Report*; FAO: Rome, Italy, 2020.
2. Liu, S.; Yang, Y.; Wang, H. Development strategy and management countermeasures of planted forests in China: Transforming from timber-centered single objective management towards multi-purpose management for enhancing quality and benefits of ecosystem services. *Acta Ecol. Sin.* **2018**, *38*, 1–10, (In Chinese with English Abstract).
3. Zhou, Q.; Li, F.; Cai, X.; Rao, X.; Zhou, L.; Liu, Z.; Lin, Y.; Fu, S. Survivorship of plant species from soil seedbank after translocation from subtropical natural forests to plantation forests. *For. Ecol. Manag.* **2019**, *432*, 741–747. [CrossRef]

4. Sujii, P.S.; Schwarcz, K.D.; Grando, C.; de Aguiar Silvestre, E.; Mori, G.M.; Brancalion, P.H.S.; Zucchi, M.I. Recovery of genetic diversity levels of a Neotropical tree in Atlantic Forest restoration plantations. *Biol. Conserv.* **2017**, *211*, 110–116. [CrossRef]
5. Mohler, C.; Bataineh, M.; Bragg, D.C.; Ficklin, R.; Pelkki, M.; Olson, M. Long-term effects of group opening size and site preparation method on gap-cohort development in a temperate mixedwood forest. *For. Ecol. Manag.* **2021**, *48*, 118616. [CrossRef]
6. Hou, L.; Zhang, Y.; Li, Z.; Shao, G.; Song, L.; Sun, Q. Comparison of soil properties, understory vegetation species diversities and soil microbial diversities between Chinese Fir plantation and close-to-natural forest. *Forests* **2021**, *12*, 632. [CrossRef]
7. Hua, F.; Bruijnzeel, L.A.; Meli, P.; Martin, P.A.; Zhang, J.; Nakagawa, S.; Miao, X.; Wang, W.; McEvoy, C.; Peña-Arancibia, J.L.; et al. The biodiversity and ecosystem service contributions and trade-offs of forest restoration approaches. *Science* **2022**, *376*, 839–844. [CrossRef]
8. Li, X.; Li, Y.; Zhang, J.; Peng, S.; Chen, Y.; Cao, Y. The effects of forest thinning on understory diversity in China: A meta-analysis. *Land Degrad. Dev.* **2020**, *31*, 1225–1240. [CrossRef]
9. Felton, A.; Lindbladh, M.; Brunet, J.; Fritz, Ö. Replacing coniferous monocultures with mixed-species production stands: An assessment of the potential benefits for forest biodiversity in northern Europe. *For. Ecol. Manag.* **2010**, *260*, 939–947. [CrossRef]
10. Bei, P.; Si, S.; Zhou, D.; Chen, J.; Zhao, R.; Mu, C. Effects of strip reform on understory plant diversity of Cupressus funebris plantations in Hilly Areas of Central Sichuan. *J. Sichuan For. Sci. Technol.* **2018**, *39*, 1–6, (In Chinese with English Abstract).
11. Zhang, W.; Yang, K.; Lyu, Z.; Zhu, J. Microbial groups and their functions control the decomposition of coniferous litter: A comparison with broadleaved tree litters. *Soil Biol. Biochem.* **2019**, *133*, 196–207. [CrossRef]
12. Sayer, E.J.; Tanner, E.V.J. Experimental investigation of the importance of litterfall in lowland semi-evergreen tropical forest nutrient cycling. *J. Ecol.* **2010**, *98*, 1052–1062. [CrossRef]
13. Neumann, M.; Ukonmaanaho, L.; Johnson, J.; Benham, S.; Vesterdal, L.; Novotný, R.; Hasenauer, H. Quantifying carbon and nutrient input from litterfall in European forests using field observations and modeling. *Glob. Biogeochem. Cycles* **2018**, *32*, 784–798.
14. Berg, B.; McClaughgerty, C.A. *Plant Litter: Decomposition, Humus Formation, Carbon Sequestration*, 3rd ed.; Springer: New York, NY, USA, 2014.
15. Paudel, E.; Dossa, G.G.; Xu, J.; Harrison, R.D. Litterfall and nutrient return along a disturbance gradient in a tropical montane forest. *For. Ecol. Manag.* **2015**, *353*, 97–106.
16. Zhang, H.; Yuan, W.; Dong, W.; Liu, S. Seasonal patterns of litterfall in forest ecosystem worldwide. *Ecol. Complex.* **2014**, *20*, 240–247.
17. Tang, J.; Cao, M.; Zhang, J.; Li, M. Litterfall production, decomposition and nutrient use efficiency varies with tropical forest types in Xishuangbanna, SW China: A 10-year study. *Plant Soil* **2010**, *335*, 271–288.
18. Acenolaza, P.G.; Zamboni, L.P.; Rodriguez, E.E.; Gallardo, J.F. Litterfall production in forests located at the Pre-delta area of the Parana River (Argentina). *Ann. For. Sci.* **2010**, *67*, 311. [CrossRef]
19. Zhu, X.; Liu, W.; Chen, H.; Deng, Y.; Chen, C.; Zeng, H. Effects of forest transition on litterfall, standing litter and related nutrient returns: Implications for forest management in tropical China. *Geoderma* **2019**, *333*, 123–134. [CrossRef]
20. Ni, X.; Lin, C.; Chen, G.; Xie, J.; Yang, Z.; Liu, X.; Yang, Y. Decline in nutrient inputs from litterfall following forest plantation in subtropical China. *For. Ecol. Manag.* **2021**, *496*, 119445.
21. Lin, N.; Bartsch, N.; Heinrichs, S.; Vor, T. Long-term effects of canopy opening and liming on leaf litter production, and on leaf litter and fineroot decomposition in a European beech (*Fagus sylvatica* L.) forest. *For. Ecol. Manag.* **2015**, *338*, 183–190. [CrossRef]
22. Wu, W.; Zhou, X.; Wen, Y.; Zhu, H.; You, Y.; Qin, Z.; Li, Y.; Huang, X.; Yan, L.; Li, H. Coniferous-broadleaf mixture increases soil microbial biomass and functions accompanied by improved stand biomass and litter production in subtropical China. *Forests* **2019**, *10*, 879.
23. Vigulu, V.; Blumfield, T.J.; Reverchon, F.; Bai, S.; Xu, Z. Nitrogen and carbon cycling associated with litterfall production in monoculture teak and mixed species teak and flueggea stands. *J. Soil Sediments* **2019**, *19*, 1672–1684. [CrossRef]
24. Tongkaemkaew, U.; Sukkul, J.; Sumkhan, N.; Panklang, P.; Brauman, A.; Ismail, R. Litterfall, litter decomposition, soil macrofauna, and nutrient contents in rubber monoculture and rubber-based agroforestry plantations. *For. Soc.* **2018**, *2*, 138–149. [CrossRef]
25. Jimenez, M.N.; Navarro, F.B. Thinning effects on litterfall remaining after 8 years and improved stand resilience in Aleppo pine afforestation (SE Spain). *J. Environ. Manag.* **2016**, *169*, 174–183. [CrossRef] [PubMed]
26. Blanco, J.A.; Imbert, J.B.; Castillo, F.J. Influence of site characteristics and thinning intensity on litterfall production in two *Pinus sylvestris* L. forests in the western Pyrenees. *For. Ecol. Manag.* **2006**, *237*, 342–352. [CrossRef]
27. Ma, Y.; Cheng, D.; Zhong, Q.; Jin, B.; Xu, C.; Hu, B. Anisotropic ratio of different fractions of Chinese forest apoplankton. *Chin. J. Plant Ecol.* **2013**, *37*, 1071, (In Chinese with English Abstract). [CrossRef]
28. Fu, C.; Yang, W.; Tan, B.; Xu, Z.; Zhang, Y.; Yang, J.; Wu, F. Seasonal dynamics of litterfall in a sub-alpine spruce-fir forest on the eastern Tibetan Plateau: Allometric scaling relationships based on one year of observations. *Forests* **2017**, *8*, 314. [CrossRef]
29. Clark, D.A.; Brown, S.; Kicklighter, D.W.; Chambers, J.Q.; Thomlinson, J.R.; Ni, J. Measuring net primary production in forests: Concepts and field methods. *Ecol. Appl.* **2001**, *11*, 356–370. [CrossRef]
30. Han, D.; Zhang, L.; Feng, M.; Wu, T.; Tie, L.; Li, W.; Wang, Y. Effects of transformation of Cypress funebrisf forest on soil microorganism, enzymatic activity and nutrient. *J. Northeast For. Univ.* **2016**, *44*, 57–62, (In Chinese with English Abstract).
31. Wang, Y.; Chen, S.; He, W.; Ren, J.; Wen, X.; Wang, Y.; Fan, C. Shrub diversity and niche characteristics in the initial stage of reconstruction of low-efficiency Cupressus funebris stands. *Forests* **2021**, *12*, 1492. [CrossRef]

32. Chen, J.; Niu, M.; Gong, G.; Zhou, D.; Zhu, Z.; Li, Y.; Zhen, S.; Xie, T.; Mu, C. Transformation of cypress monoculture by cutting in band and replanting in Sichuan central basin. *Southwest China J. Agric. Sci.* **2019**, *32*, 636–644, (In Chinese with English Abstract).

33. Gong, G.; Niu, M.; Mu, C.; Chen, J.; Li, Y.; Zhu, Z.; Zhen, S. Impacts of different thinning intensities on growth of Cupressus funebris plantation and understory plants. *Sci. Silvae Sin.* **2015**, *51*, 8–15, (In Chinese with English Abstract).

34. Li, Y.; Gong, G.; Zheng, S.; Chen, J.; Mu, C.; Zhu, Z.; Niu, M. Impact on water and soil conservation of different bandwidths in low-efficiency cypress forest transformation. *Acta Ecol. Sin.* **2013**, *33*, 934–943, (In Chinese with English Abstract).

35. Lai, J.; Zou, Y.; Zhang, J.; Peres-Neto, P. Generalizing hierarchical and variation partitioning in multiple regression and canonical analyses using the rdacca. *hp R package. Methods Ecol. Evol.* **2022**, *13*, 782–788. [CrossRef]

36. Rich, P.M. Plant allometry: The scaling of form and process. *Ecology* **1995**, *76*, 2342–2343. [CrossRef]

37. Warton, D.I.; Weber, N.C. Common slope tests for bivariate errors-in-variables models. *Biom. J.* **2002**, *44*, 161–174. [CrossRef]

38. Zhou, J.; Lang, X.; Du, B.; Zhang, H.; Liu, H.; Zhang, Y.; Shang, L. Litterfall and nutrient return in moist evergreen broad-leaved primary forest and mixed subtropical secondary deciduous broad-leaved forest in China. *Eur. J. For. Res.* **2016**, *135*, 77–86. [CrossRef]

39. Bahru, T.; Ding, Y. Effect of stand density, canopy leaf area index and growth variables on Dendrocalamus brandisii (Munro) Kurz litter production at Simao District of Yunnan Province, southwestern China. *Glob. Ecol. Conserv.* **2020**, *23*, e01051. [CrossRef]

40. Dai, Y.; Gong, F.; Yang, X.; Chen, X.; Su, Y.; Liu, L.; Sun, Q. Litterfall seasonality and adaptive strategies of tropical and subtropical evergreen forests in China. *J. Plant Ecol.* **2022**, *15*, 320–334. [CrossRef]

41. Bray, J.R.; Gorham, E. Litter production in forests of the world. *Adv. Ecol. Res.* **1964**, *2*, 101–157.

42. Roig, S.; del Río, M.; Cañellas, I.; Montero, S. Litter fall in Mediterranean Pinus pinaster Ait. *stands under different thinning regimes. For. Ecol. Manag.* **2005**, *206*, 179–190.

43. Espinosa, J.; Madrigal, J.; Pando, V.; de la Cruz, A.C.; Guijarro, M.; Hernando, C. The effect of low-intensity prescribed burns in two seasons on litterfall biomass and nutrient content. *Int. J. Wildland Fire* **2020**, *29*, 1029–1041.

44. Souza, S.R.; Veloso, M.D.; Espírito-Santo, M.M.; Silva, J.O.; Sánchez-Azofeifa, A.; Souza e Brito, B.G.; Fernandes, G.W. Litterfall dynamics along a successional gradient in a Brazilian tropical dry forest. *For. Ecosyst.* **2019**, *6*, 35.

45. Morffi-Mestre, H.; Ángeles-Pérez, G.; Powers, J.S.; Andrade, J.L.; Huechacona Ruiz, A.H.; May-Pat, F.; Dupuy, J.M. Multiple factors influence seasonal and interannual litterfall production in a tropical dry forest in Mexico. *Forests* **2020**, *11*, 1241.

46. Sun, X.; Liu, F.; Zhang, Q.; Li, Y.; Zhang, L.; Wang, J.; Wang, X. Biotic and climatic controls on the interannual variation in canopy litterfall of a deciduous broad-leaved forest. *Agric. For. Meteorol.* **2021**, *307*, 108483. [CrossRef]

47. Liu, W.; Fox, J.E.; Xu, Z. Litterfall and nutrient dynamics in a montane moist evergreen broad-leaved forest in Ailao Mountains, SW China. *Plant Ecol.* **2003**, *164*, 157–170, (In Chinese with English Abstract). [CrossRef]

48. Sun, J.; Niu, S.; Wang, J. Divergent biomass partitioning to aboveground and belowground across forests in China. *J. Plant Ecol.* **2018**, *11*, 484–492.

49. Reich, P.B.; Wright, I.J.; Cavender-Bares, J.; Craine, J.M.; Oleksyn, J.; Westoby, M.; Walters, M.B. The evolution of plant functional variation: Traits, spectra, and strategies. *Int. J. Plant Sci.* **2003**, *164*, 143–164.

50. Price, C.A.; Weitz, J.S.; Savage, V.M.; Stegen, J.; Clarke, A.; Coomes, D.A.; Dodds, P.S.; Etienne, R.S.; Kerkhoff, A.J.; McCulloh, K.; et al. Testing the metabolic theory of ecology. *Ecology* **2012**, *15*, 1465–1474.

51. Niu, K.; Choler, P.; Zhao, B.; Du, G. The allometry of reproductive biomass in response to land use in Tibetan alpine grasslands. *Funct. Ecol.* **2009**, *23*, 274–283. [CrossRef]

Article

Monthly Dynamical Patterns of Nitrogen and Phosphorus Resorption Efficiencies and C:N:P Stoichiometric Ratios in *Castanopsis carlesii* (Hemsl.) Hayata and *Cunninghamia lanceolata* (Lamb.) Hook. Plantations

Yaoyi Zhang, Jing Yang, Xinyu Wei, Xiangyin Ni and Fuzhong Wu *

Key Laboratory for Humid Subtropical Eco-Geographical Processes of the Ministry of Education, School of Geographical Sciences, Fujian Normal University, Fuzhou 350007, China
* Correspondence: wufzchina@163.com

Abstract: Trees can resorb nutrients to preserve and reuse them before leaves fall, which could efficiently adapt to environmental changes. However, the nutrient requirements of trees in different months with seasonal climate changes are often neglected. In this study, we selected plantations of an evergreen broadleaf tree (*Castanopsis carlesii* (Hemsl.) Hayata) and a coniferous tree (*Cunninghamia lanceolate* (Lamb.) Hook.) in the subtropics. The monthly dynamics of leaf nitrogen (N) and phosphorus (P) resorption efficiencies and C:N:P stoichiometric ratios were checked along a growing season from April to October 2021. Trees in both plantations exhibited efficient N and P resorption but with significant monthly variations. The N and P resorption efficiencies in the *Cunninghamia lanceolata* plantation ranged from 34.26% to 56.28% and 41.01% to 54.85%, respectively, and were highest in September. In contrast, N and P resorption efficiencies in the *Castanopsis carlesii* plantation ranged from 11.25% to 34.23% and 49.22% to 58.72%, respectively, and were highest in July. Compared with the *Cunninghamia lanceolata*, the C:N of the *Castanopsis carlesii* plantation was significantly lower, while its C:P was significantly higher in May and September. The *Castanopsis carlesii* plantation was strongly limited by P (the N:P ratios in mature leaves were higher than 20), whereas the *Cunninghamia lanceolata* plantation might be limited by both N and P (the N:P ratios in mature leaves were between 10 and 20). In addition, the statistical analyses revealed that temperature and precipitation were significantly associated with N and P resorption efficiencies, but the relationships were controlled by forest types. These findings highlight that efficient resorption of N and P may be beneficial in regulating nutrient limitation and balance in subtropical forest ecosystems. These results contribute to the understanding of N and P utilization strategies of trees and provide a theoretical basis for vegetation management in the subtropics.

Keywords: nutrient reuse strategy; nutrient limitation; ecological stoichiometric ratios; mature and senescent leaves; subtropical forest plantation

Citation: Zhang, Y.; Yang, J.; Wei, X.; Ni, X.; Wu, F. Monthly Dynamical Patterns of Nitrogen and Phosphorus Resorption Efficiencies and C:N:P Stoichiometric Ratios in *Castanopsis carlesii* (Hemsl.) Hayata and *Cunninghamia lanceolata* (Lamb.) Hook. Plantations. *Forests* **2022**, *13*, 1458. https://doi.org/10.3390/f13091458

Academic Editor: Heinz Rennenberg

Received: 30 August 2022
Accepted: 8 September 2022
Published: 10 September 2022

Publisher's Note: MDPI stays neutral with regard to jurisdictional claims in published maps and institutional affiliations.

1. Introduction

Nutrient resorption is a process by which plants transfer nutrients from senescent organs to fresh tissues [1–3]. It is an important strategy for trees to conserve and use nutrients efficiently and profoundly affects several key processes in ecosystems, such as plant nutrient uptake, ecochemical balance, and carbon cycling [4–6]. The efficiency of this nutrient transfer (nutrient resorption efficiency) can be quantified by the percentage reduction in nutrient content between mature and senescent leaves [7–9]. However, nutrients that are not transferred from senescent leaves fall to the ground with withered leaves and are recycled in the ecosystem after the withered leaves are decomposed. Therefore, the final nutrient content in senescent leaves can also be used as an indicator of nutrient resorption (nutrient resorption proficiency) [7], which reflects the extent of nutrient transfer from leaves [7,8]. However, it is well known that trees often have various nutrient requirements to satisfy the growing need with monthly climate changes [10–12], but the dynamical patterns of nutrient resorption are often neglected.

Temperature and precipitation are important drivers for regulating the nutrient utilization of trees [13–15]. Previous studies have shown that P resorption efficiency increases with increasing average annual temperature and average annual precipitation, while N resorption efficiency decreases [16]. However, other studies have suggested that N and P resorption efficiencies decrease with increasing average annual temperature or average annual precipitation [17,18]. Most of these studies were limited to large regional scales, and differences in the species selected as well as the sampling months may contribute to the uncertainty of the results. Currently, even less is known about the dynamics of leaf resorption of forest trees in response to seasonal climate changes. Given the importance of resorption for plants and ecosystems, more data are needed to clarify this issue so that we can incorporate nutrient resorption into models of plant responses to climate changes [19].

The trees with different functional types also develop various nutrient resorption strategies [8,20,21]. A meta-analysis proposed that the P resorption efficiency was higher in coniferous plantations than in broadleaf plantations, while the N resorption efficiency has no significant difference [22]. Consistent results were also observed in many natural forests [21,23]. However, a study in the karst ecosystems of Southwestern China documented that coniferous trees have a lower P resorption efficiency and a higher N resorption efficiency than broadleaf trees [24]. The differences in nutrient resorption among functional types of forests remain uncertain.

In addition, element stoichiometry focuses on the interactions and balances among elements [25–27]. Element stoichiometric balance exerts an essential role in species development, community structure, and adaptation to environmental stresses [28,29]. Specifically, C:N:P stoichiometric ratios of plants can reveal the nutrient utilization and relative nutrient limitation [30,31]. At present, C:N:P stoichiometric ratios have been evaluated at both global and regional scales, and vary with several factors such as plant functional types [2,8], soil nutrients [5,6,32], and climate changes [33,34], which can provide theoretical support for the management and conservation of forest.

Compared to other regions at the same latitude, the subtropical region of China nurtures vast humid subtropical forests [35]. In recent years, amounts of natural broadleaf evergreen forests in the region have been replaced by large areas of mono-structured plantations to meet the needs of economic development [36]. More than 60% of China's plantations are located in warm and humid subtropical regions [37]. However, N and P in subtropical forest soils are highly susceptible to leaching out of the ecosystem by rainfall [18], thus limiting plant growth. Therefore, the N and P resorption strategies of subtropical plants may be more important compared to other regions. The *Cunninghamia lanceolate* (Lamb.) Hook. is an important fast-growing tree for afforestation in subtropical regions, accounting for 17.3% of total plantations in China [38]. *Castanopsis carlesii* (Hemsl.) Hayata is one of the most typical trees of evergreen broadleaf forests in subtropical regions of China [39]. The results in studying the nutrient resorption and C:N:P stoichiometric characteristics in *Castanopsis carlesii* and *Cunninghamia lanceolata* plantations can help to understand the nutrient use strategies of subtropical forests.

Therefore, to understand the dynamics of nutrient use strategies in response to changes in temperature and precipitation in different plantations, we investigated the C, N, and P contents and stoichiometric ratios in mature and senescent leaves and nutrient resorption of *Cunninghamia lanceolata* plantation and *Castanopsis carlesii* plantation in a typical subtropical region, where both plantations have similar climates and land-use histories. We hypothesized that N and P resorption and C:N:P stoichiometric ratios in the leaves may vary significantly with forest types and are regulated by seasonal climate changes. Our objectives were to determine (i) what are the differences of nutrient resorption and C:N:P stoichiometric ratios between the two plantations, and (ii) whether and how nutrient resorption strategies were influenced by monthly temperature and precipitation dynamics.

These results can provide further insights into a better understanding of nutrient cycling and ecological processes in different forest plantations in the subtropics and provide a theoretical basis for the management of plantations.

2. Materials and Methods

2.1. Study Site

The study site was located in Fujian Sanming Forest Ecosystem National Observation and Research Station, China (26°19′ N, 117°36′ E), in the southeast of the Wuyishan National Nature Reserve and northwest of the Daiyun Mountains. The area is dominated by low hills, with an average elevation of 300 m and a slope of 25–45°. The climate is maritime subtropical monsoonal, with an average annual temperature of 19.3 °C and an annual precipitation of 1610 mm from 1960 to 2019, with about 80% of the rainfall occurring between March and August [40]. The soils were developed from granite and can be classified as Hapludults under the Ultisols order according to the United States Department of Agriculture Soil Taxonomy. The soil pH is 4.38 [37,38]. Before 1958, the area was covered by natural broadleaf forests dominated by *Castanopsis carlesii* [41]. However, as the demand for timber for construction increased, the natural forests were gradually logged and replaced by plantations (mainly *Castanopsis carlesii* and *Cunninghamia lanceolata*).

In 2011, six 20 m × 20 m plots were set up in a randomized group design in a natural forest of *Castanopsis carlesii* formed in 1976, including two treatments, a *Castanopsis carlesii* plantation and a *Cunninghamia lanceolata* plantation (Table 1), with three replicates in each plantation [40]. We investigated in 2021 when both plantations were 10 years old and in a rapid growth stage with high nutrient demand.

Table 1. The characteristics of the two plantations.

Parameters	*Castanopsis carlesii* Plantation	*Cunninghamia lanceolata* Plantation
Tree age (year)	10	10
Slope (°)	32	33
Stand density (plant·ha^{-1})	2400	2860
Average tree height (m)	8.77 ± 0.17	11.30 ± 0.37
Average breast diameter (cm)	9.93 ± 0.62	14.80 ± 0.28
Soil C (mg·g^{-1})	23.83 ± 0.56	27.02 ± 0.27
Soil N (mg·g^{-1})	1.54 ± 0.03	1.79 ± 0.04
Soil P (mg·g^{-1})	0.19 ± 0.00	0.19 ± 0.00

Values are meant ± standard errors (n = 3).

2.2. Sampling and Chemical Analysis

At the end of each month from April to October 2021, five trees with similar height and diameter at breast height were selected in each replicate sample plot to collect the mature leaves. A total of 15 trees from each plantation were collected. The green mature leaves of *Castanopsis carlesii* were collected in the *Castanopsis carlesii* plantation, while the green needles on the second or third node branches were collected in the *Cunninghamia lanceolata* plantation. The collected mature leaves were placed in moist self-sealing bags and brought back to the laboratory through an insulated box (internal temperature < 4 °C) [42]. In addition, senescent leaves were also collected at the end of each month. Three separate 0.7 m × 0.7 m nylon mesh frames were set up in each replicate sample plot in the *Castanopsis carlesii* plantation. The fresh leaf litter of *Castanopsis carlesii* in nylon mesh frames was selected as senescent leaf samples. Because of the persistence effect, the needle litter of branches of *Cunninghamia lanceolata* can retain on the trunk for many years [43], and so the newly senescent brownish-yellow needles on the trees were collected as senescent needle samples in the *Cunninghamia lanceolata* plantation. In addition, soil samples were collected at a depth of 0–20 cm in the *Castanopsis carlesii* plantation and the *Cunninghamia lanceolata* plantation to assess the chemical composition in August 2021. During the sampling period, three tipping bucket rainfall barrels were placed in a non-forested area near the sample plots for automatic recording of precipitation. A temperature logger was also placed in the non-forested area for recording air temperature (Figure 1).

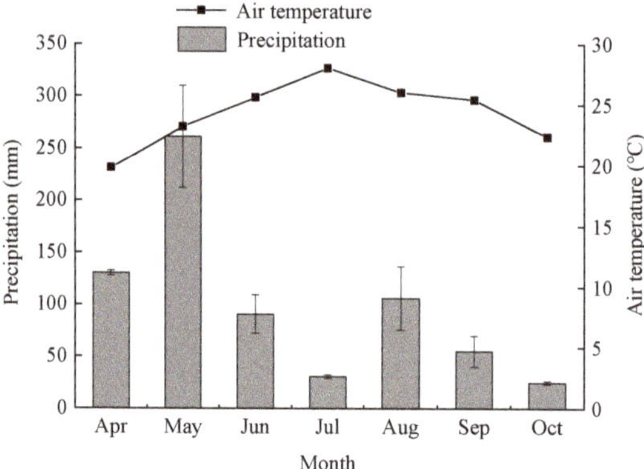

Figure 1. Monthly average air temperature and precipitation during the experiment. Each column represents the average value of three replicates ($n = 3$). Error bars represent standard errors (SE).

The collected mature and senescent leaves were dried in an oven at 65 °C for more than 72 h to a constant weight. Afterward, the dried samples were ground with a grinder (Tube Mill 100 control) and sieved through 100 mesh. The soil samples were air dried and also sieved through 100 mesh. The C and N contents of leaves and soil were determined using an elemental analyzer (Elemental Analyzer Vario EL III, Langenselbold, Germany) [37]. The P content of leaves and soil was determined using a continuous flow analyzer (SAN++, SKALAR, Holland) after the preparation of the solution, to be measured by H_2SO_4-$HClO_4$ decoction [44].

2.3. Calculation and Analysis

To eliminate the error caused by the loss of leaf mass during senescence in the calculation of the nutrient resorption efficiency (*NuRE*) during leaf senescence, we used the following formula [9,43]:

$$NuRE = \left(1 - \frac{Nu_{senescent}}{Nu_{mature}} MLCF\right) \times 100 \tag{1}$$

where Nu_{mature} and $Nu_{senescent}$ are the nutrient contents on a mass basis in mature and senescent leaves ($mg \cdot g^{-1}$), respectively; and *MLCF* is the mass loss correction factor used to compensate for the loss of leaf mass during senescence (specifically the ratio of the dry mass of senescent leaves to the dry mass of mature leaves).

The nutrient resorption proficiency (*NuRP*) is expressed directly as the nutrient contents in senescent leaves [7]. Furthermore, the amount of nutrient changes during leaf senescence was calculated by subtracting the nutrient contents of mature leaves from the nutrient contents of senescent leaves.

The C:N, C:P, and N:P ratios were determined using the C, N, and P contents in each leaf sample. The C:N:P stoichiometric ratios were calculated as mass ratios. The normality and chi-squareness of the data were checked and transformed if necessary. A two-way analysis of variance (two-way ANOVA) was used to test the effects of forest type and sampling time on leaf element contents, resorption efficiencies, and stoichiometric ratios. A one-way analysis of variance (one-way ANOVA) and Tukey's multiple comparison test were used to analyze the differences between plantations at the $P < 0.05$ level. Pearson correlation was used to determine the correlations among leaf nutrient contents, stoichiometric ratios, resorption and temperature, and precipitation. The above analyses were performed using SPSS 23.0 (IBM SPSS, Chicago, IL, USA).

3. Results

3.1. Carbon, Nitrogen, and Phosphorus Contents in Mature and Senescent Leaves

The contents of C, N, and P in mature and senescent leaves mostly differed significantly between plantations and varied with months (Figure 2). The C content in the mature leaves of the *Cunninghamia lanceolata* plantation was lower in April and May than of the *Castanopsis carlesii* plantation, but higher at subsequent times (Figure 2a). During the growing season, the N content in both mature and senescent leaves of the *Cunninghamia lanceolata* plantation was significantly lower than that of the *Castanopsis carlesii* plantation (Figure 2c,d). The P content in mature leaves of the *Cunninghamia lanceolata* plantation was significantly higher than that of the *Castanopsis carlesii* plantation in May and September, and this pattern was more pronounced in senescent leaves (Figure 2e,f).

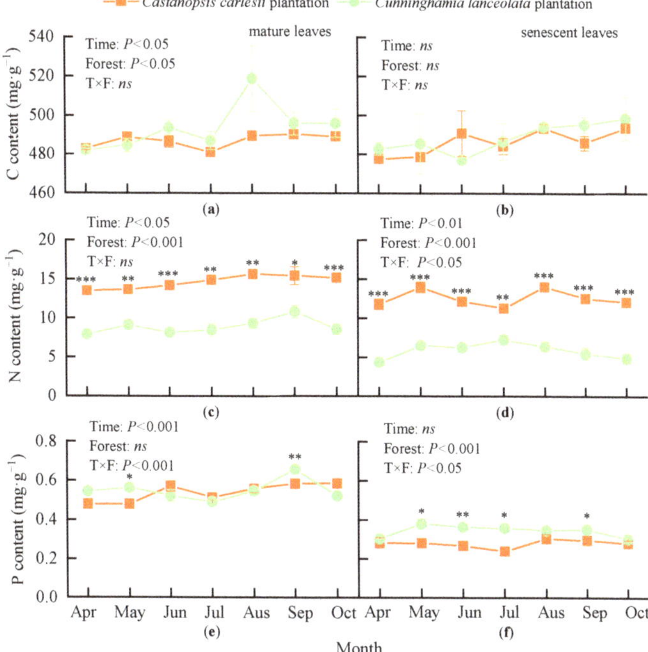

Figure 2. C, N and P contents in mature (**a,c,e**) and senescent leaves (**b,d,f**). Each scatter represents the average value of the three replicates (*n* = 3). Error bars represent standard errors (SE). The *p* values show the results from repeated measures ANOVA testing for the effect of forest type over time. *ns*: not significant. Asterisks denote significant differences between forest types: * *p* < 0.05, ** *p* < 0.01, *** *p* < 0.001.

Leaf N and P contents in both plantations decreased markedly during leaf senescence, and the dynamical patterns of decrease differed significantly between plantations (Figure 3). The N content decreased significantly more in the *Cunninghamia lanceolata* plantation than in the *Castanopsis carlesii* plantation in May, while it was significantly less than in the *Castanopsis carlesii* plantation in July (Figure 3a). In contrast, the P content decreased significantly more in the *Cunninghamia lanceolata* plantation than in the *Castanopsis carlesii* plantation in April and September, while it was significantly less than in the *Castanopsis carlesii* plantation in June, July, and October (Figure 3b).

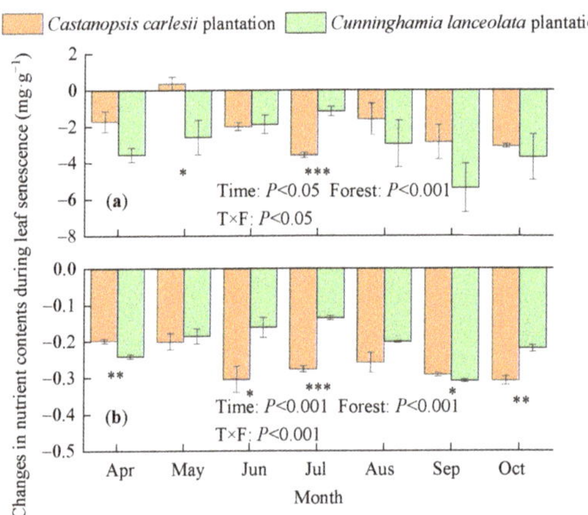

Figure 3. Changes in N (**a**) and P contents (**b**) during leaf senescence. Each column represents the average value of the three replicates ($n = 3$). Error bars represent standard errors (SE). The p values show the results from repeated measures ANOVA testing for the effect of forest type over time. Asterisks denote significant differences between forest types: * $p < 0.05$, ** $p < 0.01$, *** $p < 0.001$.

3.2. Nitrogen and Phosphorus Resorption Efficiencies in the Leaves of Castanopsis carlesii and Cunninghamia lanceolata Plantations

Dynamical patterns of N and P resorption efficiencies differed significantly between plantations (Figure 4). The N resorption efficiencies of the *Cunninghamia lanceolata* plantation and the *Castanopsis carlesii* plantation varied from 34.26% to 56.28% and from 11.25% to 34.23%, respectively (Figure 4a). The N resorption efficiency of the *Cunninghamia lanceolata* plantation was significantly lower than that of the *Castanopsis carlesii* plantation in July but higher in most months. In contrast, the P resorption efficiencies of the *Cunninghamia lanceolata* plantation and the *Castanopsis carlesii* plantation varied from 41.01% to 54.85% and from 49.22% to 58.72%, respectively (Figure 4b). The P resorption efficiency of the *Cunninghamia lanceolata* plantation was significantly lower than that of the *Castanopsis carlesii* plantation in most sampling months.

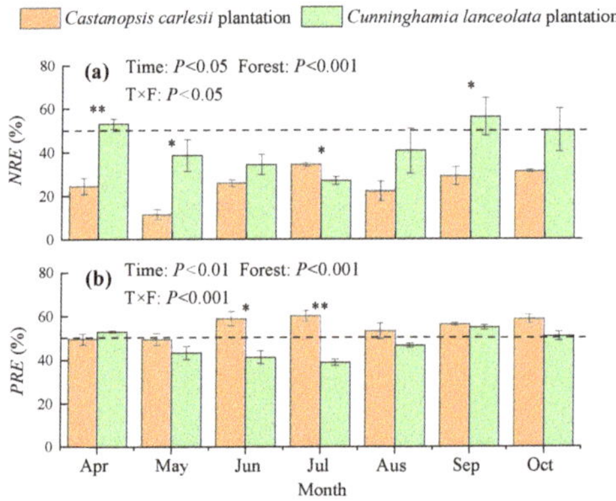

Figure 4. Leaf N (**a**) and P (**b**) resorption efficiencies. *NRE*: Nitrogen resorption efficiency, *PRE*: Phosphorus resorption efficiency. Each column represents the average value of the three replicates ($n = 3$). Error bars represent standard errors (SE). The p values show the results from repeated measures ANOVA testing for the effect of forest type over time. Asterisks denote significant differences between forest types: * $p < 0.05$, ** $p < 0.01$. The dotted line represents the 50% threshold.

3.3. Carbon, Nitrogen, and Phosphorus Stoichiometric Ratios in Mature and Senescent Leaves

The C:N, C:P, and N:P ratios in mature leaves varied from 31.39 to 61.24, from 753.09 to 1022.54, and from 14.57 to 28.92 during the investigation period, respectively (Figure 5). Forest types displayed a significant effect on C:N in mature leaves (Figure 5a). The C:N of mature leaves in the *Cunninghamia lanceolata* plantation was significantly higher than that of the *Castanopsis carlesii* plantation in all months. The C:P of mature leaves varied with time, and the variation patterns varied significantly between plantations (Figure 5c). The C:P of mature leaves in the *Cunninghamia lanceolata* plantation was significantly lower than that in the *Castanopsis carlesii* plantation in May and September, but higher in October. The N:P of mature leaves also differed significantly between plantations (Figure 5e). Compared to the *Castanopsis carlesii* plantation, the *Cunninghamia lanceolata* plantation had significantly lower mature leaf N:P.

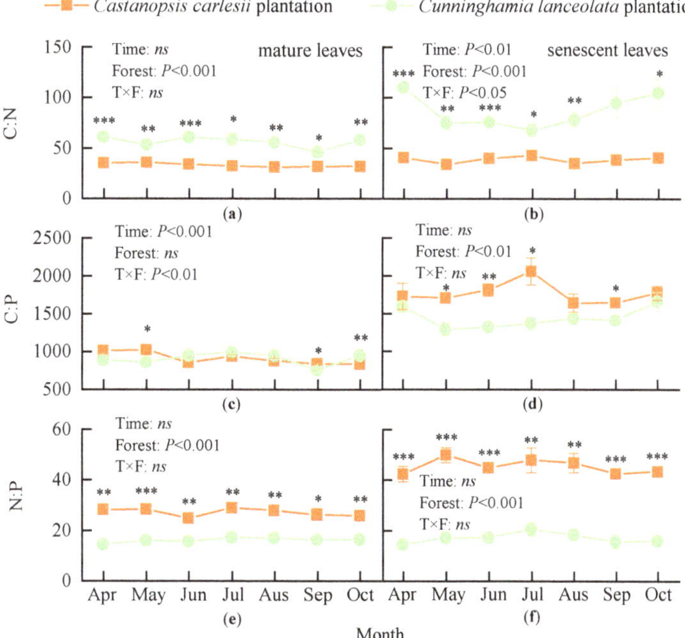

Figure 5. C:N, C:P, and N:P ratios in mature (**a,c,e**) and senescent leaves (**b,d,f**). Each scatter represents the average value of the three replicates (n = 3). Error bars represent standard errors (SE). The p values show the results from repeated measures ANOVA testing for the effect of forest type over time. *ns*: not significant. Asterisks denote significant differences between forest types: * $p < 0.05$, ** $p < 0.01$, *** $p < 0.001$.

The C:N, C:P, and N:P of senescent leaves varied from 34.45 to 110.42, from 1292.61 to 2060.64, and from 14.50 to 49.94, respectively. Forest types exhibited a significant effect on the C:N:P stoichiometric ratios in senescent leaves (Figure 5). The C:N of senescent leaves in the *Cunninghamia lanceolata* plantation was significantly higher than that of the *Castanopsis carlesii* plantation in all sampling months (Figure 5b). In contrast, C:P and N:P in senescent leaves of the *Cunninghamia lanceolata* plantation were significantly lower than that of the *Castanopsis carlesii* plantation (Figure 5d,f).

3.4. Correlation of Nutrient Contents, Resorption Efficiencies, and Stoichiometric Ratios with Climatic Factors

In the *Castanopsis carlesii* plantation, the temperature was significantly negatively correlated with senescent leaf P content and significantly positively correlated with leaf

PRE (Figure 6). Precipitation was significantly negatively correlated with leaf *NRE*, *PRE*, mature leaf N and P contents, and senescent leaf C:N and significantly positively correlated with senescent leaf N, mature leaf C:N, and C:P. In the *Cunninghamia lanceolata* plantation, the temperature was significantly negatively correlated with leaf *NRE*, *PRE*, senescent leaf C:N, and C:P, and significantly positively correlated with senescent leaf N content (Figure 7). Precipitation was significantly positively correlated with senescent leaf P content.

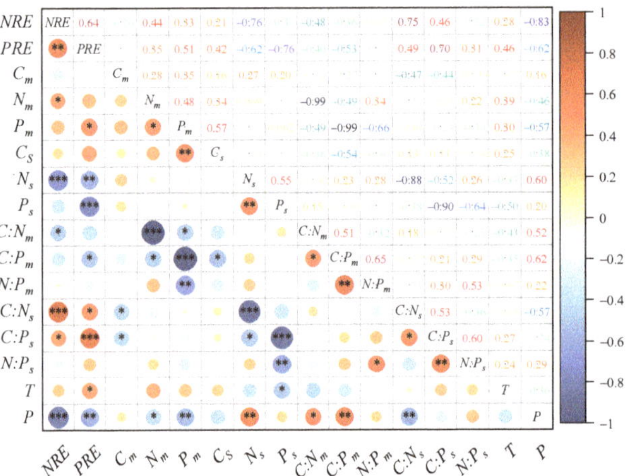

Figure 6. Correlation of nutrient contents, resorption efficiencies, and stoichiometric ratios with climatic factors in the *Castanopsis carlesii* plantation. *NRE*: N resorption efficiency; *PRE*: P resorption efficiency; C_m: C content in mature leaves; N_m: N content in mature leaves; P_m: P content in mature leaves; C_s: C content in senescent leaves; N_s: N content in senescent leaves; P_s: P content in senescent leaves. $C{:}N_m$: C:N in mature leaves; $C{:}P_m$: C:P in mature leaves; $N{:}P_m$: N:P in mature leaves; $C{:}N_s$: C:N in senescent leaves; $C{:}P_s$: C:P in senescent leaves; $N{:}P_s$: N:P in senescent leaves. *T*: Temperature; *P*: Precipitation. Asterisks indicate significant correlations among indicators. * $p < 0.05$, ** $p < 0.01$, *** $p < 0.001$. The same is below.

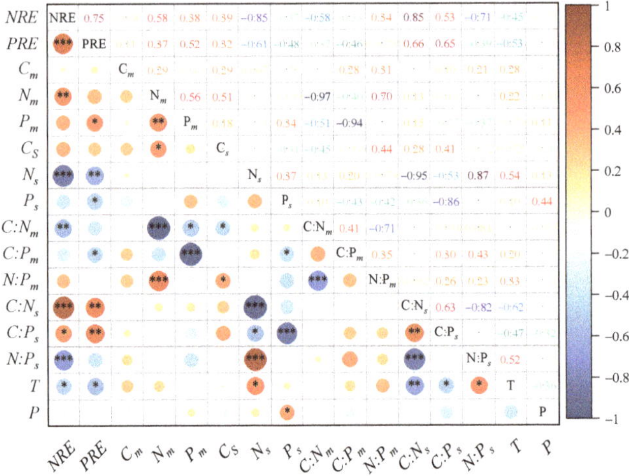

Figure 7. Correlation of nutrient contents, resorption efficiencies, and stoichiometric ratios with climatic factors in the *Cunninghamia lanceolata* plantation. Asterisks indicate significant correlations among indicators. * $p < 0.05$, ** $p < 0.01$, *** $p < 0.001$.

4. Discussion

The results here partially supported our hypothesis that the dynamical patterns of N and P resorption and C:N:P stoichiometric ratios of in leaves vary significantly with forest types and are regulated by seasonal climate changes. We found that N and P resorption efficiencies in the *Castanopsis carlesii* plantation were higher in the middle of the growing season than in the *Cunninghamia lanceolata* plantation, and lower in the early and late growing season. Moreover, the P-limited *Castanopsis carlesii* plantation exhibited better P resorption, while the *Cunninghamia lanceolata* plantation limited by both N and P showed similar levels of N and P resorption. These findings revealed that plant growth limitation could have a dominant role in controlling nutrient resorption [23,24]. We also found that temperature and precipitation were significantly associated with N and P resorption, but this relationship differed between forest types, implying that adaptive nutrient use strategies respond to the change of the external environment and ultimately to plant development and growth [16,22].

Resorption is one of the important nutrient conservation strategies to improve plant nutrient utilization, playing a key role in several ecosystem processes such as species competition, nutrient return, and decomposition of leaf litter [1,18]. We found that the N and P resorption characteristics in the two plantations were significantly different (Figure 4). Compared to the *Castanopsis carlesii* plantation, the *Cunninghamia lanceolata* plantation showed higher *NRE* and *NRP* but lower *PRE* and *PRP* (Figures 2 and 4). These results indicated that the *Castanopsis carlesii* plantation could be generally more efficient and proficient in resorbing P, while the *Cunninghamia lanceolata* plantation could be better at resorbing N. This is in agreement with findings on coniferous and broadleaf trees in subtropical karst regions [24]. The higher soil N content in the *Cunninghamia lanceolata* plantation may also be due to the fact that the needles require less N and that most of the N is obtained through resorption rather than root uptake, hence the higher N retained in the soil. In addition, *PRE* was higher than *NRE* in both plantations. This result is consistent with the average *NRE* (49.1%) and *PRE* (51.0%) of plants in Eastern China [23]. The possible reasons are the general lack of P in soils in subtropical regions and the differences in the relative amounts of N and P leached from leaves [45,46]. Therefore, subtropical trees might survive in P-deficient ecosystems by increasing P resorption.

Plant nutrient resorption may exhibit seasonal dynamics in response to environmental changes [13–15]. The dynamical patterns of N and P resorption efficiencies in response to seasonal climate changes differed significantly between the two plantations (Figure 4). In the *Castanopsis carlesii* plantation, N and P resorption efficiencies were significantly negatively correlated with precipitation, and P resorption efficiency was significantly positively correlated with temperature (Figure 6). The possible reason is that as a broad-leaved evergreen species, *Castanopsis carlesii* has a large leaf area [47]. Higher precipitation may lead to abnormal senescence of *Castanopsis carlesii* leaves and leaching of nutrients from the leaves, which results in lower nutrient resorption efficiencies [48]. In addition, it has been suggested that elevated temperatures may delay the leaf senescence in broadleaf trees, which may prolong the time to retransfer P in leaves, resulting in higher P resorption efficiency [49–51]. In the *Cunninghamia lanceolata* plantation, N and P resorption efficiencies were significantly negatively correlated with temperature, but not with precipitation (Figure 7). This indicates that the nutrient resorption of *Cunninghamia lanceolata* is more regulated by temperature than precipitation. In some studies, a higher temperature may aggravate the degree of membrane lipid peroxidation in leaf cells, thereby accelerating leaf senescence, leading to a shorter time for nutrient retransfer and lower nutrient resorption [52,53]. This may explain the negative relationship between nutrient resorption efficiencies and temperature in the *Cunninghamia lanceolata* plantation. Differences in the responses of N and P resorption efficiencies to temperature and precipitation between plantations also supports our hypothesis.

Ecological stoichiometric ratios can usually reflect the growth and nutrient utilization status of plants [25,54]. In this study, mature leaf C:N was significantly higher in the

Cunninghamia lanceolata plantation than in the *Castanopsis carlesii* plantation. This might be attributed to the higher C content as well as the lower N content in the *Cunninghamia lanceolata* plantation. Previous studies have observed that coniferous trees tend to have high amounts of resins, waxes, and tannins. These substances are rich in C, which help to increase the C content and C:N in leaves [16,24]. In addition, we found that the C:P in mature leaves varied with time, and these variation patterns varied significantly between plantations (Figure 5). Previous studies have shown that plants need more rRNA to synthesize the required proteins during high growth periods and that increased rRNA leads to an increase in P content [6,55,56]. The Mature leaf C:P in *Cunninghamia lanceolata* plantation was significantly lower in April, May, and September than in *Castanopsis carlesii* plantation, and higher in other months (Figure 5). This indicated that the growth rate of the *Cunninghamia lanceolata* plantation may be higher in the early and late growing season and lower in the middle of the growing season compared to that of the *Castanopsis carlesii* plantation. In addition, resorption causes changes in leaf nutrient contents by transferring nutrients to other active tissues before leaf drop, thus it may be an important process affecting the seasonal dynamics of leaf stoichiometric ratios [24]. In the present study, *NRE* and *PRE* were higher in the early and late growing season and lower in the middle of the growing season in the *Cunninghamia lanceolata* plantation compared to the *Castanopsis carlesii* plantation. This suggested that trees in subtropical forests might meet the nutrient demands of a high growth rate by increasing nutrient resorption.

The availability of N and P tends to limit plant growth and community composition in terrestrial ecosystems, so leaf N:P is widely used to assess the nutrient limitation of plants [28]. In this study, the leaf N:P ratios differed significantly between two plantations, but changed less over time (Figure 5), which did not support our hypothesis. The theory of dynamic equilibrium believes that organisms can control many of their characteristics, especially nutrient balance so that the internal environment will not change drastically with changes in the external environment [55]. The leaf N:P in the two plantations showed a certain degree of homeostasis, which might be the result of the long-term adaptation of different subtropical plants to the environment, or, might be related to its strong ability to maintain homeostasis [57,58]. Therefore, it is speculated that the difference in leaf N:P of trees in different plantations is more determined by the genetic characteristics of the species itself. In addition, a study proposed that the plant growth of terrestrial plants was N-limited when mature leaf N:P is less than 10, and P-limited when N:P is more than 20 [28]. Our results suggested that the *Castanopsis carlesii* plantation was strongly P-limited (average N:P of 27.23), whereas the *Cunninghamia lanceolata* plantation was both N and P-limited (average N:P of 16.24). The stronger P limitation in the *Castanopsis carlesii* plantation compared to the *Cunninghamia lanceolata* plantation may explain its more efficient P resorption. These findings implied that N and P resorption might be adapted to the nutrient limitation status of plants. In subtropical low-nutrient habitats, *Castanopsis carlesii* and *lanceolata* plantations may maintain leaf N:P relative stability as much as possible by regulating nutrient resorption. This is consistent with previous findings [58,59].

5. Conclusions

In conclusion, there were marked monthly dynamics of leaf N and P resorption and C:N:P stoichiometric ratios in both *Castanopsis carlesii* and *Cunninghamia lanceolata* plantations, but the dynamic patterns differed significantly between forest types. Due to the limitation of P, the *Castanopsis carlesii* plantation had a higher P resorption efficiency than its N resorption efficiency, and exhibited a higher N and P resorption in the middle of the growing season to meet the nutrient requirements of the rapid growth period. In contrast, the *Cunninghamia lanceolata* plantation displayed similar levels of N and P resorption due to the co-limitation of N and P and showed higher N and P resorption in the early and late growing season. Furthermore, the correlation showed that the N and P resorption efficiencies displayed strongly significant correlations with temperature and precipitation in both plantations. The results here could further clarify the differences in the dynamics of N and P utilization strategies between different functional plantations and help to understand the nutrient cycling processes in the subtropical forest ecosystems.

Author Contributions: F.W. and X.N. designed the experiments; Y.Z., J.Y., and X.W. collected and examined the samples; Y.Z. analyzed the data and drafted the manuscript; F.W. revised the manuscript. All authors have read and agreed to the published version of the manuscript.

Funding: This research was funded by the National Natural Science Foundation of China (32171641, 32101509 and 32022056).

Data Availability Statement: Not applicable.

Acknowledgments: We wish to thank the Fujian Sanming Forest Ecosystem National Observation and Research Station for their support and all those who have helped us in the field sampling and experiments.

Conflicts of Interest: The authors declare no conflict of interest.

References

1. Aerts, R. Nutrient Resorption from Senescing Leaves of Perennials: Are There General Patterns? *J. Ecol.* **1996**, *84*, 597. [CrossRef]
2. Chen, H.; Reed, S.C.; Lü, X.; Xiao, K.; Wang, K.; Li, D. Global Resorption Efficiencies of Trace Elements in Leaves of Terrestrial Plants. *Funct. Ecol.* **2021**, *35*, 1596–1602. [CrossRef]
3. Chen, X.; Chen, H.Y.H. Foliar Nutrient Resorption Dynamics of Trembling Aspen and White Birch during Secondary Succession in the Boreal Forest of Central Canada. *For. Ecol. Manag.* **2022**, *505*, 119876. [CrossRef]
4. You, C.; Wu, F.; Yang, W.; Xu, Z.; Tan, B.; Zhang, L.; Yue, K.; Ni, X.; Li, H.; Chang, C.; et al. Does Foliar Nutrient Resorption Regulate the Coupled Relationship between Nitrogen and Phosphorus in Plant Leaves in Response to Nitrogen Deposition? *Sci. Total Environ.* **2018**, *645*, 733–742. [CrossRef]
5. Drenovsky, R.E.; Pietrasiak, N.; Short, T.H. Global Temporal Patterns in Plant Nutrient Resorption Plasticity. *Glob. Ecol. Biogeogr.* **2019**, *28*, 728–743. [CrossRef]
6. Qiu, X.; Wang, H.; Peng, D.; Liu, X.; Yang, F.; Li, Z.; Cheng, S. Thinning Drives C:N:P Stoichiometry and Nutrient Resorption in *Larix Principis-Rupprechtii* Plantations in North China. *For. Ecol. Manag.* **2020**, *462*, 117984. [CrossRef]
7. Killingbeck, K.T. Nutrients in Senesced Leaves: Keys to the Search for Potential Resorption and Resorption Proficiency. *Ecology* **1996**, *77*, 1716–1727. [CrossRef]
8. Liu, C.; Liu, Y.; Guo, K.; Wang, S.; Yang, Y. Concentrations and Resorption Patterns of 13 Nutrients in Different Plant Functional Types in the Karst Region of South-Western China. *Ann. Bot.* **2014**, *113*, 873–885. [CrossRef]
9. Van Heerwaarden, L.M.V.; Toet, S.; Aerts, R. Current Measures of Nutrient Resorption Efficiency Lead to a Substantial Underestimation of Real Resorption Efficiency: Facts and Solutions. *Oikos* **2003**, *101*, 664–669. [CrossRef]
10. González-Zurdo, P.; Escudero, A.; Mediavilla, S. N Resorption Efficiency and Proficiency in Response to Winter Cold in Three Evergreen Species. *Plant Soil* **2015**, *394*, 87–98. [CrossRef]
11. Tian, D.; Kattge, J.; Chen, Y.; Han, W.; Luo, Y.; He, J.; Hu, H.; Tang, Z.; Ma, S.; Yan, Z.; et al. A Global Database of Paired Leaf Nitrogen and Phosphorus Concentrations of Terrestrial Plants. *Ecology* **2019**, *100*, e02812. [CrossRef] [PubMed]
12. Zhang, S.B.; Zhang, J.L.; Slik, J.; Cao, K.F. Leaf Element Concentrations of Terrestrial Plants across China Are Influenced by Taxonomy and the Environment. *Glob. Ecol. Biogeogr.* **2012**, *21*, 809–818. [CrossRef]
13. Kerkhoff, A.J.; Enquist, B.J.; Fagan, E.W.F. Plant Allometry, Stoichiometry and the Temperature-Dependence of Primary Productivity. *Glob. Ecol. Biogeogr.* **2010**, *14*, 585–598. [CrossRef]
14. Wang, C.G.; Zheng, X.B.; Wang, A.Z.; Dai, G.H.; Zhu, B.K.; Zhao, Y.M.; Dong, S.J.; Zu, W.Z.; Wang, W.; Zheng, Y.G.; et al. Temperature and Precipitation Diversely Control Seasonal and Annual Dynamics of Litterfall in a Temperate Mixed Mature Forest, Revealed by Long-Term Data Analysis. *J. Geophys. Res.-Biogeosci.* **2021**, *126*, e2020JG006204. [CrossRef]
15. Zhu, X.; Liu, W.; Chen, H.; Deng, Y.; Chen, C.; Zeng, H. Effects of Forest Transition on Litterfall, Standing Litter and Related Nutrient Returns: Implications for Forest Management in Tropical China. *Geoderma* **2019**, *333*, 123–134. [CrossRef]
16. Yuan, Z.Y.; Chen, H.Y.H. Global-Scale Patterns of Nutrient Resorption Associated with Latitude, Temperature and Precipitation. *Glob. Ecol. Biogeogr.* **2009**, *18*, 11–18. [CrossRef]
17. Oleksyn, J.; Reich, P.B.; Zytkowiak, R.; Karolewski, P.; Tjoelker, M.G. Nutrient Conservation Increases with Latitude of Origin in European *Pinus sylvestris* Populations. *Oecologia* **2003**, *136*, 220–235. [CrossRef]
18. Vergutz, L.; Manzoni, S.; Porporato, A.; Novais, R.F.; Jackson, R.B. Global Resorption Efficiencies and Concentrations of Carbon and Nutrients in Leaves of Terrestrial Plants. *Ecol. Monogr.* **2012**, *82*, 205–220. [CrossRef]
19. Chapin, F.S.; Matson, P.A.; Vitousek, P.M. *Principles of Terrestrial Ecosystem Ecology*; Springer: New York, NY, USA, 2011; pp. 38–40.
20. Tian, D.; Yan, Z.; Ma, S.; Ding, Y.; Luo, Y.; Chen, Y.; Du, E.; Han, W.; Kovacs, E.D.; Shen, H.; et al. Family-Level Leaf Nitrogen and Phosphorus Stoichiometry of Global Terrestrial Plants. *Sci. China Life Sci.* **2019**, *62*, 1047–1057. [CrossRef]
21. Yan, T.; Lü, X.; Zhu, J.; Yang, K.; Yu, L.; Gao, T. Changes in Nitrogen and Phosphorus Cycling Suggest a Transition to Phosphorus Limitation with the Stand Development of Larch Plantations. *Plant Soil* **2018**, *422*, 385–396. [CrossRef]

22. Jiang, D.; Geng, Q.; Li, Q.; Luo, Y.; Vogel, J.; Shi, Z.; Ruan, H.; Xu, X. Nitrogen and Phosphorus Resorption in Planted Forests Worldwide. *Forests* **2019**, *10*, 201. [CrossRef]

23. Tang, L.; Han, W.; Chen, Y.; Fang, J. Resorption Proficiency and Efficiency of Leaf Nutrients in Woody Plants in Eastern China. *J. Plant Ecol.* **2013**, *6*, 408–417. [CrossRef]

24. Pang, D.; Wang, G.; Li, G.; Sun, Y.; Liu, Y.; Zhou, J. Ecological Stoichiometric Characteristics of Two Typical Plantations in the Karst Ecosystem of Southwestern China. *Forests* **2018**, *9*, 56. [CrossRef]

25. Elser, J.J.; Fagan, W.F.; Denno, R.F.; Dobberfuhl, D.R.; Folarin, A.; Huberty, A.; Interlandi, S.; Kilham, S.S.; McCauley, E.; Schulz, K.L.; et al. Nutritional Constraints in Terrestrial and Freshwater Food Webs. *Nature* **2000**, *408*, 4. [CrossRef]

26. Hu, M.; Ma, Z.; Chen, H.Y.H. Intensive Plantations Decouple Fine Root C:N:P in Subtropical Forests. *For. Ecol. Manag.* **2022**, *505*, 119901. [CrossRef]

27. Su, H.; Cui, J.; Adamowski, J.F.; Zhang, X.; Biswas, A.; Cao, J. Using Leaf Ecological Stoichiometry to Direct the Management of *Ligularia virgaurea* on the Northeast Qinghai-Tibetan Plateau. *Front. Environ. Sci.* **2022**, *9*, 805405. [CrossRef]

28. Güsewell, S. N: P Ratios in Terrestrial Plants: Variation and Functional Significance. *New Phytol.* **2004**, *164*, 243–266. [CrossRef]

29. Yuan, Z.Y.; Chen, H.Y.H. Decoupling of Nitrogen and Phosphorus in Terrestrial Plants Associated with Global Changes. *Nat. Clim. Change* **2015**, *5*, 465–469. [CrossRef]

30. Fan, S.; Zhang, B.; Luan, P.; Gu, B.; Wan, Q.; Huang, X.; Liao, W.; Liu, J. PI3K/AKT/MTOR/P70S6K Pathway Is Involved in A β 25-35-Induced Autophagy. *BioMed Res. Int.* **2015**, *2015*, 161020. [CrossRef]

31. Yang, D.; Mao, H.; Jin, G. Divergent Responses of Foliar N:P Stoichiometry During Different Seasons to Nitrogen Deposition in an Old-Growth Temperate Forest, Northeast China. *Forests* **2019**, *10*, 257. [CrossRef]

32. Zhang, W.; Liu, W.; Xu, M.; Deng, J.; Han, X.; Yang, G.; Feng, Y.; Ren, G. Response of Forest Growth to C:N:P Stoichiometry in Plants and Soils during *Robinia pseudoacacia* Afforestation on the Loess Plateau, China. *Geoderma* **2019**, *337*, 280–289. [CrossRef]

33. Ren, H.; Kang, J.; Yuan, Z.; Xu, Z.; Han, G. Responses of Nutrient Resorption to Warming and Nitrogen Fertilization in Contrasting Wet and Dry Years in a Desert Grassland. *Plant Soil* **2018**, *432*, 65–73. [CrossRef]

34. Zhao, F.; Li, D.; Jiao, F.; Yao, J.; Du, H. The Latitudinal Patterns of Leaf and Soil C:N:P Stoichiometry in the Loess Plateau of China. *Front. Plant Sci.* **2019**, *10*, 85.

35. Yu, G.; Chen, Z.; Piao, S.; Peng, C.; Ciais, P.; Wang, Q.; Li, X.; Zhu, X. High Carbon Dioxide Uptake by Subtropical Forest Ecosystems in the East Asian Monsoon Region. *Proc. Natl. Acad. Sci. USA* **2014**, *111*, 4910–4915. [CrossRef]

36. Lin, R.; Zhang, J.; Li, S.; Yao, X.; Wang, X.; Chen, G. Overyielding of *Pinus massoniana* Mixed Forest and Its Influencing Factors. *J. Subtropic. Resour. Environ.* **2022**, *17*, 43–50.

37. Ni, X.; Lin, C.; Chen, G.; Xie, J.; Yang, Z.; Liu, X.; Xiong, D.; Xu, C.; Yue, K.; Wu, F.; et al. Decline in Nutrient Inputs from Litterfall Following Forest Plantation in Subtropical China. *For. Ecol. Manag.* **2021**, *496*, 119445. [CrossRef]

38. Lü, M.; Xie, J.; Wang, C.; Guo, J.; Wang, M.; Liu, X.; Chen, Y.; Chen, G.; Yang, Y. Forest Conversion Stimulated Deep Soil C Losses and Decreased C Recalcitrance through Priming Effect in Subtropical China. *Biol. Fertil. Soils* **2015**, *51*, 857–867. [CrossRef]

39. Lin, K.; Lyu, M.; Jiang, M.; Chen, Y.; Li, Y.; Chen, G.; Xie, J.; Yang, Y. Improved Allometric Equations for Estimating Biomass of the Three *Castanopsis carlesii* H. Forest Types in Subtropical China. *New For.* **2017**, *48*, 115–135. [CrossRef]

40. Yang, Y.; Wang, L.; Yang, Z.; Xu, C.; Xie, J.; Chen, G.; Lin, C.; Guo, J.; Liu, X.; Xiong, D.; et al. Large Ecosystem Service Benefits of Assisted Natural Regeneration. *J. Geophys. Res. Biogeosci.* **2018**, *123*, 676–687. [CrossRef]

41. Gao, Y.; Bao, Y.; Hu, W.; Li, H.; Si, Y. Soil DOM Quantity and Spectral Characteristics of Three Typical Forest Soils in Subtropical China. *J. Subtropic. Resour. Environ.* **2018**, *13*, 26–35.

42. Wright, I.J.; Westoby, M. Nutrient Concentration, Resorption and Lifespan: Leaf Traits of Australian Sclerophyll Species. *Funct. Ecol.* **2003**, *17*, 10–19. [CrossRef]

43. Zhou, L.; Li, S.; Jia, Y.; Heal, K.; He, Z.; Wu, P.; Ma, X. Spatiotemporal Distribution of Canopy Litter and Nutrient Resorption in a Chronosequence of Different Development Stages of *Cunninghamia lanceolata* in Southeast China. *Sci. Total Environ.* **2021**, *762*, 143153. [CrossRef] [PubMed]

44. Bo, F.; Zhang, Y.; Chen, H.Y.H.; Wang, P.; Ren, X.; Guo, J. The C:N:P Stoichiometry of Planted and Natural *Larix principis-rupprechtii* Stands along Altitudinal Gradients on the Loess Plateau, China. *Forests* **2020**, *11*, 363. [CrossRef]

45. Du, Y.X.; Pan, G.X.; Li, L.Q.; Hu, Z.L.; Wang, X.Z. Leaf N/P ratio and nutrient reuse between dominant species and stands: Predicting phosphorus deficiencies in Karst ecosystems, southwestern China. *Environ. Earth Sci.* **2011**, *64*, 299–309. [CrossRef]

46. Hofmeister, J.; Mihaljevic, M.; Hosek, J.; Sadlo, J. Eutrophication of deciduous forests in the Bohemian Karst (Czech Republic): The role of nitrogen and phosphorus. *For. Ecol. Manag.* **2002**, *169*, 213–230. [CrossRef]

47. Zhang, Y.; Ni, X.; Yang, J.; Tan, S.; Liao, S.; Wu, F. Nitrogen and phosphorus resorption and stoichiometric characteristics of different tree species in a mid-subtropical common-garden, China. *Chin. J. Appl. Ecol.* **2021**, *32*, 1154–1162.

48. Zhao, Z.; Wu, F.; Yang, Y.; Ni, X.; Xu, C.; Xiong, D.; Liao, S.; Yuan, J.; Tan, S.; Yue, K. Dynamics of Phosphorus Along with Stemflow and Throughfall in Middle Subtropical *Cunninghamia lanceolata* Plantations and *Castanopsis carlesii* Secondary Forests. *J. Soil Water Conserv.* **2021**, *35*, 129–134.

49. Fu, Y.H.; Piao, S.; Delpierre, N.; Hao, F.; Hanninen, H.; Liu, Y.; Sun, W.; Janssens, I.A.; Campioli, M. Larger Temperature Response of Autumn Leaf Senescence than Spring Leaf-out Phenology. *Glob. Change Biol.* **2018**, *24*, 2159–2168. [CrossRef]

50. Lawrence, B.T.; Melgar, J.C. Variable Fall Climate Influences Nutrient Resorption and Reserve Storage in Young Peach Trees. *Front. Plant. Sci.* **2018**, *9*, 1819. [CrossRef]

51. Wang, F.; Zhang, R.; Lin, J.; Zheng, J.; Hanninen, H.; Wu, J. High Autumn Temperatures Increase the Depth of Bud Dormancy in the Subtropical *Torreya grandis* and *Carya illinoinensis* and Delay Leaf Senescence in the Deciduous Carya. *Trees-Struct. Funct.* **2022**, *36*, 1053–1065. [CrossRef]
52. Gerdol, R.; Marchesini, R.; Iacumin, P. Bedrock Geology Interacts with Altitude in Affecting Leaf Growth and Foliar Nutrient Status of Mountain Vascular Plants. *J. Plant Ecol.* **2017**, *10*, 839–850. [CrossRef]
53. Jardine, K.J.; Chambers, J.Q.; Holm, J.; Jardine, A.B.; Fontes, C.G.; Zorzanelli, R.F.; Meyers, K.T.; de Souza, V.F.; Garcia, S.; Gimenez, B.O.; et al. Green Leaf Volatile Emissions during High Temperature and Drought Stress in a Central Amazon Rainforest. *Plants* **2015**, *4*, 678–690. [CrossRef] [PubMed]
54. Tian, D.; Yan, Z.; Niklas, K.J.; Han, W.; Kattge, J.; Reich, P.B.; Luo, Y.; Chen, Y.; Tang, Z.; Hu, H.; et al. Global Leaf Nitrogen and Phosphorus Stoichiometry and Their Scaling Exponent. *Natl. Sci. Rev.* **2018**, *5*, 728–739. [CrossRef]
55. Sterner, R.W.; Elser, J.J. *Ecological Stoichiometry: The Biology of Elements from Molecules to the Biosphere*; Press Princeton University Press: Princeton, NJ, USA, 2002; pp. 55–58.
56. Matzek, V.; Vitousek, P.M. N:P Stoichiometry and Protein: RNA Ratios in Vascular Plants: An Evaluation of the Growth-Rate Hypothesis. *Ecol. Lett.* **2010**, *12*, 765–771. [CrossRef] [PubMed]
57. Han, W.X.; Fang, J.Y.; Reich, P.B.; Woodward, F.I.; Wang, Z.H. Biogeography and Variability of Eleven Mineral Elements in Plant Leaves across Gradients of Climate, Soil and Plant Functional Type in China. *Ecol. Lett.* **2011**, *14*, 788–796. [CrossRef]
58. Schreeg, L.A.; Santiago, L.S.; Wright, S.J.; Turner, B.L. Stem, Root, and Older Leaf N:P Ratios Are More Responsive Indicators of Soil Nutrient Availability than New Foliage. *Ecology* **2014**, *95*, 2062–2206. [CrossRef]
59. Lambers, H.; Chapin, F.S.; Pons, T.L. *Plant. Physiological Ecology*, 2nd ed.; Springer: New York, NY, USA, 2008; pp. 47–52.

forests

MDPI

The Contributions of Soil Fauna to the Accumulation of Humic Substances during Litter Humification in Cold Forests

Yu Tan [1], Kaijun Yang [2,3], Zhenfeng Xu [1], Li Zhang [1], Han Li [1], Chengming You [1] and Bo Tan [1,*]

[1] Forestry Ecological Engineering in Upper Reaches of Yangtze River Key Laboratory of Sichuan Province, Institute of Ecology and Forestry, College of Forestry, Sichuan Agricultural University, No. 211, Huimin Road, Chengdu 611130, China

[2] CSIC, Global Ecology Unit CREAF-CSIC-UAB, 08913 Catalonia, Spain

[3] CREAF, Cerdanyola del Vallès, 08913 Catalonia, Spain

* Correspondence: bobotan1984@163.com

Citation: Tan, Y.; Yang, K.; Xu, Z.; Zhang, L.; Li, H.; You, C.; Tan, B. The Contributions of Soil Fauna to the Accumulation of Humic Substances during Litter Humification in Cold Forests. *Forests* **2022**, *13*, 1235. https://doi.org/10.3390/f13081235

Academic Editor: Choonsig Kim

Received: 20 June 2022
Accepted: 2 August 2022
Published: 4 August 2022

Publisher's Note: MDPI stays neutral with regard to jurisdictional claims in published maps and institutional affiliations.

Abstract: Litter humification is an essential process of soil carbon sequestration in forest ecosystems, but the relationship between soil fauna and humic substances has not been well understood. Therefore, a field litterbag experiment with manipulation of soil fauna was carried out in different soil frozen seasons over one year in cold forests. The foliar litter of four dominated tree species was selected as Birch (*Betula albosinensis*), Fir (*Abies fargesii* var. *faxoniana*), Willow (*Salix paraplesia*), and Cypress (*Juniperus saltuaria*). We studied the contribution of soil fauna to the accumulation of humic substances (including humic acid and fulvic acid) and humification degree as litter humification proceeding. The results showed that soil fauna with litter property and environmental factor jointly determined the accumulation of humic substances (humic acid and fulvic acid) and humification degree of four litters. After one year of incubation, the contribution rates of soil fauna to the accumulation of humic substances were 109.06%, 71.48%, 11.22%, and −44.43% for the litter of fir, cypress, birch, and willow, respectively. Compared with other stages, both growing season and leaf falling stage could be favorable to the contributions of soil fauna to the accumulation of humic substances in the litter of birch, fir, and cypress rather than in willow litter. In contrast, the contribution rates of soil fauna to humification degree were −49.20%, −7.63%, −13.27%, and 12.66% for the litter of fir, cypress, birch, and willow, respectively. Statistical analysis indicated that temperature changes at different sampling stages and litter quality exhibited dominant roles in the contributions of soil fauna on the accumulation of humus and litter humifiaction degree in the cold forests. Overall, the present results highlight that soil fauna could play vital roles in the process of litter humification and those strengths varied among species and seasons.

Keywords: soil fauna; humic substances; humic acid; fulvic acid; humification degree

1. Introduction

Humus has been known as the major component of soil organic carbon, which can improve soil fertility and soil physical and chemical structure [1–3]. Litter humification is an important pathway for the formation of soil humus and plays an important role in maintaining soil fertility and carbon sequestration in nature ecosystems [4–7]. Soil fauna are not only decomposers during litter humification, but also act as boosters throughout the formation of soil organic matter by producing and transferring humic substances [8–11]. The roles of soil fauna on the process of decomposition have been wildly documented, and the effect of soil fauna on litter decomposition can be elucidated from different functional groups and feeding habits [12–14]. However, their impacts on litter humification have not been well understood. Therefore, there is still a lack of unified cognition on the role of soil fauna in the accumulation of humus during litter humification.

Generally, the litter humification by soil fauna works through nesting, constructing shuttle, foraging fragmentation, and regulating microorganism's structure [15]. Additionally, these feces and dead bodies of soil fauna are the main sources of soil humus [16,17]. Moreover, some species of soil fauna can weaken microbial activities and feed humic substances, displaying negative effects on the accumulation of humic substances [18]. Including other factors, microclimate and litter quality are widely considered as vital constraints regulating the relationship between soil fauna and litter humification [19–21]. Humic substances are more sensitive and unstable in cold forests than those are in subtropic/tropical forests, while stable temperature and humidity are more likely to promote the formation of humic substances [22–24]. The feeding of soil fauna may be limited under low temperature conditions, thereby affecting the formation of humic substances [25]. Nevertheless, litter quality and its interaction with living conditions of soil fauna could control the humus accumulation by regulating the community structure of soil fauna [22], leading to currently inconsistent online information.

The cold forest on the eastern edge of the Qinghai–Tibet Plateau is an important part of the ecological barrier in the upper reaches of the Yangtze River [26]. The forests have served as prominent strategic positions in regional climate regulation, biodiversity conservation, water conservation, and river runoff regulation [27,28]. However, the cold forests are also fragile and irreversible, since they could be affected by geology and long-term snow cover and seasonal freeze–thaw cycles [29–31]. As an important indicator of soil fertility, soil humus is one of the essential factors in keeping forest soil productivity [32,33]. However, the processes of litter humification and its relationships with soil fauna have not been well addressed in the cold regions. Therefore, a field incubation has carried out to investigate the contributions of soil fauna in the accumulation of humus for the foliar litter of four dominated tree species. The following hypothesis is proposed: (1) soil fauna could promote the process of humidification depending on species and key times in this region; (2) the individual density of soil fauna is related to the degree of humification of leaf litter.

2. Materials and Methods

2.1. Site Description

The study site was conducted at the Long-term Research Station of Alpine Forest Ecosystem of Sichuan Agricultural University (31°14′–31°19′ N, 102°53′–102°57′ E, 2458–4619 m a.s.l), which is located in the eastern Tibet Plateau [34,35]. The average annual temperature is 2 °C, and the average annual rainfall in this site is 850 mm. The seasonal freeze–thaw period is from November to April and lasts for 5 to 6 months each year. The experimental sites are located at two elevations, where the higher elevation (3593 m) is a primary natural forest that is dominated with Willow (*Salix paraplesia* Schneid) and Cypress (*Juniperus saltuaria* (Rehd. et Wils) Cheng et W. T. Wang), whereas the other site (3028 m) is a plantation that dominated with Birch (*Betula albosinensis* Burkill) and Fir (*Abies fargesii* var. *faxoniana* (Rehder & E. H. Wilson) Tang S. Liu) trees. The soil is a Cambic Umbrisol (IUSS Working Group WRB, 2006), and more information on soil properties can be found in Tan et al. (2020) [36].

2.2. Litterbag Experiment and Sample Collection

A primary natural forest and a cypress plantation were selected in the study site. Fresh senesced foliar litters were collected through a 5 m × 5 m litter collector at each site in October 2013. All litter materials were air-dried at room temperature (25 °C) for 15 days and then weighted 10 ± 0.05 g for each sample that was placed in a 20 cm × 20 cm litterbag with one type of foliar litter. Litterbags with two mesh sizes were used to exclude and permit the access of soil fauna into the litterbags to determine fauna effects on litter humification [37]. Specifically, there were two types of litterbags, with both types having the bottoms of the litterbags with a mesh size of 0.04 mm, but the tops had two mesh sizes (0.04 mm and 3.00 mm): 0.04 mm for the treatment to exclude the entry of soil fauna (fauna exclusion) and 3.00 mm for the control to permit the entry of macro-, meso- and microfauna [36,38]. A total of 360 litterbags (2 meshes × 4 species × 5 times × 3 replicates × 3 plots) were randomly distributed on the soil surface (with a distance between litter bags of about 5 cm) just after the litterfall peak in November 2013.

The temperature of litterbag was measured using DS1923-F5 Recorders (iButton DS1923-F5, Maxim/Dallas Semiconductor, Sunnyvale, CA, USA) every 2 h at each site and used to calculate the daily average (DAT) and total accumulated temperature (TAT, the accumulation of daily average temperature in the litterbag during each stage) during the whole experimental periods. The average daily temperature in the alpine forests was showed in Figure 1. The trend of the daily average temperature of the primary forest and the plantation forest is basically the same, but from the growing season to the leaf falling stage period, the daily average temperature of the plantation forest is higher than that of the primary forest. The water content of litter was dried into oven for 48 h until its weight was constant, then it was recorded.

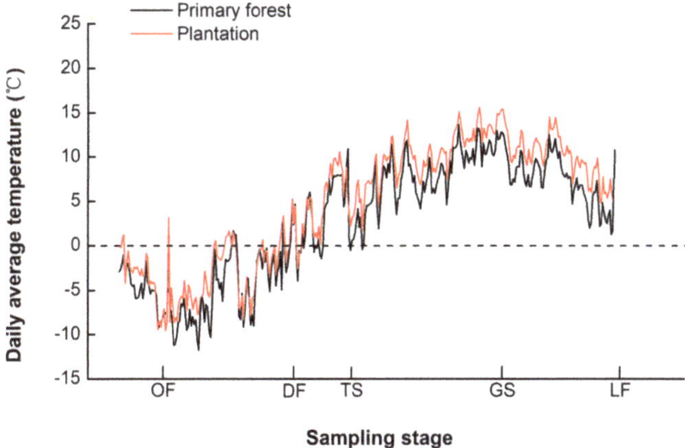

Figure 1. The average daily temperature in the alpine forests of different sampling stages. OF: onset of freezing, DF: deep freezing stage, TS: thawing stage, GS: growing season, LF: leaf falling stage.

According to our previous research and long-term sequential temperature observations [39], litterbags were collected in different stages: onset of freezing stage (OF, December 2013), deep freezing stage (DF, March 2014), thawing stage (TS, April 2014), growing season (GS, August 2014), and leaf falling stage (LS, October 2014). Thereafter, 9 litterbags filled with each litter species were randomly collected at each sampling site at every stage. A part of the recovered litterbags was used for soil fauna collection and the determination of matrix quality. The other part was air-dried to remove impurities, then oven-dried at 65 °C for 48 h (DHJ 2450B, Sheng Yuan Instrument, Zhengzhou, China) and weighed (Figure 2). The air-dried fallen leaves were crushed, sieved (0.25 mm mesh), and stored in a kraft paper bag for the extraction experiment of humus.

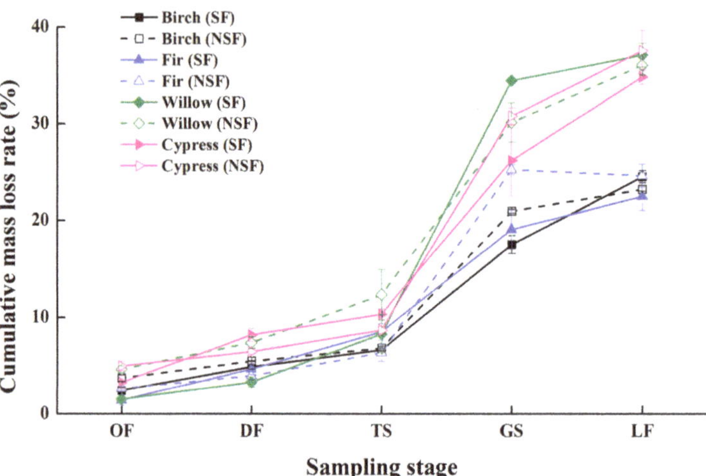

Figure 2. Cumulative mass loss rates of four litter (excluded (NSF) and non-excluded soil fauna (SF)) at different sampling stages (mean ± standard error, n = 3). OF: onset of freezing, DF: deep freezing stage, TS: thawing stage, GS: growing season, LF: leaf falling stage.

2.3. Soil Fauna Collection and Chemistry Analysis

The soil fauna in the litterbags were extracted by the Tullgren funnel method over a period of 48 h at 30 °C, as previously described [37]. All extracted soil fauna were counted and classified under a microscope (Leica MZ 125, Leica Microsystems GmbH, Wetzlar, Germany). After one year of exposure, we collected a total number of 2005 soil fauna individuals in the litter bag, among which Collembola (*Isotomidae* and *Campodeoidea*, etc.) and mites (*Oribatida* and *Prostigmata*, etc.) were the dominant groups, accounting for 49% and 46% of the total number of soil fauna, respectively (Figure 3, Supplementary Materials Table S1).

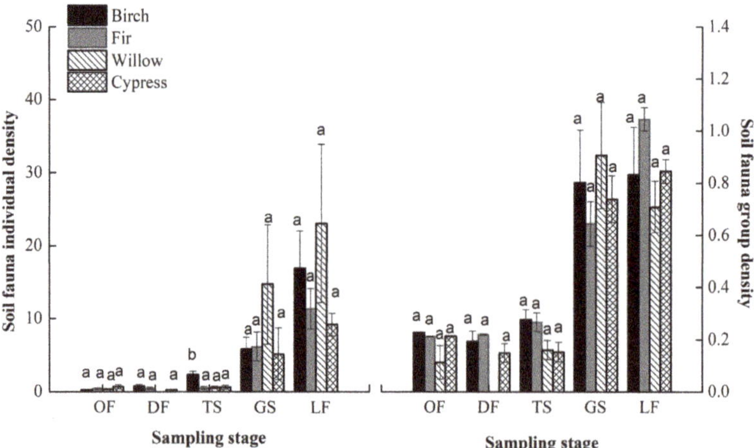

Figure 3. Individual and group density of soil fauna in different sampling stage (mean ± standard error, n = 3). OF: onset of freezing, DF: deep freezing stage, TS: thawing stage, GS: growing season, LF: leaf falling stage. The letters a and b mean in the sampling stage with different letters differ significantly, *p* < 0.05.

NaOH method was used to extracted humic substances [40]. 0.5 g of litter sample was added to 100 mL of 0.1 mol/L NaOH + $Na_4P_2O_7 \cdot 10\,H_2O$ mixed solutions, shaken for 10 min, and put in a water bath with a constant temperature at 100 °C for 1 h. The liquid was filtered and used in part to measure humic substances (HS). We took out another 20 mL, added 0.5 mol/L H_2SO_4 to pH 3, and put it in a water bath with a constant temperature at 80 °C for 30 min. The liquid was taken out and filtered, and then the precipitated part was dissolved with 0.05 mol/L NaOH solution for the determination of humic acid (HA). The humic substances and the humic acid test solution were both filtered with a 0.45 μm filter membrane, and then measured with TOC analyzer (multi N/C 2100, Analytik Jena, Thüringen, Germany).

2.4. Data Calculation and Statistical Analyses

Soil fauna individual density (D_i) and group density (D_g) were calculated as [37]:

$$\text{Individual density } (D_i) = N_i/M$$

$$\text{Group density } (D_g) = N_g/M$$

where, N_i, N_g, and M represent the individual number of soil fauna, the group number of soil fauna, and the mass residue of litter, respectively.

Fulvic acid concentration (FA), humification degree (HD), the contribution rate of soil fauna to the accumulation of humic substances, humic and fulvic acid, and humification degree were calculated as [30,36]:

$$\text{FA } (g/Kg) = HS - HA$$

$$\text{HD } (\%) = HS/OC \times 100\%$$

$$C_{HS}\ (\%) = (HS_{(SFt)} \times M_{(SFt)} - HS_{(NSFt)} \times M_{(NSFt)})/HS_{(NSFt)} \times M_{(NSFt)} \times 100\%$$

$$C_{HA}\ (\%) = (HA_{(SFt)} \times M_{(SFt)} - HA_{(NSFt)} \times M_{(NSFt)})/HA_{(NSFt)} \times M_{(NSFt)} \times 100\%$$

$$C_{FA}\ (\%) = (FA_{(SFt)} \times M_{(SFt)} - FA_{(NSFt)} \times M_{(NSFt)})/FA_{(NSFt)} \times M_{(NSFt)} \times 100\%$$

$$C_{HD}\ (\%) = (HD_{(SFt)} \times M_{(SFt)} - HD_{(NSFt)} \times M_{(NSFt)})/HD_{(NSFt)} \times M_{(NSFt)}$$

where HS and HA represent the concentrations of humic substances and humic acid, OC represents organic carbon concentration of litter. C_{HS}, C_{HA}, C_{FA}, and C_{HD} represent the contribution rate of soil fauna to the accumulation of humic substances, humic and fulvic acid, and humification degree, respectively. $HS_{(SFt)}$, $HA_{(SFt)}$, $FA_{(SFt)}$, and $HD_{(SFt)}$ represent the accumulation of humic substances, humic and fulvic acid, and humification degree under non-excluded soil fauna at the stage t, respectively. $HS_{(NSFt)}$, $HA_{(NSFt)}$, $FA_{(NSFt)}$, and $HD_{(NSFt)}$ represent the accumulation of humic substances, humic and fulvic acid, and humification degree under excluded soil fauna at the stage t, respectively. $M_{(SFt)}$ and $M_{(NSFt)}$ represent the mass residue of excluded soil fauna and non-excluded soil fauna at the stage t.

One-way ANOVA was used to analyze individual density and group density of soil fauna and the contribution rate of soil fauna to humic substances, humic acid, fulvic acid, and humification degree accumulation. Multivariate analysis of variance was used to analyze the individual density and group density of soil fauna, accumulation of humic substances, humic acid, fulvic acid, and degree of humification with species time, and their interaction. Pearson correlation analysis was used to calculate the correlation parameters of the contribution rate of soil fauna to the accumulation of humic substances (HS), humic acid (HA), fulvic acid (FA), and humification degree (HD) to carbon, nitrogen, phosphorus, lignin (L), cellulose (CE), total accumulated temperature (TAT), daily average temperature (DAT), litter water content (LWC), soil fauna individual density (D_i), and group density (D_g) at different stages. Linear regression analysis was used for the individual and group density of soil fauna and the contribution rate of soil fauna to the degree of litter humification. All significant differences were set at a level of $p = 0.05$. All statistical analyses were performed with SPSS 20.0 (IBM SPSS Statistics Inc., Chicago, IL, USA).

3. Results

3.1. The Contributions of Soil Fauna on Humic Substances Accumulation

Soil fauna showed significant effects on the accumulation of humic substance, but these strengths differed among sampling time and species (Figure 4 and Table 1). Over one year incubation, the contribution rates of soil fauna to humic substances were 109.06%, 71.48%, 11.22%, and −44.43% for fir, cypress, birch, and willow, respectively. Pearson correlation showed that soil fauna group density was significantly correlated with the contribution rates of soil fauna to humic substances at OF stage ($r = -0.59$, $p < 0.05$, Table 2). At the stage of OF, soil fauna in willow litter promoted the accumulation of humic substances, while it inhibited the accumulation of humic substances in fir and birch litters. Moreover, soil fauna improved the accumulation of humic substances at both the GS and LF stages in fir, birch, and cypress litters, but negative effects of soil fauna on the accumulation of humic substances in willow were observed during both stages.

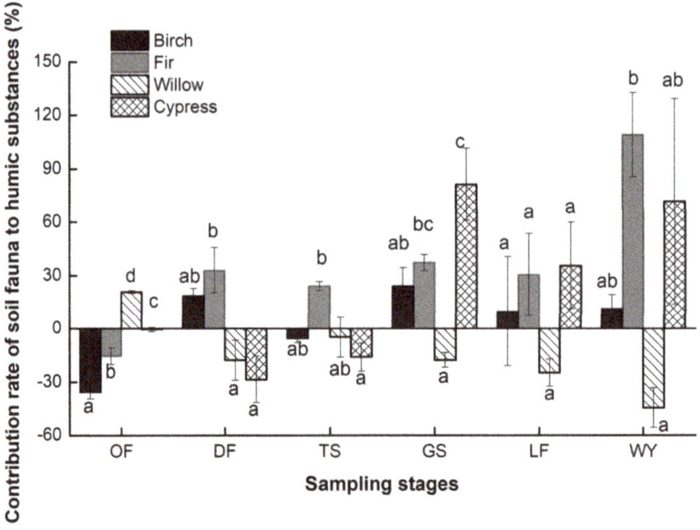

Figure 4. The contribution rate of soil fauna to the accumulation of four litter humic substance in different stages (mean ± standard error, n = 3). OF: onset of freezing stage, DF: deep freezing stage, TS: thawing stage, GS: growing season, LF: leaf falling stage, WY: whole year. The letters a–c mean in the sampling stage with different letters differ significantly, $p < 0.05$.

Table 1. F values for multivariate analysis of variance for individual (D_i) and group density (D_g) of soil fauna, the contribution rate of humic substances (HS), humic acid (HA), fulvic acid (FA), and humification degree (HD) by species and time.

Factor	D_i	D_g	HS	HA	FA	HD
Species (S)	2.04	0.67	5.38 **	23.79 ***	73.71 ***	2.470
Time (T)	12.23 ***	54.46 ***	5.43 **	9.63 ***	35.88 ***	10.04 ***
S × T	0.73	1.13	4.75 ***	26.73 ***	36.65 ***	8.09 ***

** $p < 0.01$. *** $p < 0.001$.

Table 2. Pearson correlation analysis of contribution rate of soil fauna to humic substances (HS), humic acid (HA), fulvic acid (FA) and humification degree (HD) and C, N, P, lignin (L), cellulose (CE), total accumulated temperature (TAT), daily average temperature (DAT), litter water content (LWC), soil fauna individual density (D_i), and group density (D_g) at different stages. OF: onset of freezing stage, DF: deep freezing stage, TS: thawing stage, GS: growing season, LF: leaf falling stage.

Time	Factor	C	N	P	L	CE	TAT	DAT	LWC	D_i	D_g
OF	HS	−0.72 **	0.64 *	0.77 **	−0.52	0.68 *	−0.85 **	−0.85 **	−0.47	0.173	−0.59 *
	HA	−0.42	0.72 **	0.67 *	−0.22	0.63 *	−0.48	−0.48	−0.37	−0.33	−0.39
	FA	−0.69 *	0.62 *	0.76 **	−0.55	0.69 *	−0.86 **	−0.86 **	−0.45	0.21	−0.52
	HD	−0.32	0.63 *	0.14	0.02	0.06	−0.04	−0.04	−0.56	−0.33	−0.47
DF	HS	−0.46	−0.16	−0.63 *	0.43	−0.66 *	0.83 **	0.83 **	−0.51	0.51	0.49
	HA	−0.33	−0.57	−0.74 **	0.59 *	−0.75 **	0.89 **	0.89 **	0.08	0.45	0.48
	FA	−0.54	−0.14	−0.69 *	0.37	−0.71 *	0.82 **	0.82 **	−0.55	0.59	0.43
	HD	−0.29	0.75 **	0.04	−0.07	0.01	−0.17	−0.17	−0.31	0.24	0.35
TS	HS	0.41	−0.02	−0.26	0.09	0.29	0.54	0.54	−0.64 *	−0.17	0.44
	HA	0.46	−0.21	−0.53	0.25	0.34	0.76 **	0.76 **	−0.86 **	−0.01	0.69 *
	FA	−0.02	0.32	0.11	−0.4	0.03	0.17	0.17	−0.41	−0.51	0.08
	HD	−0.13	−0.12	0.05	0.35	−0.23	−0.28	−0.28	0.49	0.61 *	−0.11
GS	HS	−0.21	−0.77 **	0.72 **	−0.47	0.34	−0.01	−0.01	−0.38	−0.35	−0.33
	HA	0.32	−0.26	0.01	0.57	−0.62 *	0.39	0.39	0.54	−0.16	−0.01
	FA	−0.48	−0.37	0.33	−0.71 **	0.61 *	−0.21	−0.21	−0.59 *	−0.18	−0.34
	HD	0.02	0.23	−0.01	−0.13	−0.24	−0.44	−0.44	−0.23	0.14	0.32
LF	HS	−0.03	−0.56	−0.33	−0.12	0.54	0.18	0.18	−0.49	−0.49	0.13
	HA	0.56	−0.16	0.21	0.02	0.51	−0.04	−0.04	−0.59 *	−0.28	0.42
	FA	−0.2	−0.71 **	−0.48	0.44	0.60 *	0.75 **	0.75 **	−0.34	−0.31	0.62 *
	HD	0.05	−0.67 *	−0.45	−0.55	0.63 *	−0.08	−0.08	−0.65 *	−0.23	0.02

* $p < 0.05$; ** $p < 0.01$.

3.2. The Contributions of Soil Fauna to Humic Acid Accumulation

The contribution of soil fauna to accumulation of humic acid was significantly affected by species ($F = 23.79$), sampling time ($F = 9.63$), and their interaction ($F = 27.53$, all $p < 0.001$, Table 1). Overall, the contribution rate of soil fauna to the accumulation of humic acid showed the order as fir (111.87%) > birch (91.24%) > cypress (−34.83%) > willow (−49.9%). Soil fauna group density was significantly correlated with the contribution rate of soil fauna to humic acid at TS stage ($r = 0.69$, $p < 0.05$, Table 2). Significant increases in the accumulation of humic acid contributed by soil fauna were detected in willow litter rather than in other species at OF stage (Figure 5). In contrast, the contributions of soil fauna to humic acid exhibited the same trend with that soil fauna promoted the accumulation of humic acid in the litters of birch and fir but inhibited them in the litters of willow and cypress at DF and TS stages. Soil fauna promoted the accumulation of humic acid in birch litter at GS and LF stages, while limiting the accumulation in other litters.

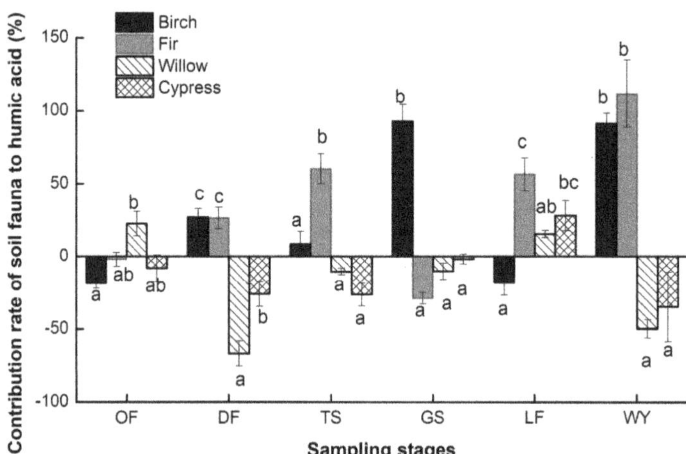

Figure 5. The contribution rate of soil fauna to the accumulation of four litter humic acid in different stages (mean ± standard error, n = 3). OF: onset of freezing stage, DF: deep freezing stage, TS: thawing stage, GS: growing season, LF: leaf falling stage, WY: whole year. The letters a–c mean in the sampling stage with different letters differ significantly, $p < 0.05$.

3.3. The Contributions of Soil Fauna to Fulvic Acid Accumulation

The contribution rate of soil fauna to the accumulation of fulvic acid showed the order as, fir (220.24%) > cypress (35.43%) > willow (−28.24%) > birch (−55.64%). Soil fauna promoted the accumulation of fulvic acid in the litter of willow and cypress but limited the accumulations of fulvic acid in other two litters at OF stage (Figure 6), which showed the opposite side at OF stage. Soil fauna promoted the accumulation of fulvic acid in the litter of fir and willow and inhibited the accumulation in the litter of birch and cypress at TS stage. At the stage of GS, soil fauna promoted the accumulation of fulvic acid in the litters of fir and cypress but inhibited the accumulation in the litters of birch and willow. Soil fauna only inhibited the accumulation of fulvic acid in willow litter at LF stage, whereas soil fauna contributed 40.42% and 128.10% increase of fulvic acid in the litters of birch and fir, respectively.

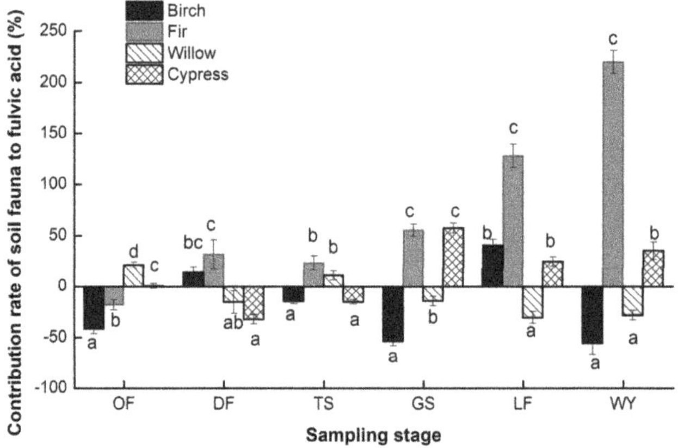

Figure 6. The contribution rate of soil fauna to the accumulation of four litter fulvic acid in different stages (mean ± standard error, n = 3). OF: onset of freezing stage, DF: deep freezing stage, TS: thawing stage, GS: growing season, LF: leaf falling stage, WY: whole year. The letters a–d mean in the sampling stage with different letters differ significantly, $p < 0.05$.

3.4. Contributions of Soil Fauna to Litter Humification Degree

Sampling time and the interaction of species with sampling time significantly affected the contribution rate of soil fauna on litter humification degree ($F = 10.04$ and $F = 8.09$, all $p < 0.001$, Table 1). The whole year contribution rate of soil fauna to humification degree was 12.66% for willow litter, and positive contribution appeared at both OF and DF stages (Figure 7). Contrarily, soil fauna showed -13.27%, -49.20%, and -7.63% of contribution rates to humification degree in the litter of birch, fir, and cypress in the whole year. However, the contribution rates of soil fauna to humification degree were 14.67%, 6.55%, and 48.79% in the litter of birch, fir, and cypress at LF stage.

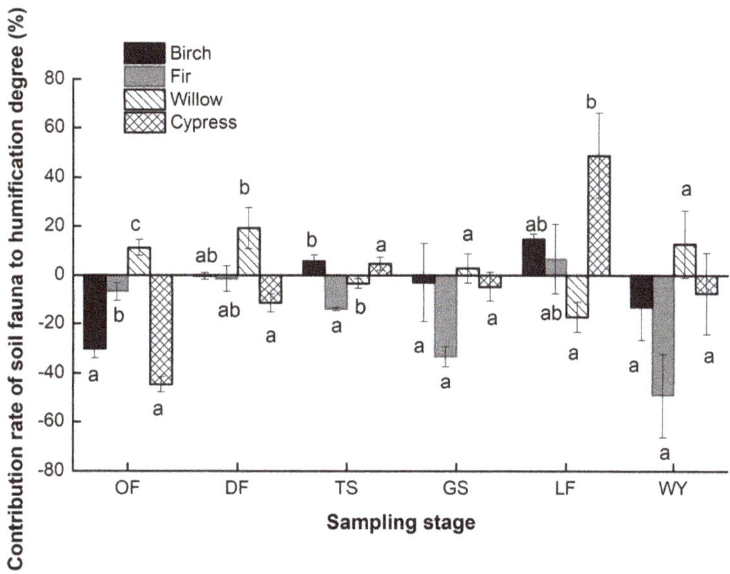

Figure 7. The contribution rate of soil fauna to four litter humification degree in different stages (mean ± standard error, n = 3). OF: onset of freezing stage, DF: deep freezing stage, TS: thawing stage, GS: growing season, LF: leave falling stage. OF: onset of freezing stage, DF: deep freezing stage, TS: thawing stage, GS: growing season, LF: leaf falling stage, WY: whole year. The letters a–c mean in the sampling stage with different letters differ significantly, $p < 0.05$.

The linear regressions show the correlation of individual and group density of soil fauna with cumulative contribution rate of soil fauna to the litter humification degree differed among species (Figure 8). Soil fauna significantly influenced the humification degree of the litter of birch, willow, and cypress, while it had no significant effect for fir litter. There was a significant correlation of individual density of soil fauna with humification degree for the litter of birch ($p < 0.05$, $R^2 = 0.27$, Figure 8a) and cypress ($p < 0.01$, $R^2 = 0.37$, Figure 8d). Additionally, there were significant correlations of group density for the litter of willow ($p < 0.05$, $R^2 = 0.28$, Figure 8c) and cypress ($p < 0.05$, $R^2 = 0.28$, Figure 7d).

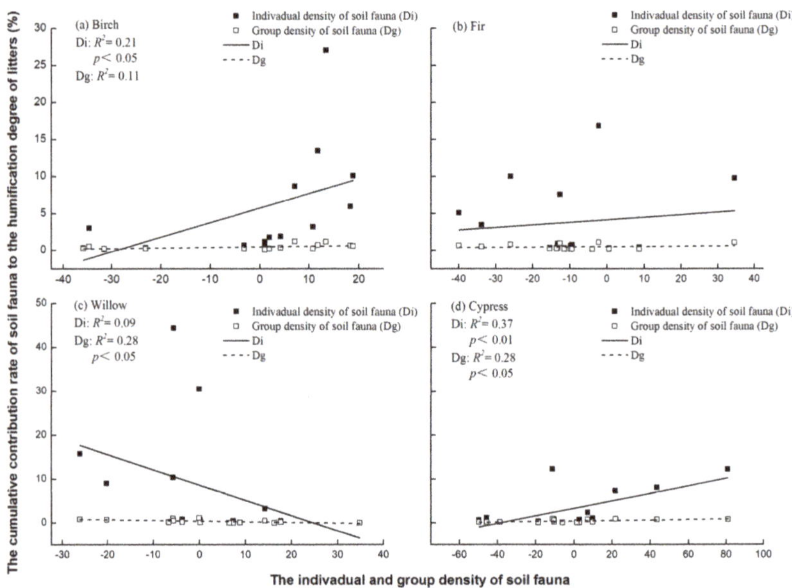

Figure 8. Linear regression analysis between the cumulative contribution rate of soil fauna to the humification degree of four kinds of litters and the individual and group density of soil fauna.

4. Discussion

Following our first hypothesis, we verified that the process of humidification promoted by soil fauna differed on species. Except for willow, we found soil fauna significantly promoted the accumulation of humic substances in litters of birch, fir, and cypress (Figure 4), indicating soil fauna could promote the process of humidification depending on species [37,41]. Additionally, we found soil fauna individual density significantly related to humification degree besides cypress (Figure 8), possibly by the presence of high concentration of volatile organic compounds in cypress, which could inhibit the activity of soil decomposer [42]. Overall, our result clearly demonstrated that soil fauna promotes the process of litter humification but varied by species and environmental factor.

Species, sampling, stage and their interactions had significant effects on the humic substances accumulations as litter humification proceeding in the alpine forests (all $p < 0.001$, Table 1). The positive effect of soil fauna on the accumulation of humic substances is high in the stages of GS and LF, which is in accordance with the densities of individual and group of soil fauna, indicating that soil fauna is favorable to the accumulations of humic substance of birch, fir, and cypress. Our result showed that soil fauna plays vital roles during the period of accumulation of humic substances, agreement with previous studies [19,43]. However, soil fauna (*Prostigmata* and *Oribatida*, etc.) also feed on humic substances and reduce the content of humic acid through the intestines, thereby converting organic matter [16], but these activities are varied among seasons and resulting in the capability of producing humic substances are difference. Interestingly, soil fauna had a positive effect on the accumulation of humic substance of willow litter in each stage aside from the stage of OF, while inhabited other foliar litters (Figure 4). The presence of scavenging soil fauna was found in birch, fir, and cypress litter, but not in willow litter (Supplementary Materials Table S1), so the humic substances might be highly consumed and decomposed by both soil microbes and scavenging fauna [8,10]. Moreover, we found the overall contribution rate of soil fauna to humic substance in coniferous litters (fir and cypress) is higher than that in broadleaf litters (birch and willow), which is related to litter quality. As higher contents of lignin in coniferous litters are more conducive to the formation of humus than broadleaf litters [43,44].

The contribution rate of litter humic substance response to soil fauna differs among sampling time, which is affected by litter qualities and environmental factors (Table 2). For fir litter, soil fauna had positive effects on both humic acid and fulvic acid, indicating a net accumulation of humic substances on the whole year, which was influenced by litter substrate quality [43]. Moreover, temperature and humidity can affect the biodiversity of soil fauna [44], and the growing season increases with soil fauna density and individual density, resulting in the contribution rate of soil fauna on fulvic acid being much higher than humic acid, as the contribution of soil fauna to humic acid was negative in the stage of GS (Figure 5). More importantly, after one-year incubation, the contribution rate of litter humic acid to soil fauna seems related to elevation, as positive contribution of birch and fir litters appeared in the plantation and negative values of willow and cypress litter in the primary forest. Differently, the contribution rate of litter fulvic acid to soil fauna was related to species, which showed negative contributions in coniferous litters and positive values in broadleaf litters. The divergent responses of humic acid and fulvic acid to soil fauna in different litter species highlighted that environmental factors and litter quality jointly controlled the content of humic substances [4,20,44].

Generally, environment exerts significant effects in the early decomposition stages due to physical effects [39]. When litter components that are susceptible to physical effects are consumed, the influence of the environment on litter decay is weakened. Otherwise, environmental factor directly affected the individual density and group density of soil fauna in each stage (Table 2), thus affecting the process of litter humification. We found the contribution rates of humic acid and fulvic acid to soil fauna reached maximum values in litters of birch, fir, and cypress at the stages of GS and LF. The individual density and group density of soil fauna both reached their peak values under the influence of temperature in these two stages, so their contribution rate reached the maximum. Moreover, Larionova et al. (2017) found there was an optimal temperature (12 °C) for the formation of humic substances [24]. In our site, the mean air temperature is 11 °C in the growing season, which provides good conditions to promote the humic substances of the fallen leaves accumulate rapidly. The annual contribution rate of soil fauna to the accumulation of humic acid in the plantation forest is higher than that of primary forest, and the contribution rate of the fulvic acid accumulation in coniferous litter is higher than that in broad-leaved litter, which indicates that the formation of humic acid by soil fauna is mainly affected by environmental factors, while the formation of fulvic acid is mainly affected by the quality of the matrix.

Over one-year field incubation, only willow litter showed a positive effect on the humification degree by soil fauna, and other litters were negative values (Figure 6). The different responses of humification degree to soil fauna in litters appeared significantly in the stages of OF and DF (Figure 7). The positive effects in willow litter at the stages of OF (16%) and DF (22%) indicate that soil fauna also play an important function, improving humification degrees even in cold winter. Previous studies have suggested a nonnegligible process of litter humification in winter, but their humification degree depends on litter quality [4]. Otherwise, the humification degree is positively correlated with the individual and group density of soil fauna (Figure 8), but this relationship is also controlled by litter quality and type [8,20]. We found the individual density of soil fauna in the litter of birch and cypress significantly correlated with humification degree, and the group density for willow and cypress litters was significantly correlated with the humification degree. In contrast, the fir litter showed no significant effect to litter humification mediated by soil fauna.

For litters of birch, fir and cypress, the annual contributions of soil fauna to the humification degree are negative, ranging from −48% to −8% after a year field incubation (Figure 7), indicating soil fauna may not be a key factor controlling the humification degree of these litters; other factors, such as initial litter quality and environmental factory, are also important [8,11,45]. The contribution rate of soil fauna to humification degrees reached the maximum values in the stage of LF (Figure 7), which are in accordance with the patterns of individual density and group density (Figure 3 and Supplementary Materials Table S1). Other studies also showed that the population of soil fauna has a positive correlation with the humification degree of litter [4,45].

5. Conclusions

The accumulation of humic substances (including humic acid and fulvic acid) and litter humification degree were jointly determined by soil fauna, species, and environmental factors. Soil fauna is favorable to the accumulation of humic substances in growing and leaf falling seasons for the litter of birch, fir, and cypress rather than willow litter. Moreover, the accumulations of humic acid and fulvic acid were promoted by soil fauna in fir litter, but these accumulations were inhibited in willow litter. Additionally, soil fauna showed positive effects on the humification degree in willow litter, but negative effects on the other three litter. Overall, our study gives a deep insight in the contribution of soil fauna on litter humification, which could provide useful data in soil carbon sequestration in the cold forests.

Supplementary Materials: The following supporting information can be downloaded at: https://www.mdpi.com/article/10.3390/f13081235/s1, Table S1: The average densities and eating patterns of soil fauna in the four studied litters under different stages.

Author Contributions: Formal analysis, Z.X. and L.Z.; Investigation, K.Y. and H.L.; Methodology, C.Y.; Supervision, B.T.; Writing—original draft, Y.T. All authors have read and agreed to the published version of the manuscript.

Funding: This research was funded by the National Natural Science Foundation of China (31870602, 32071745, 31901295 and 32001165), the Program of Sichuan Excellent Youth Sci-Tech Foundation (2020JDJQ0052) and the Program of Sichuan Applied Basic Research Foundation (2021YJ0340).

Institutional Review Board Statement: Not applicable.

Informed Consent Statement: Not applicable.

Data Availability Statement: The study did not report any data.

Conflicts of Interest: The authors declare no conflict of interest.

References

1. Andreetta, A.; Cecchini, G.; Bonifacio, E.; Comolli, R.; Vingiani, S.; Carnicelli, S. Tree or soil? Factors influencing humus form differentiation in Italian forests. *Geoderma* **2016**, *264*, 195–204. [CrossRef]
2. Andreetta, A.; Cecchini, G.; Carnicelli, S. Forest humus forms in Italy: A research approach. *Appl. Soil Ecol.* **2018**, *123*, 384–390. [CrossRef]
3. Alekseev, I.; Kraev, G.; Shein, A.; Petrov, P. Soil Organic Matter in Soils of Suburban Landscapes of Yamal Region: Humification Degree and Mineralizing Risks. *Energies* **2022**, *15*, 2301. [CrossRef]
4. Ni, X.Y.; Yang, W.Q.; Li, H.; Xu, L.X.; He, J.; Tan, B.; Wu, F.Z. The responses of early foliar litter humification to reduced snow cover during winter in an alpine forest. *Can. J. Soil Sci.* **2014**, *94*, 453–461. [CrossRef]
5. Ponge, J.F. Plant–soil feedbacks mediated by humus forms: A review. *Soil Biol. Biochem.* **2013**, *57*, 1048–1060. [CrossRef]
6. Yang, J.P.; Zhang, Y.; Liang, Z.Y.; Yue, K.; Fu, C.K.; Ni, X.Y.; Wu, F.Z. The optical properties in alkali-soluble fractions extracted from newly shed litters in a subalpine forest. *J. Soils Sediments* **2020**, *20*, 1276–1284. [CrossRef]
7. Zhang, Y.; Zhang, X.J.; Wen, J.; Wang, Y.N.; Zhang, N.; Jia, Y.H.; Su, S.M.; Wu, C.X.; Zeng, X.B. Exogenous fulvic acid enhances stability of mineral-associated soil organic matter better than manure. *Environ. Sci. Pollut. Res.* **2021**, *29*, 9805–9816. [CrossRef]
8. Wang, W.; Yang, W.Q.; Tan, B.; Liu, R.; Wu, F.Z. Contributions of soil fauna to litter decomposition in subtropical evergreen broad-leaved forests in Sichuan basin. *Ecol. Environ. Sci.* **2013**, *24*, 3354–3360.
9. Zanella, A.; Bolzonella, C.; Lowenfels, J.; Ponge, J.F.; Bouché, M.; Saha, D. Humusica 2, article 19: Techno humus systems and global change–conservation agriculture and 4/1000 proposal. *Appl. Soil Ecol.* **2018**, *122*, 271–296. [CrossRef]
10. Tan, B.; Yin, R.; Zhang, J.; Xu, Z.F.; Liu, Y.; He, S.Q.; Zhang, L.; Li, H.; Wang, L.X.; Liu, S.N.; et al. Temperature and Moisture Modulate the Contribution of Soil Fauna to Litter Decomposition via Different Pathways. *Ecosystems* **2020**, *24*, 1142–1156. [CrossRef]
11. Huang, Y.M.; Yang, X.; Zhang, D.J.; Zhang, J. The effects of gap size and litter species on colonization of soil fauna during litter decomposition in *Pinus massoniana* plantations. *Appl. Soil Ecol.* **2020**, *155*, 103611. [CrossRef]

12. Tardy, V.; Spor, E.; Mathieu, O.; Eque, J.; Maron, P.A. Shifts in microbial diversity through land use intensity as drivers of carbon mineralization in soil. *Soil Biol. Biochem.* **2015**, *90*, 204–213. [CrossRef]

13. Gessner, M.O.; Swan, C.M.; Dang, C.K.; Mckie, B.G.; Bardgett, R.D.; Wall, D.H.; Hattenschwiler, S. Diversity meets decomposition. *Trends Ecol. Evol.* **2010**, *25*, 372–380. [CrossRef] [PubMed]

14. Steinwandter, M.; Seeber, J. The buffet is open: Alpine soil macro-decomposers feed on a wide range of litter types in a microcosm cafeteria experiment. *Soil Biol. Biochem.* **2020**, *144*, 107786. [CrossRef]

15. Fujii, S.; Berg, M.P.; Cornelissen, J.H.C. Living Litter: Dynamic Trait Spectra Predict Fauna Composition. *Trends Ecol. Evol.* **2020**, *35*, 886–896. [CrossRef] [PubMed]

16. Komarov, A.; Chertov, O.; Bykhovets, S.; Shaw, C.; Nadporozhskaya, M.; Frolov, P.; Shashkov, M.; Shanin, V.; Grabarnik, G.; Priputina, I.; et al. Romul_Hum model of soil organic matter formation coupled with soil biota activity. I. Problem formulation, model description, and testing. *Ecol. Modell.* **2017**, *345*, 113–124. [CrossRef]

17. Nakatsuka, H.; Karasawa, T.; Ohkura, T.; Wagai, R. Soil faunal effect on plant litter decomposition in mineral soil examined by two in-situ approaches: Sequential density-size fractionation and micromorphology. *Geoderma* **2020**, *357*, 113910. [CrossRef]

18. Hobbie, S.N.; Li, X.; Basen, M.; Stingl, U.; Brune, A. Humic substance-mediated Fe(III) reduction by a fermenting Bacillus strain from the alkaline gut of a humus-feeding scarab beetle larva. *Syst. Appl. Microbiol.* **2012**, *35*, 226–232. [CrossRef]

19. Ni, X.Y.; Yang, W.Q.; Tan, B.; Li, H.; He, J.; Xu, L.X.; Wu, F.Z. Forest gaps slow the sequestration of soil organic matter: A humification experiment with six foliar litters in an alpine forest. *Sci. Rep.* **2016**, *6*, 19744. [CrossRef] [PubMed]

20. Zhou, Y.; Wang, L.F.; Chen, Y.M.; Zhang, J.; Liu, Y. Litter stoichiometric traits have stronger impact on humification than environment conditions in an alpine treeline ecotone. *Plant Soil.* **2020**, *453*, 545–560. [CrossRef]

21. Zhou, Y.; Wang, L.F.; Chen, Y.M.; Zhang, J.; Xu, Z.F.; Guo, L.; Wang, L.X.; You, C.M.; Tan, B.; Zhang, L.; et al. Temporal dynamics of mixed litter humification in an alpine treeline ecotone. *Sci. Total Environ.* **2022**, *803*, 150122. [CrossRef] [PubMed]

22. Liu, Y.; Wang, L.F.; He, R.L.; Chen, Y.M.; Xu, Z.F.; Tan, B.; Zhang, L.; Xiao, J.J.; Zhu, P.; Chen, L.H.; et al. Higher soil fauna abundance accelerates litter carbon release across an alpine forest-tundra ecotone. *Sci. Rep.* **2019**, *9*, 10561. [CrossRef] [PubMed]

23. Koven, C.D.; Hugelius, G.; Lawrence, D.M.; Wieder, W.R. Higher climatological temperature sensitivity of soil carbon in cold than warm climates. *Nat. Clim. Chang.* **2017**, *7*, 817–822. [CrossRef]

24. Larionova, A.A.; Maltseva, A.N.; Lopes de Gerenyu, V.O.; Kvitkina, A.K.; Bykhovets, S.S.; Zolotareva, B.N.; Kudeyarov, V.N. Effect of temperature and moisture on the mineralization and humification of leaf litter in a model incubation experiment. *Eurasian Soil Sci.* **2017**, *50*, 422–431. [CrossRef]

25. Gongalsky, K.B.; Persson, T.; Pokarzhevskii, A.D. Effects of soil temperature and moisture on the feeding activity of soil animals as determined by the bait-lamina test. *Appl. Soil Ecol.* **2008**, *39*, 84–90. [CrossRef]

26. Zhou, C.X.; He, X.Y.; Zhao, Y.H.; Hu, Y.M.; Chan, Y.; Zhou, Q.X. Landscape Changes from 1974 to 1995 in the Upper Minjiang River Basin, China. *Pedosphere* **2006**, *16*, 398–405.

27. Cui, Y.X.; Bing, H.J.; Fang, L.C.; Wu, Y.H.; Yu, J.L.; Shen, G.T.; Mao, J.; Wang, X.; Zhang, X.C. Diversity patterns of the rhizosphere and bulk soil microbial communities along an altitudinal gradient in an alpine ecosystem of the eastern Tibetan Plateau. *Geoderma* **2019**, *338*, 118–127. [CrossRef]

28. Li, J.B.; Shen, Z.N.; Li, C.H.; Kou, Y.P.; Wang, Y.S.; Tu, B.; Zhang, S.H.; Li, X.Z. Stair-Step Pattern of Soil Bacterial Diversity Mainly Driven by pH and Vegetation Types Along the Elevational Gradients of Gongga Mountain, China. *Front. Microbiol.* **2018**, *9*, 569. [CrossRef]

29. Kreyling, J. Winter climate change: A critical factor for temperate vegetation performance. *Ecology* **2010**, *91*, 1939–1948. [CrossRef]

30. Ni, X.Y.; Yang, W.Q.; Tan, B.; He, J.; Xu, L.Y.; Li, H.; Wu, F.Z. Accelerated foliar litter humification in forest gaps: Dual feedbacks of carbon sequestration during winter and the growing season in an alpine forest. *Geoderma* **2015**, *241–242*, 136–144.

31. Millar, C.I.; Stephens, S. Climate Change and Forests of the Future: Managing in the Face of Uncertainty. *Ecol. Appl.* **2007**, *17*, 2145–2151. [CrossRef]

32. Radea, C.; Arianoutsou, M. Soil Micro-and Macrofauna in Mediterranean Pine and Mixed Forests. In *Pines and Their Mixed Forest Ecosystems in the Mediterranean Basin*; Springer: Cham, Switzerland, 2021; pp. 379–394.

33. Makarova, V.; Mykhaylov, A. Soil Fertility as a Criterion of Value Classification of Agricultural Land. *Mod. Econ.* **2020**, *23*, 108–113. [CrossRef]

34. Zhuang, L.Y.; Yang, W.Q.; Wu, F.Z.; Tan, B.; Zhang, L.; Yang, K.J.; He, R.Y.; Li, Z.J.; Xu, Z.F. Diameter-related variations in root decomposition of three common subalpine tree species in southwestern China. *Geoderma* **2018**, *311*, 1–8. [CrossRef]

35. Liu, Q.; Zhuang, L.Y.; Yin, R.; Ni, X.Y.; You, C.M.; Yue, K.; Tan, B.; Liu, Y.; Zhang, L. Root diameter controls the accumulation of humic substances in decomposing root litter. *Geoderma* **2019**, *348*, 68–75. [CrossRef]

36. Tan, B.; Yin, R.; Yang, W.Q.; Zhang, J.; Xu, Z.F.; Liu, Y.; He, S.Q.; Zhou, W.; Zhang, L.; Li, H.; et al. Soil fauna show different degradation patterns of lignin and cellulose along an elevational gradient. *Appl. Soil Ecol.* **2020**, *155*, 103673. [CrossRef]

37. Tan, Y.; Yang, W.Q.; Ni, X.Y.; Tan, B.; Yue, K.; Cao, R.; Liao, S.; Wu, F.Z. Soil fauna affects the optical properties in alkaline solutions extracted (humic acid-like) from forest litters during different phenological periods. *Can. J. Soil Sci.* **2019**, *99*, 195–207. [CrossRef]

38. Liao, S.; Ni, X.Y.; Yang, W.Q.; Li, H.; Wang, B.; Fu, C.K.; Xu, Z.F.; Tan, B.; Wu, F.Z. Water, Rather than Temperature, Dominantly Impacts How Soil Fauna Affect Dissolved Carbon and Nitrogen Release from Fresh Litter during Early Litter Decomposition. *Forests* **2016**, *7*, 249. [CrossRef]

39. Wu, F.Z.; Peng, C.H.; Zhu, J.X.; Zhang, J.; Tan, B.; Yang, W.Q. Impact of changes in freezing and thawing on foliar litter carbon release in alpine/subalpine forests along an altitudinal gradient in the eastern Tibetan Plateau. *Biogeosciences* **2014**, *11*, 6471–6481.

40. Adani, F.; Ricca, G. The contribution of alkali soluble (humic acid-like) and unhydrolyzed-alkali soluble (core-humic acid-like) fractions extracted from maize plant to the formation of soil humic acid. *Chemosphere* **2004**, *56*, 13–22. [CrossRef]

41. Bal, L. Morphological investigation in two moder-humus profiles and the role of the soil fauna in their genesis. *Geoderma* **1970**, *4*, 5–36. [CrossRef]

42. Werner, S.; Polle, A.; Brinkmann, N. Belowground communication: Impacts of volatile organic compounds (VOCs) from soil fungi on other soil-inhabiting organisms. *Appl. Microbiol. Biotechnol.* **2016**, *100*, 8651–8665. [CrossRef]

43. Ono, K.; Hiradate, S.; Morita, S.; Ohse, K.; Hirai, K. Humification processes of needle litters on forest floors in Japanese cedar (*Cryptomeria japonica*) and Hinoki cypress (*Chamaecyparis obtusa*) plantations in Japan. *Plant Soil.* **2010**, *338*, 171–181. [CrossRef]

44. Garcia-Palacios, P.; Maestre, F.T.; Kattge, J.; Wall, D.H. Climate and litter quality differently modulate the effects of soil fauna on litter decomposition across biomes. *Ecol. Lett.* **2013**, *16*, 1045–1053. [CrossRef] [PubMed]

45. Szanser, M.; Ilieva-Makulec, K.; Kajak, A.; Górska, E.; Kusińska, A.; Kisiel, M.; Olejniczaka, I.; Russelbe, S.; Sieminiakf, D.; Wojewodaa, D. Impact of litter species diversity on decomposition processes and communities of soil organisms. *Soil Biol. Biochem.* **2011**, *43*, 9–19. [CrossRef]

Article

Effects of Forest Gaps on *Abies faxoniana* Rehd. Leaf Litter Mass Loss and Carbon Release along an Elevation Gradient in a Subalpine Forest

Han Li [1,2], Ting Du [2], Yulian Chen [2], Yu Zhang [2], Yulian Yang [1], Jiaping Yang [1], Qing Dong [1], Li Zhang [2] and Qinggui Wu [1,*]

[1] Ecological Security and Protection Key Laboratory of Sichuan Province, Mianyang Normal University, Mianyang 621006, China; lihansc@sicau.edu.cn (H.L.); yangyulian2015@163.com (Y.Y.); iamyyyjp@163.com (J.Y.); zhaohrchina@163.com (Q.D.)

[2] Forestry Ecological Engineering in the Upper Reaches of the Yangtze River Key Laboratory of Sichuan Province, Institute of Ecology & Forestry, Sichuan Agricultural University, Chengdu 611130, China; 18292896853@163.com (T.D.); chenyulian09@163.com (Y.C.); wbcdmm123@163.com (Y.Z.); 14046@sicau.edu.cn (L.Z.)

* Correspondence: qgwu30@mtc.edu.cn; Tel.: +86-816-2579941

Abstract: Changes in the microenvironment induced by forest gaps may affect litter decomposition, yet it is unclear how the gap effects respond to altitudinal and seasonal differences. Here, a four-year litterbag decomposition experiment along an elevation gradient (3000, 3300, 3600 m) was conducted in an *Abies faxoniana* Rehd. subalpine forest of southwestern China, to assess the potential seasonal effects of forest gaps (large: ≈ 250 m^2, middle: ≈ 125 m^2, small: ≈ 40 m^2 vs. closed canopy) on litter mass loss and carbon release at different elevations. We found that the *A. faxoniana* litter mass loss and carbon release reached 50~53 and 58~64% after four years of decomposition, respectively. Non-growing seasons (November to April) had a greater decline than the growing seasons (May to October). Litter in the forest gaps exhibited significantly higher mass loss than that under the closed canopy, and the decomposition constant (k) exhibited a gradually declining trend from large gaps, middle gaps, small gaps to closed canopy. Moreover, more significant differences of gap on both carbon content and release were observed at the 3600 m site than the other two elevations. Our findings indicate that (i) a rather high mass loss and carbon release during the decomposition of *A. faxoniana* litter was observed at high elevations of the subalpine forest subjected to low temperatures in the non-growing seasons and (ii) there were stimulative effects of forest gaps on litter mass loss and carbon release in early decomposition, especially in the non-growing seasons, driven by fewer freeze–thaw cycles when compared to the closed canopy, which diminished at the end of the experiment. The results will provide crucial ecological data for further understanding how opening gaps as a main regeneration method would induce changes in carbon cycling in subalpine forest ecosystems.

Keywords: forest gap; litter decomposition; carbon release; elevation; subalpine forest

Citation: Li, H.; Du, T.; Chen, Y.; Zhang, Y.; Yang, Y.; Yang, J.; Dong, Q.; Zhang, L.; Wu, Q. Effects of Forest Gaps on *Abies faxoniana* Rehd. Leaf Litter Mass Loss and Carbon Release along an Elevation Gradient in a Subalpine Forest. *Forests* **2022**, *13*, 1201. https://doi.org/10.3390/f13081201

Academic Editor: Francisco B. Navarro

Received: 17 June 2022
Accepted: 26 July 2022
Published: 29 July 2022

Publisher's Note: MDPI stays neutral with regard to jurisdictional claims in published maps and institutional affiliations.

1. Introduction

Litter decomposition is a crucial component of carbon and nutrient turnover, and determines the carbon balance in forest ecosystems [1,2]. In the past decades, research into this complex ecological process has grown steadily since the development of the litterbag technique [3,4], which reveals that ambient temperature and moisture influence litter decomposition rate in the early stage [5]. Forest gaps are a main natural regeneration method, caused by pests, wildfires, natural stem breakage, human deforestation and other disturbances, and are widely distributed in forest ecosystems [6], which may induce changes in the microenvironment such as: temperature, precipitation, sunlight exposure, snow coverage and further affecting litter mass loss and carbon release during the decomposition processes [7]. Previous studies suggested forest gaps affected decomposition rates in a different way, which were documented to be greater, weaker or the same compared to that under closed canopies [8–10]. Furthermore, litter decomposition rates and nutrient release were reported to be stimulated, inhibited and insignificantly related to increasing gap sizes across multiple forests ecosystems [11–13]. Such uncertainties may limit our understanding of how gaps affect decomposition-induced carbon cycling in forest ecosystems.

The altitudinal gradient across a relatively smaller spatial scale could mirror the large-scale environmental conditions along climatic gradients, leading to changes in soil microbial biomass, enzyme activity and other driving factors for decomposition [14–17]. In general, the atmospheric and soil temperatures gradually decrease with increasing elevation, accompanied by deeper seasonal snowpack as well as a longer coverage period in higher elevations [18,19]. However, there were divergent results of the variations in soil biological indicators with an increasing elevation, such as promoted or inhibited enzyme activities and soil microbial quotient [20–22], which may directly affect litter decomposition, implying that the decomposition pattern with increasing elevation is still uncertain. Theoretically, getting away from the canopy interception would help the floor in forest gap obtain more precipitation and sunlight [7,23,24], further favoring the abundant biological community [25,26]. Therefore, the effects of forest gap size on litter decomposition along an elevation gradient need to be further assessed to strengthen our uncertain knowledge of this critical carbon-cycling ecological process.

Litter decomposition has obvious seasonal dynamics, which are more remarkable in high-elevation regions with distinct growing and non-growing seasons driven mainly by soil temperature and moisture [27]. Consequently, the strength of forest gap effects on litter mass loss and carbon release in response to an increasing elevation might be highly season-dependent. The non-growing season tends to be a critical period for the decomposition of newly shed litter [28,29]. Snowpack accumulated in forest gaps offers insulated protection against the outer extremely low temperature and frequent freeze–thaw cycles to maintain preferable microenvironmental conditions for decomposers [30], and this induces intensive leaching losses attributed to the strong snowmelt in early spring. Both functions would promote litter decomposition in winter [31,32]. Moreover, enhanced photodegradation and eluviation from sunlight exposure and heavy rain wash would also stimulate the litter decomposition within gaps when compared to that under closed canopy [33,34]. In subalpine regions, relevant information is scarce, especially when discussed in conjunction with different elevation effects.

Subalpine forests in the eastern Qinghai–Tibetan Plateau contain shallow soil thus making litter decomposition a crucial role in ecological processes, such as maintaining carbon balance in the area [35]. In elevations from 3000 to 3700 m, *Abies faxoniana* Rehd. is the dominant tree species and 84% of forest gaps are less than 240 m^2 in these forests according to our previous investigation [36]. A four-year litterbag decomposition experiment was conducted in an *A. faxoniana* subalpine forest of southwestern China, to assess the potential effects of forest gaps (large: $\approx$250 m^2, middle: $\approx$125 m^2, small: $\approx$40 m^2 vs. closed canopy) on litter mass loss and carbon release in decomposing *A. faxoniana* litter. More specifically, we evaluated the responses of decomposition for three elevations (3000, 3300, 3600 m) to forest gaps at different critical stages for four years. We hypothesized that (i) the *A. faxoniana* leaf litter mass loss and carbon release were lower under closed canopies than in forest gaps; (ii) the gap effects on the mass loss and carbon release would vary with gap size and elevation, and diminish gradually along with the decomposition; (iii) temperature was the key driver of the gap and elevation induced changes in litter decomposition. The objective of this study was to understand the seasonal effects of forest gap on litter decomposition along elevational differences, which could provide crucial ecological data for further understanding how the opening of gaps would induce changes in carbon cycling in subalpine forest ecosystems.

2. Materials and Methods

2.1. Site Description and Experimental Design

The study was conducted at the Long-term Research Station of Alpine Forest Ecosystems (31°14′–31°19′ N, 102°53′–102°57′ E, 2458~4619 m *a.s.l.*), located in Li County, southwestern China. It is a transitional area between the Tibetan Plateau and Sichuan Basin with remarkable changes in climate, vegetation and soil types along the elevation gradient [37,38]. The mean annual temperature and precipitation are 2~4 °C and 850 mm, respectively [39]. Winter generally starts from late October to the next April, with seasonal snowpack accumulated on the forest floor of up to approximately 50 cm and frequent soil freeze–thaw cycles [9]. *Abies faxoniana* Rehd., *Betula albo-sinensis* and *Sabina saltuaria* are the dominant trees species, and *Salix paraplesia*, *Rhododendron Lapponicum* and *Fargesia spathacea* are the dominant shrubs. The forest soil is classified as Cambisols [40]. The concentrations of total C, N, and P in soil (0~20 cm) were 363, 6.6 and 1.6 g·kg^{-1}, respectively [33].

Based on the previous investigation, forest gaps caused by human deforestation and natural stem breakage were widely distributed in the study area [36]. In September 2011, we selected *A. faxoniana* forests with similar slopes (24°~34°) and aspects (NE 38°~45°) at 3000 m (31°19′ N, 102°56′ E), 3300 m (31°17′ N, 102°56′ E) and 3600 m (31°15′ N, 102°53′ E)as our three study sites. For the 3000 m site, the dominant species are *Betula albosinensi*, *A. faxoniana*, *Berberis diaphana*, *Fragaria orientalis*, *Thalictrum aquilegifolium*, *Polygonum viviparum* etc., the annual precipitation and temperature are 825 mm and 3.6 °C, the forest soil is classified as Cambisols, and the organic layer is 15 ± 2 cm. For the 3300 m site, the dominant species are *A. faxoniana*, *Betula albosinensis*, *Sorbus*, *Hippophae rhamnoides*, *Parasenecio forrestii*, *Fragaria orientalis* etc. the annual precipitation and temperature are 825 mm and 2.8 °C, the forest soil is classified as Cambisols, and the organic layer is 12 ± 2 cm. For the 3600 m site, the dominant species are *A. faxoniana*, *Prunus tatsienensis*, *Rhododendron lapponicum*, *Salix paraplesia*, *Fargesia spathacea*, *Fargesia spathacea* etc., the annual precipitation and temperature are 825 mm and 1.6 °C, the forest soil is classified as Cambisols, and the organic layer is 12 ± 2 cm. At each site, three plots of each gap size were established: three gap sizes were large gaps (≈250 m^2), middle gaps (≈125 m^2) and small gaps (≈40 m^2), with 9 gap plots in total; another three plots (10 m×10 m) under the ambient closed canopy were set up as the control. These forests gaps have all existed for about 25 years, and all the 36 plots were located apart with over 50 m apart from each other. Other detailed information is shown in Table S1.

2.2. Litterbag Experiment

Foliar litter of *A. faxoniana* in this study area was collected using a litter trap method in late autumn (October 2011). Only freshly fallen leaves (without signs of galls, herbivory or atypical coloration) were collected and air-dried for two weeks. Nylon litterbags (20 cm × 20 cm, mesh size: 1 mm top vs. 0.5 mm bottom) were used to fill 10 g of dried litter. Litterbags were placed on the top of existing litter layer in each plot on November 21, 2011. In total, we applied 3 elevations × 4 gap sizes × 3 replicates × 3 bags × 16 samplings = 1728 litterbags. The snow depth in each plot was obtained by averaging the direct measurements from 9 random locations, and the measurements were performed immediately after each sampling event using a ruler. The temperatures of the litter layers were recorded every 2 h by iButton DS1923-F5 recorders (Maxim Inc., Sunnyvale, CA, USA), which were placed in the litterbags.

Based on previous field investigations, the growing season was from April to October every year [31]. In order to explore litter decay processes comprehensively, the bags were retrieved at the end of each period: these were in December (OF, onset of freezing stage), February (DP, deep freezing stage), April (TS, thawing stage), June (EGS, early snow-free stage), August (MGS, middle snow-free stage) and October (LGS, late snow-free stage) of the initial two decomposing years (2011–2013). For the latter two years (2014–2015), bags were only retrieved in April (NGS, non-growing season) and October (GS, growing season), which were defined according to the phenological characteristics of the vegetation in the study area. After carefully removing the roots, mosses and foreign materials from the litter, each sample was oven-dried at 65 °C for 48 h to a constant weight to determine the litter's remaining mass. Litter carbon content was determined using the dichromate oxidation–ferrous sulfate titration method [41].

2.3. Data Analysis

The mean temperature and frequency of the freeze–thaw cycles of each forest gap were calculated in each decomposition stage using the every 2 h temperature data in situ. Freezing and thawing in the non-growing season are important physical processes that impact below-ground ecological functions in alpine forests. We calculated the freeze–thaw cycles per day during each decomposition period; one freeze–thaw cycle was defined as the temperature increasing above 0 °C or decreasing below 0 °C for 3 h or more, followed by a decrease below 0 °C or an increase above 0 °C for at least 3 h [42].

The litter mass loss (*ML*, %) and carbon release (*CR*, %) were calculated as follows:

$$ML_t = (M_0 - M_t)/M_0 \times 100\%$$

$$CR_t = (C_{t\text{-}1} \times M_{t\text{-}1} - C_t \times M_t)/(C_0 \times M_0) \times 100\%$$

where M_0 (g) and M_t (g) are the litter's remaining mass in the litterbags at the initial and sampling times, respectively; C_0 (g·kg^{-1}), C_t (g·kg^{-1}) $C_{t\text{-}1}$ (g·kg^{-1}) are the litter carbon content at the initial, sampling and previous sampling times, respectively. All analyses were conducted in triplicate.

The Olson exponential decay model [43] was used to fit the relationship of litter remaining mass and decomposition year:

$$M_t = M_0 \cdot e^{-kt}$$

where M_0 is the litter initial dry mass; e is a natural constant; k is the litter decomposition constant; t is the decomposition year; M_t is the litter's remaining mass at t.

In this study, gap size (with three sizes and closed canopy) was a main-plot factor, elevation (with three elevations) was a sub-plot factor, and decomposition time (with sixteen stages) was a sub-subplot factor.

A two-way ANOVA (with the LSD post hoc test) was used to examine the differences of litter decomposition constant (k) generated from the Olson exponential decay model, mass loss, carbon content and release both among the gap sizes and elevations. Linear mixed models were conducted to analyze the effects of gap size, elevation, decomposition time and their interaction on litter mass loss, carbon content and release; the bags within gap sizes were set as random effects. All data were subjected to a normality test and homogeneity of variance test prior to the analysis of variance. Multiple non-linear regressions were conducted to evaluate the links of the litter's remaining mass and carbon content from the scales of gap size and elevation. Spearman correlation analysis was performed to describe the relationship between mean temperature, freeze–thaw cycles and litter mass loss, carbon release in the non-growing season, growing season and the whole four decomposition years. All statistical analyses were performed using SPSS 27.0 (SPSS Inc., Chicago, IL, USA) and Prism GraphPad 8.0 (GraphPad Software Inc. San Diego, CA, USA).

3. Results

3.1. Temperature Characteristics and Snow Depth

Temperature characteristics of litter layer fluctuated consistently in different forest gaps and elevations are shown in Table 1. For the non-growing and growing seasons, a higher mean temperature was observed in forest gaps when compared to that under the closed canopy in all the three elevations. Furthermore, the highest mean temperatures were shown in the large gap in the 3000 m, while it was found in the middle or small gaps in the 3300 m and 3600 m plots. Freeze–thaw cycles' frequency was lower in large and middle gaps at the 3600 m and 3300 m sites, while a higher cycle frequency was found at the 3000 m site. During the non-growing season (November to April), the snow depth on the forest soil was consistently ranked in the following two orders: large gap > middle gap > small gap > closed canopy, 3600 m > 3300 m > 3000 m. At the OF and DF of the first two decomposing years, the accumulated snow was deeper than for other stages.

Table 1. Seasonal environmental factors of in the study sites.

Environmental Factor	Stage	3000 m				3300 m				3600 m			
		Large Gap	Middle Gap	Small Gap	Closed Canopy	Large Gap	Middle Gap	Small Gap	Closed Canopy	Large Gap	Middle Gap	Small Gap	Closed Canopy
Mean temperature (°C)	OF	−1.02	−1.14	−2.27	−1.98	−1.76	−0.98	−1.22	−1.37	0.78	−0.23	−0.51	−0.70
	DF	−2.45	−2.26	−4.11	−4.55	−2.69	−1.05	−1.29	−2.85	−2.35	−1.81	−1.59	−4.21
	TS	6.25	4.63	5.04	3.91	4.58	5.26	5.06	4.50	0.05	2.94	2.87	0.34
	EGS	12.63	12.21	11.78	9.77	11.72	10.64	11.31	9.65	9.17	9.60	7.22	6.47
	MGS	15.28	12.92	14.50	13.08	13.61	13.60	11.98	12.47	16.11	12.02	11.54	12.11
	LGS	9.32	9.19	8.48	8.08	8.10	9.07	10.01	8.01	7.34	5.57	6.27	7.25
	NGS	0.79	0.42	−0.57	−1.04	−0.04	0.97	0.69	0.01	−0.93	−0.28	−0.35	−1.76
	GS	12.38	11.38	11.55	10.29	11.09	11.10	11.12	10.05	10.50	8.77	8.11	8.34
Frequency of freeze-thaw cycles (time·d^{-1})	OF	0.92	0.59	0.51	0.50	0.64	0.84	0.78	0.91	0.36	0.41	0.91	0.81
	DF	0.70	0.72	0.45	0.43	0.07	0.61	0.89	0.74	0.41	0.40	0.76	0.76
	TS	0.41	0.51	0.47	0.40	0.47	0.36	0.51	0.68	0.09	0.45	0.48	0.65
	EGS	—	—	—	—	0.03	0.01	0.01	0.07	0.09	0.03	0.05	0.03
	MGS	—	—	—	—	—	—	—	—	—	—	—	—
	LGS	0.03	0.02	0.03	0.01	0.05	0.07	0.03	0.01	0.13	0.30	0.22	0.05
	NGS	0.63	0.53	0.48	0.46	0.35	0.61	0.70	0.75	0.30	0.36	0.68	0.75
	GS	0.01	0.01	0.01	0.01	0.03	0.03	0.01	0.04	0.08	0.12	0.08	0.05
Snow depth (cm)	OF	10.58	8.42	5.75	0.17	20.50	16.42	9.67	0.25	31.67	24.17	11.58	0.25
	DF	11.92	9.42	6.75	0.50	28.33	19.75	8.67	0.50	42.92	29.92	12.25	0.50
	TS	6.17	4.00	2.88	0.33	7.88	5.08	3.58	0.13	12.79	8.79	6.42	0.25

Notes: OF, onset of freezing stage; DF, deep freezing stage; TS, thawing stage; EGS, early snow-free stage; MGS, middle snow-free stage; LGS, late snow-free stage; NGS, non-growing season; GS, growing season.

3.2. Litter Mass Loss and Decomposition Constant k

After the four decomposition years, the *A. faxoniana* litter mass loss reached 50~53%, and it was significantly affected by gap size, elevation, and decomposition time, as well as their interactions (Table 2). Specifically, the litter in the forest gaps exhibited significantly higher mass loss at the 3300 m and 3600 m sites than that under the closed canopy, especially in the late decomposition period (Figure 1b,c). However, litter at the 3000 m site showed significantly higher mass loss under the closed canopy when compared to those in the forest gaps (Figure 1a). Furthermore, there were higher litter mass loss rates during the non-growing season than during the growing season (Figure 2a,b). Although there were no significant differences, the mass loss was ranked in the order: large gap > middle gap > small gap in the non-growing season, but this showed a opposite trend in the growing season. For the four years, the mass loss was significantly lower at the 3000 m site than that at the 3300 m site. Besides, the decomposition k exhibited a gradually increasing trend from the large gap, middle gap, small gap to closed canopy plots at the 3000 m site, while that at 3300 m and 3600 m sites showed a totally opposite decreasing trend (Figure 1d).

Table 2. Results (F-values) of linear mixed effects models testing the effects of elevation (3000, 3300, 3600 m), gap size (large, middle, small), decomposition time (sampled times) and their interactions on mass loss and carbon content and release in in decomposing *A. faxoniana* litter. ** $p < 0.001$.

Sources of Variance	df	Mass Loss		Carbon Content		Carbon Release	
		F-Value	*p*-Value	F-Value	*p*-Value	F-Value	*p*-Value
Elevation	2	56.5 **	<0.001	1465.2 **	<0.001	56.5 **	<0.001
Gap	3	39.7 **	<0.001	137.9 **	<0.001	39.7 **	<0.001
Time	15	2513.6 **	<0.001	232.5 **	<0.001	2513.6 **	<0.001
Elevation × Gap	6	49.4 **	<0.001	68.8 **	<0.001	49.4 **	<0.001
Elevation × Time	30	6.6 **	<0.001	30.6 **	<0.001	6.6 **	<0.001
Gap × Time	45	2.7 **	<0.001	10.5 **	<0.001	2.7 **	<0.001
Elevation × Gap × Time	90	4.8 **	<0.001	8.9 **	<0.001	4.8 **	<0.001

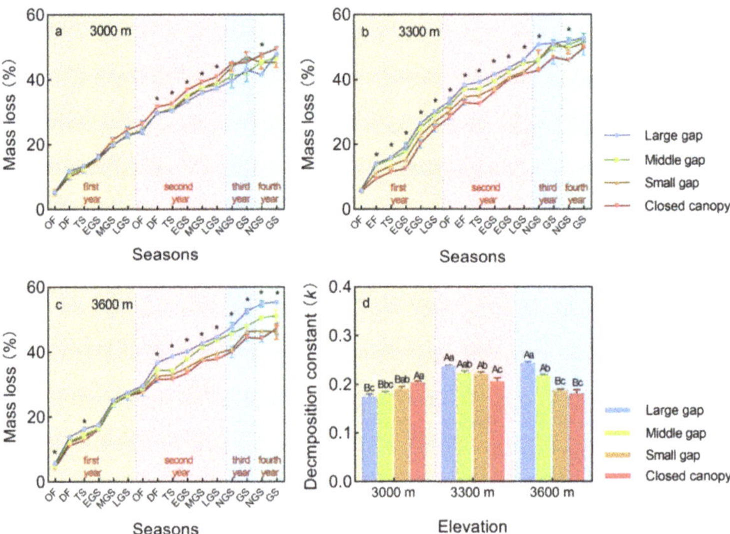

Figure 1. Mass loss and decomposition constant (*k*) of *A. faxoniana* litter in forest gaps (large gap, middle gap, small gap, closed canopy) along an elevation gradient (3000, 3300, 3600 m) at different decomposition stages in a subalpine forest of southwestern China. (**a**) Mass loss at 3000 m; (**b**) Mass loss at 3300 m; (**c**) Mass loss at 3600 m; (**d**) Decomposition constant k. Mean ± SE, n = 3. OF, onset of freezing stage; DF, deep freezing stage; TS, thawing stage; EGS, early snow-free stage; MGS, middle snow-free stage; LGS, late snow-free stage; NGS, non-growing season; GS, growing season. Asterisk above lines indicates significant differences among forest gaps (*p* < 0.05), lowercase letters above columns indicate significant differences among forest gaps (*p* < 0.05), capital letters above columns indicate significant differences among elevations (*p* < 0.05).

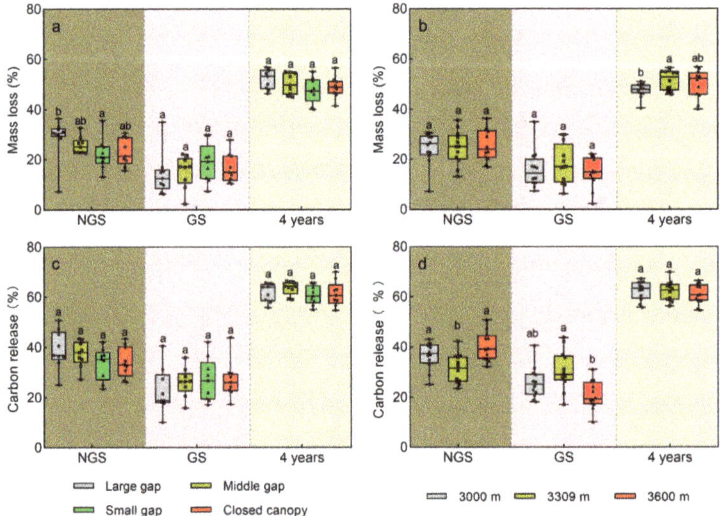

Figure 2. Effects of forest size and elevation on mass loss and carbon release in decomposing *A. faxoniana* litter. (**a**) Mass loss on forest gap scale; (**b**) Mass loss on elevation scale; (**c**) Carbon release on forest gap scale; (**d**) Carbon release on elevation scale. Mean ± SE, n = 3. NGS, the non-growing seasons in the four decomposing years; GS, the growing seasons in the four decomposing years. Lowercase letters indicate significant differences among forest gaps or elevations (*p* < 0.05).

3.3. Litter Carbon Content

During the four decomposition years, the *A. faxoniana* litter content gradually decreased by 18~29%, and it was significantly affected by gap size, elevation, decomposition time as well as their interactions (Figure 3, Table 2). Specifically, obvious declines were observed in the first non-growing season and the second growing season during the first two years. When compared with the closed canopy, higher carbon contents were shown in forest gaps at the 3000 m and 3300 m sites. However, at the 3000 m site, the highest carbon content was measured under the closed canopy in the first decomposition year, and then exhibited the following order: large gap > closed canopy > middle/small gap. Additionally, the non-linear regression indicates that the carbon content significantly declined with decomposition processing, meanwhile the decomposed litter with less remaining mass had a higher carbon content at the 3300 and 3600 m sites than that at 3000 m site (Figure 4a,b).

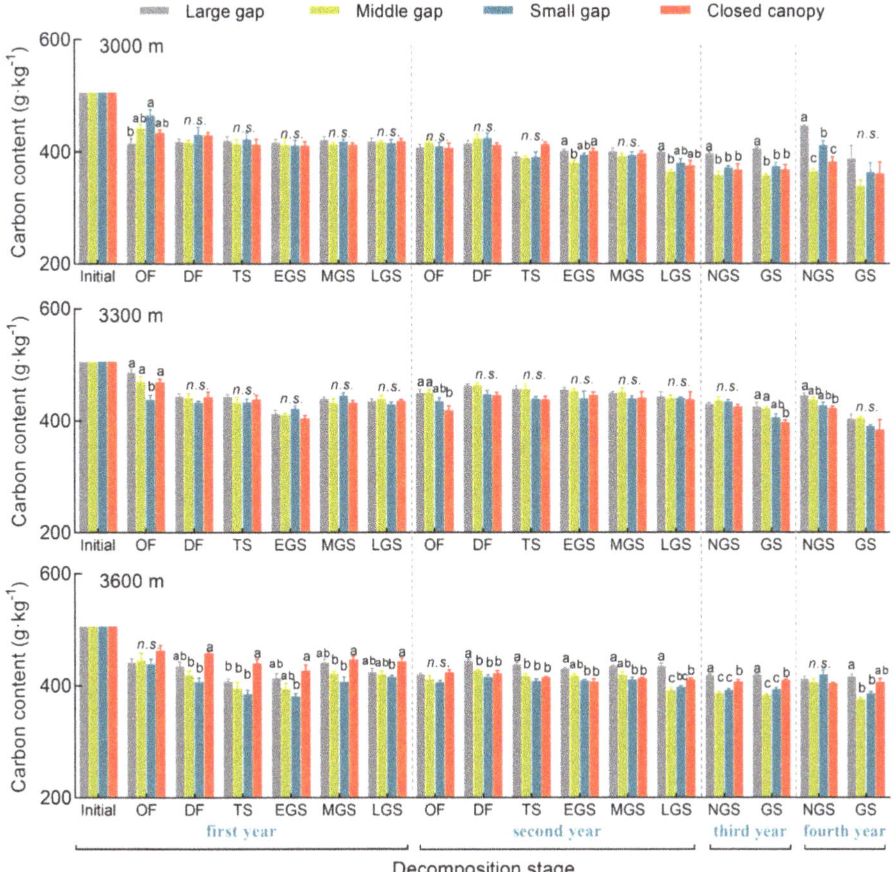

Figure 3. Carbon content of *A. faxoniana* litter in forest gaps (large gap, middle gap, small gap, closed canopy) along an elevation gradient (3000, 3300, 3600 m) at different decomposition stages in a subalpine forest of southwestern China. Mean ± SE, n = 3. OF, onset of freezing stage; DF, deep freezing stage; TS, thawing stage; EGS, early snow-free stage; MGS, middle snow-free stage; LGS, late snow-free stage; NGS, non-growing season; GS, growing season. Lowercase letters indicate significant differences among forest gaps ($p < 0.05$), *n.s.* indicates no significant differences ($p > 0.05$).

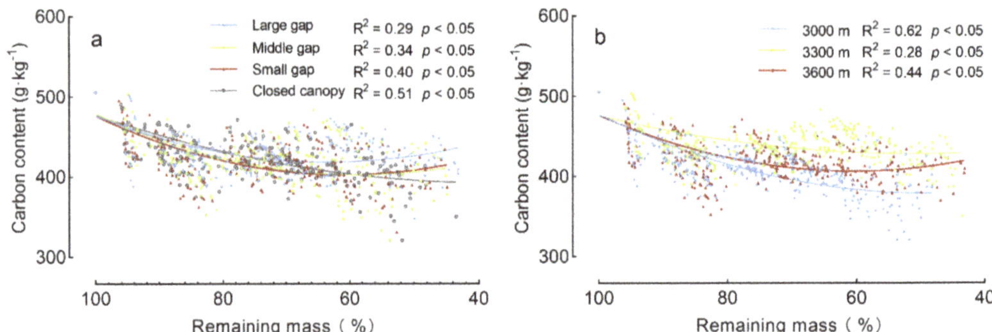

Figure 4. Carbon content versus remaining mass in decomposing *A. faxoniana* litter. (**a**) Forest gap scale; (**b**) Elevation scale. R^2 and p values from linear regression are shown in each panel.

3.4. Litter Carbon Release

After the four decomposition years, 58~64% of carbon was released from the *A. faxoniana* litter; carbon release was not affected by the gap size (Figure 5), but within the time period, it was significantly affected by gap size, elevation, decomposition time as well as their interactions (Table 2). When compared to the 3000 m and 3300 m sites, the carbon release variation was greater at the 3600 m site. When compared with closed canopy, more carbon was released from decomposing litter in forest gaps at the 3300 m and 3600 m sites. However, less carbon was released from decomposing litter in the large/small gasp at the 3000 m site. At the end of decomposition, there were no significant differences among forest gaps and the closed canopy at all sites, and significant differences were only observed in the first two years. Under the closed canopy, less carbon was released from decomposing litter at the 3300 and 3600 m sites in the first year but more carbon was released at the 3000 and 3300 m sites in the second year. Carbon released in the first two years accounted for 81~86% of the total four-year release, while that of the combined non-growing seasons' release across the four years accounted for 45~63%. There were significant differences in litter carbon release in the combined non-growing seasons across the four years among the gap sizes at the 3600 m site, but there were no significant differences in the combined growing seasons at all the sites. Further, there was a higher litter carbon release during the non-growing season than during the growing season (Figure 2c,d). The effects of gap size on carbon release were not significant, but at 3300 m they were significant higher and lower in the non-growing season and growing season, respectively.

3.5. Key Drivers of Litter Mass Loss and Carbon Release

Based on the Spearman correlation analysis, litter mass loss and carbon release were both significantly negative correlated with freeze–thaw cycles in the non-growing season, and carbon release was significantly negative correlated with freeze–thaw cycles for the whole four decomposition years. Besides, the litter carbon release was also significantly negative correlated with the positive accumulated temperature and the mean temperature in the non-growing season (Table 3).

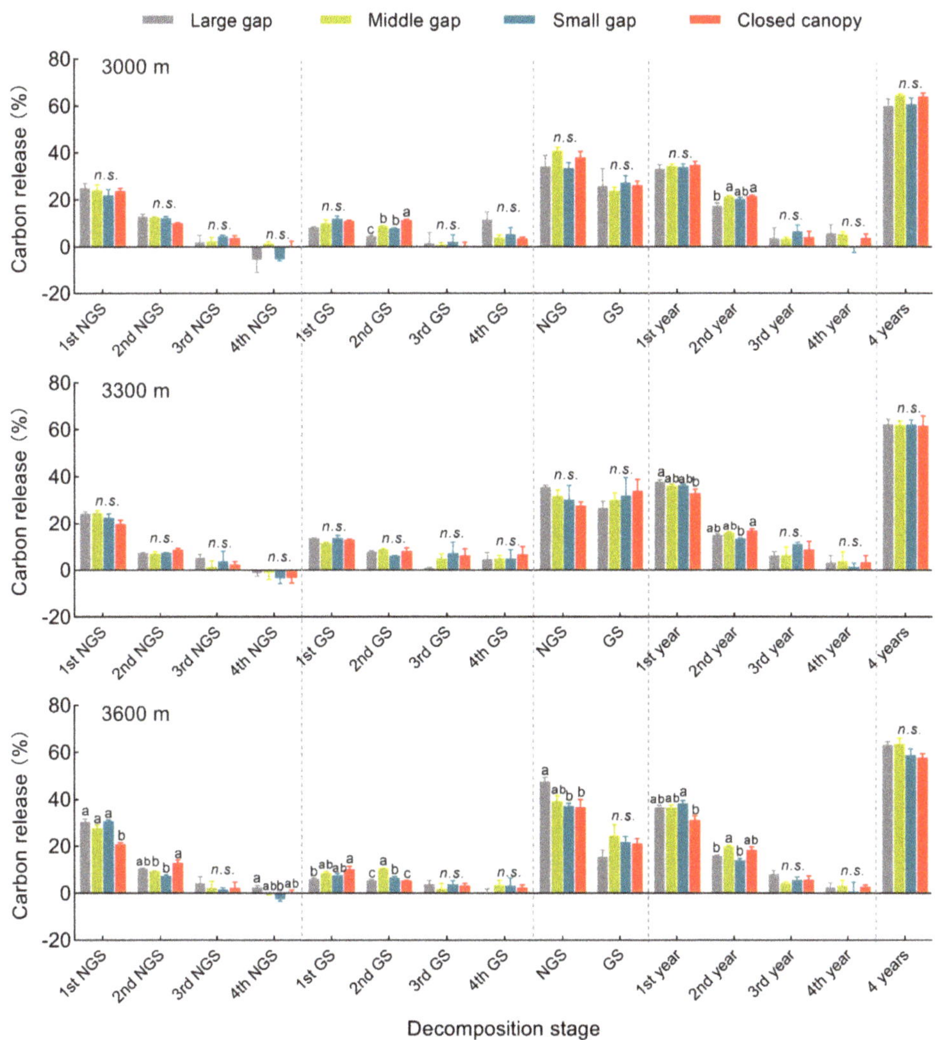

Figure 5. Carbon release percentage of *A. faxoniana* litter in forest gaps (large gap, middle gap, small gap, closed canopy) along an elevation gradient (3000, 3300, 3600 m) at different decomposition stages in a subalpine forest of southwestern China. Mean $\pm$ SE, n = 3. 1st NGS, non-growing season in the first decomposing year; 2nd NGS, non-growing season in the second decomposing year; 3rd NGS, non-growing season in the third decomposing year; 4th NGS, non-growing season in the fourth decomposing year; 1st GS, growing season in the first decomposing year; 2nd GS, growing season in the second decomposing year; 3rd GS, growing season in the third decomposing year; 4th GS, growing season in the fourth decomposing year; NGS, the non-growing seasons in the four decomposing years; GS, the growing seasons in the four decomposing years. Lowercase letters indicate significant differences among forest gaps ($p < 0.05$), *n.s.* indicates no significant differences ($p > 0.05$).

Table 3. Spearman correlation coefficient between mass loss, carbon release and environmental factors in the non-growing season, growing season and the 4 total decomposition years. Significant correlations are indicated in a bold font, with * $p < 0.05$, ** $p < 0.001$.

	Parameter	Freeze–Thaw Cycles	Mean Temperature
	Non-growing season	−0.548 **	−0.055
Mass loss	Growing season	-	0.042
	4 years	−0.273	0.070
	Non-growing season	−0.473 **	−0.315 *
Carbon release	Growing season	-	0.144
	4 years	−0.328 *	0.127

4. Discussion

After four years of decomposition, 50~53 and 58~64% of mass and carbon were lost (Figures 1 and 3), which were similar to other pine tree species in other temperate forests [44,45], and more than 50% of the losses were observed in the 3600 m sites during the non-growing seasons (Figure 2), indicating rather high mass loss and carbon release in decomposition. *A. faxoniana* litter was present at higher elevations of the subalpine forest subjected to low temperatures in the non-growing seasons. Forest gaps formed by natural stem breakage were commonly distributed in high-latitude and high-altitude ecosystems, which would modify the temperature, precipitation, snow coverage and radiation conditions within the gap areas when compared to the closed canopies [10,46]. Our results partially support the first hypothesis that forest gaps increased the mass loss and release in the decomposing *A. faxoniana* leaf litter, but only in the 3300 m and 3600 m sites. For the decomposition constant (*k*), we found that higher decomposition rates were in the gaps when compared to the closed canopy in both the 3300 and 3600 m sites. A previous study has demonstrated that there was more accumulated snow coverage within gaps, helping microbes maintain high activities through the isolated protection in winter, as well as the more intense precipitation and strong radiation attributed to the opening canopy increasing the leaching loss and photodegradation in the growing season, thereby promoting the litter mass loss [8,47], which was consist with our results. However, the gap-stimulated effects were not observed in the 3000 m site, exhibiting a higher decomposition rate under the closed canopy. According to the environmental factors presented in Table 1, we recorded relatively small differences in the snow depth among different gaps and the closed canopy; besides, there was a lower frequency of freeze–thaw cycles under closed canopy than in gaps in the non-growing seasons, which was the opposite to the other two elevations. Therefore, a likely explanation was that the lower variables of environmental conditions among the gaps and closed canopy compared to the two high elevations resulted in different effects of gaps on the decomposition rate. Additionally, the Spearman correlation analysis showed that both mass loss and carbon release were significantly negatively related to the freeze–thaw cycles (Table 3), implying that the effects of gaps on litter decomposition were more correlated to the freeze–thaw cycles in winter. It has been reported that soil freezing would modify carbon release from plant litter in the subalpine forests of this region [9,48], and our study addressed that freeze–thaw cycle variations were a key driver changing the effects of gaps on litter decomposition at different elevations of high-altitude forest ecosystems, which also supported our third hypothesis.

In this study, the effects of forest gaps were not always in accordance with the gap size gradient. Previous studies have revealed that nitrogen mineralization in forest gaps increased in the first decomposition year and then decreased later [49,50]; thus, the soil microorganisms contributing to the degradation of carbon-associated compounds at the early decomposition stage would benefit from increasing the nitrogen availability [46,51]. However, carbon release did not consistently present a general pattern along with the gap size gradient (Figure 2c). In general, although deep accumulated snowpack on the forest floor within gaps decreased the contribution of soil freezing to the physical breakdown of litter plant-derived materials [52,53], higher biological activities were still maintained under the snow cover [8,54]. Moreover, intense hydrological leaching during the snowmelt period also contributes to the labile carbon releasing from newly shed litter [55,56]. These parts of carbon, in turn, provide available carbon and nutrients for microbial utilization and thus affect the decomposition processes [50,57]. When compared with the lower elevation sites, the effect of forest gaps was greater in the high elevation site (3600 m) for both carbon content and release from *A. faxoniana* litter. Previous studies revealed that cold regions are proven to be more sensitive to environmental disturbances, and the relatively lower temperatures in the high elevation site are at much higher risk of promoting soil freezing, which would physically favor the decomposition of the *A. faxoniana* litter [9]. Meanwhile, snow coverage in forest gaps has also been documented to accelerate carbon-associated compound (both labile and recalcitrant substrate) degradation [58]. Therefore, the stimulative effects of forest gaps on mass loss and carbon release were more responsive to higher elevation in subalpine forests.

The stimulative effects of forest gap on litter decomposition were more remarkable in the non-growing seasons when compared to the growing seasons. Carbon content decreased over time at the three elevations as decomposition proceeded, particularly for the non-growing seasons (Figure 2). Moreover, a rather higher mass loss and carbon release in *A. faxoniana* litter was observed in subalpine forests subjected to low temperatures in the non-growing seasons. These results suggest that carbon release greatly varies between non-growing and growing seasons in the subalpine forest. When carbon release in the non-growing seasons throughout the four decomposition years was combined, we observed that the large gaps significantly enhanced the carbon release at the 3600 m site, suggesting that deep snow coverage at the high-elevation forest promoted carbon release from the *A. faxoniana* litter [59]. Forest gaps only exhibited significant effects at the early litter decomposition stage (the first two decomposition years), but such a gap influence was dissimilar, with the increasing carbon release in the first decomposition year and decreasing in the second year. Previous studies indicated that the difference between forest gaps was greater at the late stage of decomposition [23], which is attributed to the disturbances of the original environment resulting from the formation of forest gaps. However, in this study, the gap influence on carbon release disappeared by the end of four decomposition years, which supported our second hypothesis. Another likely explanation was that 56, 50 and 49% of the initial carbon was lost in the first two decomposition years at all the three sites, and the remaining carbon compounds were primarily recalcitrant components with a resistant structure, which decay quite slowly and are not able to be affected by environmental alteration easily [45].

5. Conclusions

Collectively, our study found that (i) rather high mass loss and carbon release in decomposition *A. faxoniana* litter was observed in high elevations of the subalpine forest subjected to low temperatures in the non-growing seasons and (ii) there were stimulative effects of forest gaps on litter decomposition in early decomposition, especially in the non-growing seasons, driven by fewer freeze–thaw cycles when compared to the closed canopy, which diminished at the end of the experiment though. Overall, these findings provide new insights into how opening gaps as a main regeneration method at different elevations would induce changes in carbon cycling in subalpine forests. We therefore suggest that future models of litter decomposition may benefit from including forest gaps and emphasizing non-growing seasons to make more accurate predictions. However, the biological mechanisms involved in the decomposition processes may need further investigation, such as considering the microbial and faunal strategy reactions to the forest gap formation.

Supplementary Materials: The following supporting information can be downloaded at: https://www.mdpi.com/article/10.3390/f13081201/s1, Table S1: Geographic information of forest gaps.

Author Contributions: Conceptualization, H.L. (Han Li), Q.W. and L.Z.; software, T.D., Y.C. and J.Y.; validation, T.D., Y.C., Y.Z. and Q.D.; data curation, H.L. (Han Li); writing—original draft preparation, H.L. (Han Li) and Q.W.; writing—review and editing, H.L. (Han Li), L.Z. and Q.W.; visualization, H.L. (Han Li), Y.Y. and L.Z.; funding acquisition, H.L. (Han Li), L.Z. and Q.W. All authors have read and agreed to the published version of the manuscript.

Funding: This research was funded by the National Natural Science Foundation of China (31901295, 32071747 and 32001165), the Program of Sichuan Applied Basic Research Foundation (2022NSFSC0997, 2022NSFSC0083, 2022NSFSC0087 and 2022NSFSC1173), the Open Fund of Ecological Security and Protection Key Laboratory of Sichuan Province Mianyang Normal University (ESP1902), and Research Fund of Mianyang Normal University (QD2020A18).

Institutional Review Board Statement: Not applicable.

Informed Consent Statement: Written informed consent has been obtained from the patient to publish this paper.

Data Availability Statement: Not applicable.

Acknowledgments: We are grateful to the Long-term Research Station of Alpine Forest Ecosystems and the Collaborative Innovation Center of Ecological Security in the Upper Reaches of the Yangtze River.

Conflicts of Interest: The authors declare no conflict of interest. The founding sponsors had no role in the design of the study; in the collection, analyses, or interpretation of data; in the writing of the manuscript, and in the decision to publish the results.

References

1. Ssmaneh, T.; Shamsollah, A.; Jahangir, K.; Shaban, S. Effects of Tree Species Composition on Soil Properties and Invertebrates in A Deciduous Forest. *Arab. J. Geosci.* **2019**, *12*, 368.
2. Prescott, C.E.; Grayston, S.J.; Helmisaari, H.S.; Kaštovská, E.; Körner, C.; Lambers, H.; Meier, I.C.; Millard, P.; Ostonen, I. Surplus Carbon Drives Allocation and Plant–Soil Interactions. *Trends Ecol. Evol.* **2020**, *35*, 1110–1118. [CrossRef] [PubMed]
3. Prescott, C.E. Litter Decomposition: What Controls It and How Can We Alter It to Sequester More Carbon in Forest Soils? *Biogeochemistry* **2010**, *101*, 133–149. [CrossRef]
4. Berg, B. Decomposition Patterns for Foliar Litter—A Theory for Influencing Factors. *Soil Biol. Biochem.* **2014**, *78*, 222–232. [CrossRef]
5. Aerts, R. The Freezer Defrosting: Global Warming and Litter Decomposition Rates in Cold Biomes. *J. Ecol.* **2006**, *94*, 713–724. [CrossRef]
6. Schliemann, S.A.; Bockheim, J.G. Methods for Studying Treefall Gaps: A Review. *For. Ecol. Manag.* **2011**, *261*, 1143–1151. [CrossRef]
7. Ritter, E. Litter Decomposition and Nitrogen Mineralization in Newly Formed Gaps in a Danish Beech (*Fagus sylvatica*) Forest. *Soil Biol. Biochem.* **2005**, *37*, 1237–1247. [CrossRef]
8. González, G.; Lodge, D.J.; Richardson, B.A.; Richardson, M.J. A Canopy Trimming Experiment in Puerto Rico: The Response of Litter Decomposition and Nutrient Release to Canopy Opening and Debris Deposition in a Subtropical Wet Forest. *For. Ecol. Manag.* **2014**, *332*, 32–46. [CrossRef]
9. Gliksman, D.; Haenel, S.; Osem, Y.; Yakir, D.; Zangy, E.; Preisler, Y.; Grünzweig, J.M. Litter decomposition in Mediterranean pine forests is enhanced by reduced canopy cover. *Plant Soil* **2018**, *422*, 317–329. [CrossRef]
10. Tan, B.; Zhang, J.; Yang, W.Q.; Yin, R.; Xu, Z.F.; Liu, Y.; Zhang, L.; Li, H.; You, C.M. Forest gaps retard carbon and nutrient release from twig litter in alpine forest ecosystems. *Eur. J. For. Res.* **2020**, *139*, 53–65. [CrossRef]
11. Prescott, C.E.; Blevins, L.L.; Staley, C.L. Effects of clear-cutting on decomposition rates of litter and forest floor in forests of British Columbia. *Can. J. For. Res.* **2020**, *30*, 1751–1757. [CrossRef]
12. Sariyildiz, T. Effects of Gap-Size Classes on Long-Term Litter Decomposition Rates of Beech, Oak and Chestnut Species at High Elevations in Northeast Turkey. *Ecosystems* **2008**, *11*, 841–853. [CrossRef]
13. Obojes, N.; Meurer, A.; Newesely, C.; Tasser, E.; Oberhuber, W.; Mayr, S.; Tappeiner, U. Water Stress Limits Transpiration and Growth of European Larch up to the Lower Subalpine Belt in an Inner-Alpine Dry Valley. *New Phytol.* **2018**, *220*, 460. [CrossRef] [PubMed]
14. Cao, R.; Yang, W.; Chang, C.; Wang, Z.; Wang, Q.; Jiang, Y.; Li, H.; Tan, B. Soil Microbial Biomass Carbon and Freeze-Thaw Cycles Drive Seasonal Changes in Soil Microbial Quotient Along a Steep Altitudinal Gradient. *J. Geophys. Res. Biogeosci.* **2021**, *126*, e2021JG006325. [CrossRef]
15. Cao, R.; Yang, W.; Chang, C.; Wang, Z.; Wang, Q.; Li, H.; Tan, B. Differential Seasonal Changes in Soil Enzyme Activity along an Altitudinal Gradient in an Alpine-Gorge Region. *Appl. Soil Ecol.* **2021**, *166*, 104078. [CrossRef]

16. Withington, C.L.; Sanford, R.L. Decomposition Rates of Buried Substrates Increase with Altitude in the Forest-Alpine Tundra Ecotone. *Soil Biol. Biochem.* **2007**, *39*, 68–75. [CrossRef]

17. Fierer, N.; Mccain, C.M.; Meir, P.; Zimmermann, M.; Rapp, J.M.; Silman, M.R.; Knight, R. Microbes Do Not Follow the Elevational Diversity Patterns of Plants and Animals. *Ecology* **2011**, *92*, 797–804. [CrossRef] [PubMed]

18. Körner, C. The Use of "altitude" in Ecological Research. *Trends Ecol. Evol.* **2007**, *22*, 569–574. [CrossRef] [PubMed]

19. Gutiérrez-Girón, A.; Díaz-Pinés, E.; Rubio, A.; Gavilán, R.G. Both Altitude and Vegetation Affect Temperature Sensitivity of Soil Organic Matter Decomposition in Mediterranean High Mountain Soils. *Geoderma* **2015**, *237–238*, 1–8. [CrossRef]

20. Ma, H.P.; Yang, X.L.; Guo, Q.Q.; Zhang, X.J.; Zhou, C.N. Soil Organic Carbon Pool along Different Altitudinal Level in the Sygera Mountains, Tibetan Plateau. *J. Mt. Sci.* **2016**, *13*, 476–483. [CrossRef]

21. He, X.; Hou, E.; Liu, Y.; Wen, D. Altitudinal Patterns and Controls of Plant and Soil Nutrient Concentrations and Stoichiometry in Subtropical China. *Sci. Rep.* **2016**, *6*, 24261. [CrossRef] [PubMed]

22. Denslow, J.S.; Ellison, A.M.; Sanford, R.E. Treefall Gap Size Effects on Above- and below-Ground Processes in a Tropical Wet Forest. *J. Ecol.* **1998**, *86*, 597–609. [CrossRef]

23. Zhang, Q.; Liang, Y. Effects of Gap Size on Nutrient Release from Plant Litter Decomposition in a Natural Forest Ecosystem. *Can. J. For. Res.* **1995**, *25*, 1627–1638. [CrossRef]

24. Zhang, Q.; Zak, J.C. Potential Physiological Activities of Fungi and Bacteria in Relation to Plant Litter Decomposition along a Gap Size Gradient in a Natural Subtropical Forest. *Microb. Ecol.* **1998**, *35*, 172–179. [CrossRef] [PubMed]

25. Yin, R.; Qin, W.; Zhao, H.; Wang, X.; Cao, G.; Zhu, B. Climate Warming in an Alpine Meadow: Differential Responses of Soil Faunal vs. Microbial Effects on Litter Decomposition. *Biol. Fertil. Soils* **2022**, *58*, 509–514. [CrossRef]

26. Tan, B.; Yin, R.; Zhang, J.; Xu, Z.; Liu, Y.; He, S.; Zhang, L.; Li, H.; Wang, L.; Liu, S.; et al. Temperature and Moisture Modulate the Contribution of Soil Fauna to Litter Decomposition via Different Pathways. *Ecosystems* **2021**, *24*, 1142–1156. [CrossRef]

27. Li, H.; Wu, F.; Yang, W.; Xu, L.; Ni, X.; He, J.; Tan, B.; Hu, Y. Effects of Forest Gaps on Litter Lignin and Cellulose Dynamics Vary Seasonally in an Alpine Forest. *Forests* **2016**, *7*, 27. [CrossRef]

28. Blok, D.; Elberling, B.; Michelsen, A. Initial stages of tundra shrub litter decomposition may be accelerated by deeper winter snow but slowed down by spring warming. *Ecosystems* **2016**, *19*, 155–169. [CrossRef]

29. Gong, L.; Chen, X.; Zhang, X.N.; Yang, X.D.; Cai, Y.J. Schrenk spruce leaf litter decomposition varies with snow depth in the Tianshan Mountains. *Sci. Rep.* **2020**, *10*, 19556. [CrossRef] [PubMed]

30. Baptist, F.; Yoccoz, N.G.; Choler, P. Direct and Indirect Control by Snow Cover over Decomposition in Alpine Tundra along a Snowmelt Gradient. *Plant Soil* **2010**, *328*, 397–410. [CrossRef]

31. He, W.; Wu, F.; Yang, W.; Tan, B.; Zhao, Y.; Wu, Q.; He, M. Lignin Degradation in Foliar Litter of Two Shrub Species from the Gap Center to the Closed Canopy in an Alpine Fir Forest. *Ecosystems* **2016**, *19*, 115–128. [CrossRef]

32. Wu, Q. Short- and Long-Term Effects of Snow-Depth on Korean Pine and Mongolian Oak Litter Decomposition in Northeastern China. *Ecosystems* **2020**, *23*, 662–674. [CrossRef]

33. Ni, X.; Yang, W.; Liao, S.; Li, H.; Tan, B.; Yue, K.; Xu, Z.; Zhang, L.; Wu, F. Rapid Release of Labile Components Limits the Accumulation of Humic Substances in Decomposing Litter in an Alpine Forest. *Ecosphere* **2018**, *9*, e02434. [CrossRef]

34. Wang, Q.W.; Pieristè, M.; Liu, C.; Kenta, T.; Robson, T.M.; Kurokawa, H. The Contribution of Photodegradation to Litter Decomposition in a Temperate Forest Gap and Understorey. *New Phytol.* **2021**, *229*, 2625–2636. [CrossRef] [PubMed]

35. Wu, F.; Yang, W.; Zhang, J.; Deng, R. Litter Decomposition in Two Subalpine Forests during the Freezeethaw Season. *Acta Oecologica* **2010**, *36*, 135–140. [CrossRef]

36. Wu, Q.; Wu, F.; Yang, W.; Tan, B.; Yang, Y.; Ni, X.; He, J. Characteristics of Gaps and Disturbance Regimes of the Alpine Fir Forest in Western Sichuan. *Chin. J. Appl. Environ. Biol.* **2013**, *19*, 922–928. [CrossRef]

37. Zhu, J.; Yang, W.; He, X. Temporal Dynamics of Abiotic and Biotic Factors on Leaf Litter of Three Plant Species in Relation to Decomposition Rate along a Subalpine Elevation Gradient. *PLoS ONE* **2013**, *8*, e62073. [CrossRef]

38. Xu, Z.; Zhu, J.; Wu, F.; Liu, Y.; Tan, B.; Yang, W. Effects of Litter Quality and Climate Change along an Elevational Gradient on Litter Decomposition of Subalpine Forests, Eastern Tibetan Plateau, China. *J. For. Res.* **2016**, *27*, 505–511. [CrossRef]

39. Zhu, J.; He, X.; Wu, F.; Yang, W.; Tan, B. Decomposition of Abies Faxoniana Litter Varies with Freeze-Thaw Stages and Altitudes in Subalpine/Alpine Forests of Southwest China. *Scand. J. For. Res.* **2012**, *27*, 586–596. [CrossRef]

40. IUSS Working Group. *WRB World Reference Base for Soil Resources 2014. International Soil Classification System for Naming Soils and Creating Legends for Soil Maps*; IUSS Working Group: Vienna, Austria, 2014.

41. Reed, S.; Martens, D. *Methods of Soil Analysis, Part 3: Chemical Methods*; John Wiley & Sons: Hoboken, NJ, USA, 1996.

42. Konestabo, H.S.; Michelsen, A.; Holmstrup, M. Responses of Springtail and Mite Populations to Prolonged Periods of Soil Freeze-Thaw Cycles in a Sub-Arctic Ecosystem. *Appl. Soil Ecol.* **2007**, *36*, 136–146. [CrossRef]

43. Olson, J.S. Energy Storage and the Balance of Producers and Decomposers in Ecological Systems. *Ecology* **1963**, *44*, 322–331. [CrossRef]

44. Qualls, R.G. Long-Term (13 Years) Decomposition Rates of Forest Floor Organic Matter on Paired Coniferous and Deciduous watersheds with Contrasting Temperature Regimes. *Forests* **2016**, *7*, 231. [CrossRef]

45. Rovira, P.; Vallejo, V.R. Labile and Recalcitrant Pools of Carbon and Nitrogen in Organic Matter Decomposing at Different Depths in Soil: An Acid Hydrolysis Approach. *Geoderma* **2002**, *107*, 109–141. [CrossRef]

46. Bauhus, J.; Vor, T.; Bartsch, N.; Cowling, A. The Effects of Gaps and Liming on Forest Floor Decomposition and Soil C and N Dynamics in a Fagus Sylvatica Forest. *Can. J. For. Res.* **2004**, *34*, 509–518. [CrossRef]
47. Bagnato, S.; Marziliano, P.A.; Sidari, M.; Mallamaci, C.; Marra, F.; Muscolo, A. Effects of gap size and cardinal directions on natural regeneration, growth dynamics of trees outside the gaps and soil properties in European beech forests of southern Italy. *Forests* **2021**, *12*, 1563. [CrossRef]
48. Fuzhong, W.; Changhui, P.; Jianxiao, Z.; Jian, Z.; Bo, T.; Wanqin, Y. Impacts of Freezing and Thawing Dynamics on Foliar Litter Carbon Release in Alpine/Subalpine Forests along an Altitudinal Gradient in the Eastern Tibetan Plateau. *Biogeosciences* **2014**, *11*, 6471–6481. [CrossRef]
49. Li, H.; Wu, F.; Yang, W.; Xu, L.; Ni, X.; He, J.; Tan, B.; Hu, Y.; Justin, M.F. The Losses of Condensed Tannins in Six Foliar Litters Vary with Gap Position and Season in an Alpine Forest. *iForest* **2016**, *9*, 910. [CrossRef]
50. Tamura, M.; Tharayil, N. Plant Litter Chemistry and Microbial Priming Regulate the Accrual, Composition and Stability of Soil Carbon in Invaded Ecosystems. *New Phytol.* **2014**, *203*, 110–124. [CrossRef]
51. Hobara, S.; Osono, T.; Hirose, D.; Noro, K.; Hirota, M.; Benner, R. The Roles of Microorganisms in Litter Decomposition and Soil Formation. *Biogeochemistry* **2014**, *118*, 471–486. [CrossRef]
52. Scharenbroch, B.C.; Bockheim, J.G. Gaps and Soil C Dynamics in Old Growth Northern Hardwood-Hemlock Forests. *Ecosystems* **2008**, *11*, 426–441. [CrossRef]
53. Ni, X.; Yang, W.; Li, H.; Xu, L.; He, J.; Tan, B.; Wu, F. The Responses of Early Foliar Litter Humification to Reduced Snow Cover during Winter in an Alpine Forest. *Can. J. Soil Sci.* **2014**, *94*, 453–461. [CrossRef]
54. Groffman, P.M.; Hardy, J.P.; Fisk, M.C.; Fahey, T.J.; Driscoll, C.T. Climate Variation and Soil Carbon and Nitrogen Cycling Processes in a Northern Hardwood Forest. *Ecosystems* **2009**, *12*, 927–943. [CrossRef]
55. Bokhorst, S.; Metcalfe, D.B.; Wardle, D.A. Reduction in Snow Depth Negatively Affects Decomposers but Impact on Decomposition Rates Is Substrate Dependent. *Soil Biol. Biochem.* **2013**, *62*, 157–164. [CrossRef]
56. Zhao, Y.; Wu, F.; Yang, W.; Tan, B.; He, W. Variations in Bacterial Communities during Foliar Litter Decomposition in the Winter and Growing Seasons in an Alpine Forest of the Eastern Tibetan Plateau. *Can. J. Microbiol.* **2015**, *62*, 35–48. [CrossRef] [PubMed]
57. Aerts, R.; Callaghan, T.V.; Dorrepaal, E.; van Logtestijn, R.S.P.; Cornelissen, J.H.C. Seasonal Climate Manipulations Have Only Minor Effects on Litter Decomposition Rates and N Dynamics but Strong Effects on Litter P Dynamics of Sub-Arctic Bog Species. *Oecologia* **2012**, *170*, 809–819. [CrossRef] [PubMed]
58. Guenet, B.; Danger, M.; Abbadie, L.; Lacroix, G. Priming Effect: Bridging the Gap between Terrestrial and Aquatic Ecology. *Ecology* **2010**, *91*, 2850–2861. [CrossRef] [PubMed]
59. Kreyling, J.; Haei, M.; Laudon, H. Snow Removal Reduces Annual Cellulose Decomposition in a Riparian Boreal Forest. *Can. J. Soil Sci.* **2013**, *93*, 427–433. [CrossRef]

*forests*

Article

Impact of Moso Bamboo (*Phyllostachys edulis*) Expansion into Japanese Cedar Plantations on Soil Fungal and Bacterial Community Compositions

Haifu Fang [1], Yuanqiu Liu [1,2], Jian Bai [1], Aixin Li [1], Wenping Deng [1,2], Tianjun Bai [2], Xiaojun Liu [1], Meng Lai [1], Yan Feng [2], Jun Zhang [2], Qin Zou [2], Nansheng Wu [3,*] and Ling Zhang [1,2,*]

[1] Jiangxi Provincial Key Laboratory of Silviculture, College of Forestry, Jiangxi Agricultural University, Nanchang 330045, China; fanghaifu11@163.com (H.F.); liuyq404@163.com (Y.L.); q751737790@163.com (J.B.); lax18317921522@163.com (A.L.); deng_wen_ping@126.com (W.D.); liuxiaojun.lxj@163.com (X.L.); laimeng21@163.com (M.L.)
[2] Lushan National Nature Reserve, Lushan National Observation and Research Station of Chinese Forest Ecosystem, Jiujiang 332900, China; baitianjun@139.com (T.B.); 18970261990@139.com (Y.F.); zhangjun8919330@163.com (J.Z.); 13970246138@139.com (Q.Z.)
[3] National Innovation Alliance of *Choerospondias axillaris*, Nanchang 330045, China
[*] Correspondence: rensh111@126.com (N.W.); lingzhang09@126.com (L.Z.)

Citation: Fang, H.; Liu, Y.; Bai, J.; Li, A.; Deng, W.; Bai, T.; Liu, X.; Lai, M.; Feng, Y.; Zhang, J.; et al. Impact of Moso Bamboo (*Phyllostachys edulis*) Expansion into Japanese Cedar Plantations on Soil Fungal and Bacterial Community Compositions. *Forests* 2022, 13, 1190. https://doi.org/10.3390/f13081190

Academic Editors: Fuzhong Wu, Zhenfeng Xu and Wanqin Yang

Received: 19 June 2022
Accepted: 21 July 2022
Published: 27 July 2022

Publisher's Note: MDPI stays neutral with regard to jurisdictional claims in published maps and institutional affiliations.

Simple Summary: Moso bamboo (*Phyllostachys edulis*) expansion caused substantial changes in the ecosystem process. Changes in plant–soil chemical characteristics and microbial community compositions have not been thoroughly studied. To understand changes in forest ecosystem process and the underlining microbial community compositions, we studied changes in plant–soil chemical characteristics and microbial community compositions. The results showed that moso bamboo expansion into Japanese cedar plantations altered litter C:N and divergently affected soil fungal and bacterial community compositions. Specifically, moso bamboo expansion decreased soil organic carbon, total nitrogen, litter carbon, and the carbon to nitrogen ratio. Moso bamboo expansion also increased soil NH_4^+-N and pH, while it decreased fungi OTUs at the phyla, class, order, family, and genus levels. The expansion of moso bamboo into Japanese cedar substantially altered soil-fungal- and bacterial-community structure, which might have implications for changes in the ecosystem element-cycling process.

Abstract: Moso bamboo expansion is common across the world. The expansion of moso bamboo into adjacent forests altered plant and soil characteristics. While the community structure of soil fungi and bacteria plays an important role in maintaining the function of forest ecosystems, changes in microbial community compositions remain unclear, limiting our understanding of ecological process changes following moso bamboo expansion. To explore changes in the community structure of soil fungi and bacteria in Japanese cedar plantations experiencing expansion of moso bamboo, Illumina NovaSeq high-throughput sequencing technology was used to elucidate changes in soil microbial communities as well as alteration in litter and soil chemical characteristics. The results showed that moso bamboo expansion decreased content of soil organic carbon, total nitrogen, litter carbon, and the carbon to nitrogen ratio as well as the number of bacterial operational taxonomic units (OTUs) at the genus level, the α-diversity Simple index, and the abundance of *Acidobacteria*, *Chloroflexi*, and *Gemmatimonadetes*. Moso bamboo expansion also increased soil NH_4^+-N, pH, while it decreased fungi OTUs at the phyla, class, order, family, and genus level. The expansion of moso bamboo into Japanese cedar substantially altered soil fungal and bacterial community structure, which might have implications for changes in the ecosystem element-cycling process. In the forest ecosystem and expansion management of moso bamboo, the types and different expansion stages of moso bamboo should be paid attention to, in the assessment of ecological effects and soil microbial structure.

Keywords: litter decomposition; soil organic carbon; microbial community composition; plant invasion ecology

1. Introduction

Biological invasion is a global problem, resulting in major changes to ecosystem diversity and stability [1,2]. The introduction of exotic plants into new habitats within a certain period and with limited space can cause environmental degradation, resource shortages, natural disasters, and other phenomena, thus altering the original ecological balance and reducing biological diversity [2–5]. When alien plants are introduced into a new environment, they exhibit strong adaptability and growth advantages due to changes in the external environmental factors [1,6,7], ultimately causing changes to the soil carbon and nitrogen cycle and the microbial structure.

Bamboo is widely regarded as one of the most economically useful species in the 21st century and is mainly distributed in the tropical and subtropical regions of the Asia-Pacific, Africa, and Latin America [8]. Moso bamboo (*Phyllostachys edulis*) is a very important forest resource in southern China [9]. Moso bamboo has the biological characteristics of rapid growth and a strong reproductive capacity. Its rhizomes are also characterized as being strongly aggressive and being able to gradually spread into the adjacent forest vegetation, resulting in an expansion phenomenon in the stand, thus forming a mixed forest or even a pure bamboo forest [10].

The expansion of moso bamboo can seriously threaten the adjacent forest vegetation and lead to the death of the surrounding forest vegetation, thus altering the vegetation community structure and reducing biodiversity [9]. The expansion of moso bamboo reduces soil organic matter carbon [8,11], nitrogen transform rates [10,12], and the changes to soil microorganisms mainly depend on the original soil type [9,13,14]. The expansion of moso bamboo also causes significant changes in the emissions of the soil greenhouse gases nitrous oxide and carbon dioxide [15–19].

Soil microorganisms are an important component of forest ecosystems and play a significant role in the mineralization of soil organic matter and nutrient cycling [20]. Fungi and bacteria are the main components of the soil microbial community, and their abundance and diversity are directly affected by soil properties and environmental factors [21], which are of great significance for energy conversion and material circulation [22]. Different environmental conditions and vegetation types can change the structure and function of the soil microbial community [23,24]. Changes in stand types have different effects on the community structure, root exudates, litter quantity and quality, and nutrient availability. Therefore, in recent years, soil microbial diversity has become a pertinent issue in the field of ecology [25]. Fang et al. [26] studied changes in the soil nitrogen cycle, as affected by Japanese cedar and moso bamboo, by adding biological inhibitors. However, there are few studies on soil microbial diversity under different expanding stages.

The expansion of moso bamboo into surrounding forests will reduce species diversity and alter the vegetation type [27]. Vegetation-species composition plays an important role in soil microbial-community structure [28,29]. The expansion of moso bamboo leads to alterations in the forest-species composition, which has a direct impact on soil microbial-community structure, in turn affecting plant development, plant-community composition, and ecosystem functioning. Here, soil microbial communities' compositions as well as litter and soil carbon and nitrogen status, during different expanding stages of moso bamboo, were studied. We predicted that (1) litter and soil carbon and nitrogen status differ with expanding stages of moso bamboo; (2) soil microbial-community compositions altered differently between microbial and fungal communities, in response to the expansion of moso bamboo. We expected clearly separated changes in both biotic compositions and abiotic characteristics along the expanding stages of moso bamboo.

2. Materials and Methods

2.1. Study Area

This study area was in Lushan Nature Reserve in Jiangxi province, encompassing a total area of 30.2 km^2. The region has a subtropical monsoon climate, an annual average rainfall of 2070 mm, an annual average temperature of 11.6 °C, a maximum temperature of 31.1 °C, and a minimum temperature of -16.7 °C. The average annual fog days are 191 days, and the frost periods are 150 days. In recent years, many moso bamboo forests have been expanding into the surrounding Japanese cedar forests, in areas over 800 m above sea level. Moreover, the number of moso bamboo plants has increased sharply, which greatly affects the forest landscape pattern and the stability of the forest ecosystem in the protected area.

2.2. Experimental Design and Sample Collection

Complete randomized experimental design was used in this study. Japanese cedar forests with different mixing rate of moso bamboo were selected to represent the expanding stage of moso bamboo into Japanese cedar plantations. Specifically, the research area was divided into four treatments, including Japanese cedar forests (115°57′17″ E, 29°32′42″ N), altitude 980 m; mixed 1 (30% moso bamboo and 70% Japanese cedar) (115°57′21″ E, 29°32′46″ N), altitude 980 m; mixed 2 (60% moso bamboo and 40% Japanese cedar) (115°57′12″ E, 29°32′44″ N), altitude 960 m; and moso bamboo forests (115°57′12″ E, 29°32′44″ N), altitude 950 m. Each treatment including four replications, and each of them were separated by at least 500 m in the studied area. In July 2020, soils were collected from four 20 × 20 m sample plots. A five-point sampling method was adopted, and a soil drill (diameter of 5 cm) was used to take soil samples from the 0–20-cm soil layer, which were transported back to the laboratory in an icebox. The soil samples from the five points were mixed evenly by plots, the visible stones and plant residues (such as roots, stems and leaves) were carefully removed from the fresh samples, and the samples were then passed through a 2-mm sieve for the experiment. At the same time, the litter of Japanese cedar and moso bamboo were collected and brought back to the laboratory.

2.3. Soil and Litter Chemical Characteristic Analyses

We removed part of the soil sample for air drying, one part for the determination of soil pH, and another part for the determination of soil total organic carbon (TOC) and total nitrogen (TN), after screening with a pore sieve of 0.149 mm. Some of the remaining fresh samples were placed at 4 °C and measured for available nitrogen (AN = NH_4^+-N + NO_3^--N) soon after, while the other samples were placed in a -80 °C refrigerator for microbiological measurements [30]. Soil pH was calculated using the electric electrode method (water: soil = 2.5:1) (Metter Toledo, Shanghai, China). Moso bamboo and Japanese cedar litter was dried at 60 °C and weighed, and the litter carbon (LC) and litter total nitrogen (LN) contents were determined after passing through a 0.149-mm aperture sieve. TOC and LC were determined by the potassium dichromic (H_2SO_4–$K_2Cr_2O_7$) method; and soil TN and LN, NO_3^--N, and NH_4^+-N were determined by automatic analyzer (Smart Chem 200, Westco, Rome, Italy) after being extracted by 2 mol L^{-1} KCl solution [30].

2.4. DNA Extraction and Gene Sequencing

DNA (Deoxyribonucleic acid) was extracted from 0.5 g of fresh soil samples, and the DNA concentration and purity (0.8% agarose gel) were monitored. The primers used for bacterial 16S rRNA were 338F: 5′-ACTCCTACGGGAGGCAGCA-3′, 806R: 5′-GGACTACHVGGGTWTCTAAT-3′ for the V4 region. The soil fungal ITS1 region was detected by 18S rRNA, and PCR amplification was performed. The primer sequences were ITS5F: 5′-GGAAGTAAAAGTCGTAACAAGG-3′ and ITS2R: 5′-GCTGCGTTCTTCATCGATGC-3′. Approximately 2 μL of template DNA, 1 μL of each forward and reverse primer, and 15 μL of Phusion High-Fidelity PCR Master Mix (New England Biolabs, Ipswich, MA USA) were used for the PCR reactions. The thermal-cycling program was as follows: after the components required for the PCR (Polymerase Chain Reaction) reaction were configured, the template DNA was fully denatured at 98 °C for 30 s on the PCR instrument, following which the amplification cycle was entered. In each cycle, the template was denatured by maintaining it at 98 °C for 15 s, following which the temperature was reduced to 50 °C for 30 s, allowing for the primer and template to be fully annealed. At 72 °C for 30 s, the primer was extended on the template to synthesize DNA, completing a cycle. This cycle was repeated 25 to 27 times, resulting in a large accumulation of amplified DNA fragments. Finally, the product was kept at 72 °C for 5 min to complete the extension and was stored at 4 °C. The amplification results were electrophoresed by 2% agarose gel. The target fragment was cut and then recovered with an Axygen gel recovery kit. The samples were sequenced using the Illumina NovaSeq platform for high-throughput sequencing by Shanghai Personalbio Technology (Shanghai, China).

2.5. Statistical Analyses

One-way analysis of variance was used to examine the effects of moso bamboo expansion on soil chemical properties, microbial-community compositions, diversity and species abundance, and litter chemical properties. Post hoc Student's *t*-tests were used to examine significant differences among treatment in multiple comparisons. All analyses were performed by JMP 9.0 (SAS Institute, Cary, NC, USA).

3. Results

3.1. Results and Analysis

3.1.1. Changes in Soil and Litter Chemical Characteristics

Different stand types had different effects on the chemical properties of the soil. In terms of soil pH and NH_4^+-N, moso bamboo and mixed 2 were significantly higher than Japanese cedar and mixed 1 (Table 1). In terms of soil NO_3^--N, there were significant differences among Japanese cedar, mixed, and moso bamboo, and the trend of mixed 1 > moso bamboo > mixed 2 > Japanese cedar indicated that soil NO_3^--N in the expansion area tended to increase during moso bamboo expansion. In terms of soil organic carbon and nitrogen, Japanese cedar and mixed 1 were significantly higher than moso bamboo and mixed 2, indicating that the contents of soil TOC and TN tended to decrease during the expansion of moso bamboo. In terms of litter organic carbon and the litter carbon–nitrogen ratio, Japanese cedar was significantly higher than moso bamboo. There was no significant difference in litter nitrogen (Table 1).

Table 1. Chemical properties (means ± se) of soil (Japanese cedar, mixed 1, mixed 2, and moso bamboo) and litter (Japanese cedar and moso bamboo). *p* values based on one-way ANOVA are shown. Means with the same letter indicate no significant difference in post hoc tests. Results with significant *p* values are shown in bold.

Variable	TOC (g kg^{-1})	TN (g kg^{-1})	TOC:TN	pH	NH_4^+-N (mg kg^{-1})	NO_3^--N (mg kg^{-1})
Soil						
Japanese cedar	72.7 (4.5) A	3.0 (0.3) AB	24.1 (0.7) A	4.5 (0.0) C	11.9 (0.3) C	2.7 (0.1) D
Mixed 1	73.9 (1.8) A	4.0 (0.3) A	18.8 (1.1) A	4.4 (0.0) C	11.6 (0.1) C	7.2 (0.1) A
Mixed 2	55.4 (5.0) B	2.8 (0.2) B	19.9 (2.3) A	5.1 (0.0) B	17.0 (0.2) A	3.9 (0.1) C
Moso bamboo	55.8 (1.9) B	2.6 (0.2) B	21.4 (2.2) A	5.2 (0.0) A	13.7 (0.1) B	4.3 (0.1) B
p	**0.0082**	**0.0212**	0.2364	**<0.0001**	**<0.0001**	**<0.0001**
Litter						
Japanese cedar	510.5 (11.3) A	14.6 (0.4) A	35.1 (1.0) A			
Moso bamboo	399.7 (7.7) B	15.5 (0.3) A	25.9 (1.0) B			
p	**<0.0001**	0.138	**<0.0001**			

Notes: NH_4^+-N, ammonium nitrogen; NO_3^--N, nitrate nitrogen; TOC, total organic carbon; TN, total nitrogen; TOC:TN: total carbon:nitrogen ratio. Capital letters next to means in the same column indicated significantly different within soil or litter.

3.1.2. Changes in Fungal-Community Structure

The high-throughput sequencing results showed that there were significant differences in the number of soil fungi operational taxonomic units (OTUs) in the four vegetation types at the phylum, class, order, family, and genus levels (Table 2, Figure 1). At the phylum level, the numbers of OTUs in mixed 2 were significantly higher than in Japanese cedar. At the level of class and order, mixed 2 and moso bamboo were significantly higher than mixed 1 and Japanese cedar. At the family level, there were significant differences between Japanese cedar, mixed, and moso bamboo in the order of moso bamboo > mixed 2 > mixed 1 > Japanese cedar. At the genus level, mixed 1, mixed 2, and moso bamboo were significantly higher than Japanese cedar. This indicated that the number of OTUs in the soil fungal of moso bamboo exhibited an increasing trend.

Table 2. Effect of moso bamboo expansion on the number of OTUs of soil microbial fungi and bacteria in ANOVAs. Results with significant *p* values are shown in bold.

OTUs	Fungal			Bacterial		
	DF	**F**	*p*	**DF**	**F**	*p*
Phylum	3	10.6	**0.0037**	3	0.1	0.9663
Class	3	30.4	**0.0001**	3	1.6	0.2719
Order	3	40.4	**<0.0001**	3	7.6	**0.0101**
Family	3	12.7	**0.0021**	3	5.3	**0.0261**
Genus	3	6.9	**0.013**	3	5.2	**0.028**

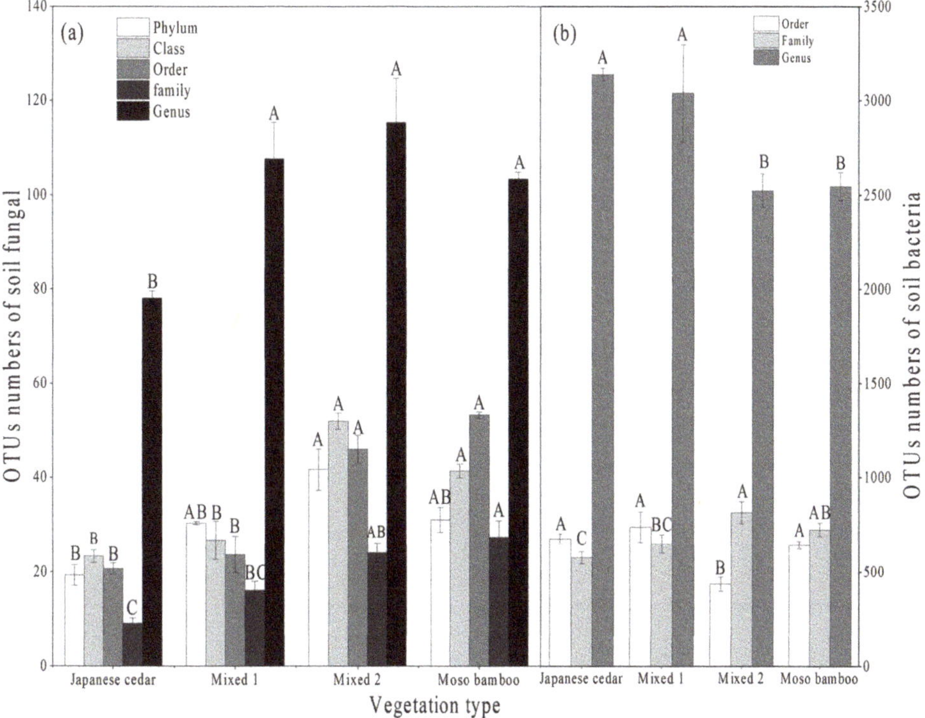

Figure 1. Effect of different vegetation types on the number of soil fungal and bacterial OTUs (means ± se): (**a**) number of soil fungal OTUs, (**b**) number of soil bacterial OTUs. Different letters above columns in the same color indicate significantly different.

The changes in the biodiversity of the soil fungal communities in the four vegetation types differed. There was no significant difference in the diversity index of the soil fungal community (goods coverage, Pielou's evenness, and simple), but there were significant differences in the Chao1, observed species, and Shannon indexes. Additionally, in terms of the Chao1 and observed species diversity indexes, mixed 2 and moso bamboo were significantly higher than Japanese cedar. In terms of the α-diversity Shannon index, moso bamboo was higher than Japanese cedar (Table 3, Figure 2).

Table 3. Effects of moso bamboo expansion on soil microbial fungi and bacteria α-diversity in ANOVAs. Results with significant *p* values are shown in bold.

α-Diversity	Fungal			Bacterial		
	DF	F	*p*	DF	F	*p*
Chao1	3	10.8	**0.0035**	3	2.4	0.3054
Goods coverage	3	1.3	0.3398	3	2.8	0.1068
Observed species	3	14.4	**0.0014**	3	2.1	0.1762
Pielou-e	3	2.6	0.1286	3	0.3	0.8124
Shannon	3	8.0	**0.0088**	3	1.8	0.2212
Simpson	3	2.0	0.2008	3	29.6	**0.0001**

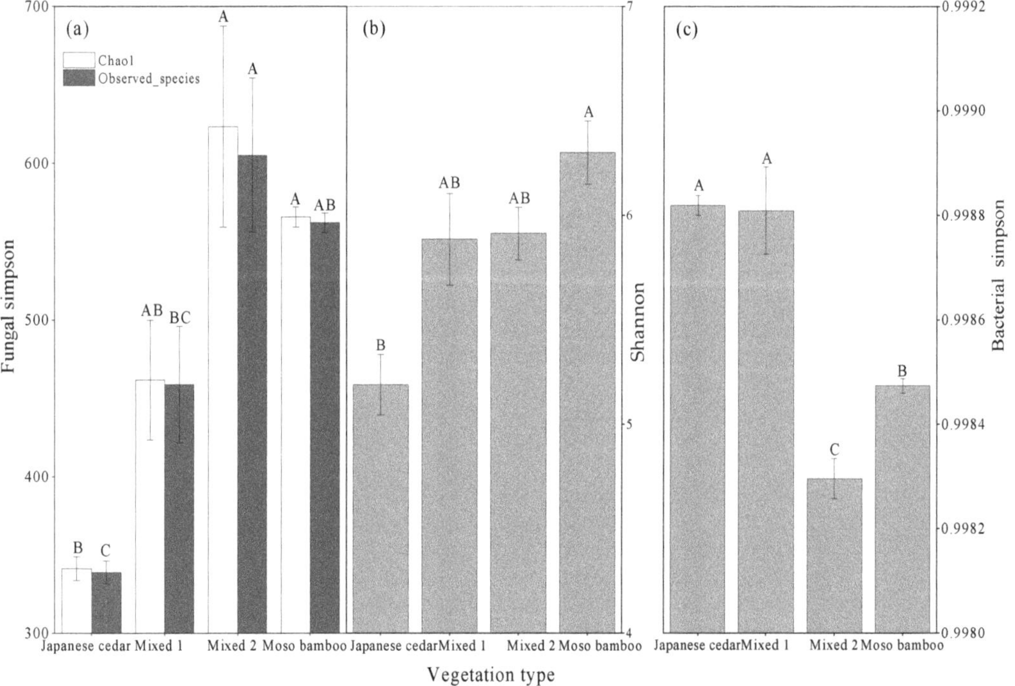

Figure 2. Effect of different vegetation types on soil fungal and bacterial α-diversity index (means ± se). (**a**) Fungal a diversity Chao1 and observed species index, (**b**) fungal a diversity Shannon index, (**c**) bacterial a diversity Simpson index. Different letters above columns in the same color indicate significantly different.

The relative abundances of the soil fungal species at the phylum level of the four vegetation types differed. *Basidiomycota* and *Mucoromycota* were significantly higher in moso bamboo than in Japanese cedar and mixed 1. However, among *Ascomycota* species, Japanese cedar and mixed 1 were significantly higher than mixed 2 and moso bamboo (Figure 3).

3.1.3. Changes in Bacterial Community Structure

The results of the high-throughput sequencing showed that there were significant differences in the number of soil bacterial OTUs in the four vegetation types at the order, family, and genus levels, but no significant differences at the phylum and class levels (Table 3, Figure 1). At the order level, Japanese cedar, mixed 1, and moso bamboo possessed significantly more OTUs than mixed 2. At the family level, moso bamboo and mixed 2 were significantly higher than Japanese cedar. At the genus level, Japanese cedar and mixed 1 were significantly higher than moso bamboo and mixed 2.

The diversity of soil bacteria in the four vegetation types differed. There was no significant difference in the soil bacterial-community diversity indexes (Chao1, goods coverage, observed species, Pielou's evenness, Shannon). There was a highly significant difference in the simple diversity index, in the order of Japanese cedar and mixed 1 > moso bamboo > mixed 2 (Table 2, Figure 2), indicating a downward trend in the soil bacterial diversity with moso bamboo expansion.

The relative abundances of the soil bacteria of the four vegetation types at the phylum level differed greatly. The three species of *Acidobacteria*, *Chloroflexi*, and *Gemmatimonadetes* were significantly higher in Japanese cedar and mixed 1 than in moso bamboo. Among the three species of *Proteobacteria*, *Verrucomicrobia*, and *Bacteroidetes*, the four vegetation types differed significantly in the order of mixed 2 > moso bamboo > Japanese cedar (Figure 3).

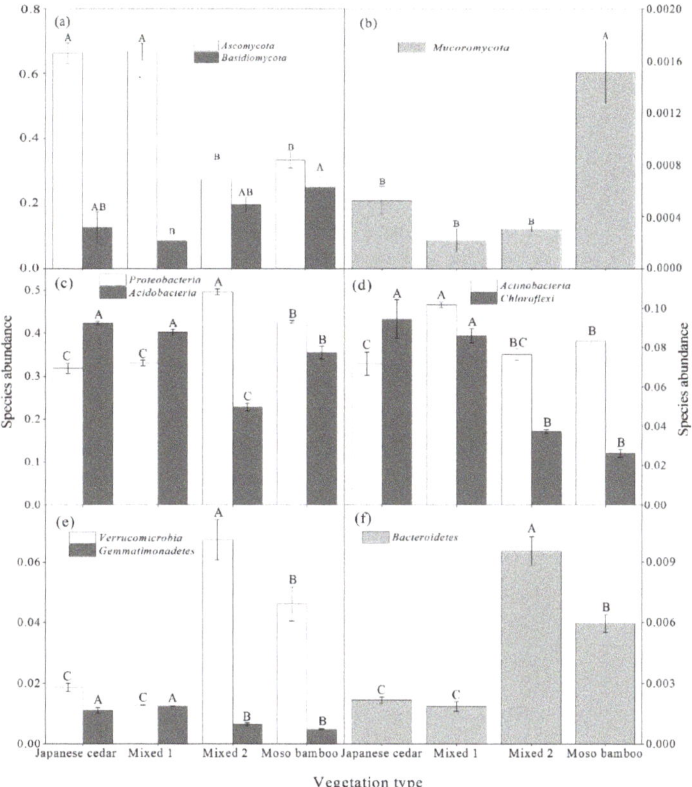

Figure 3. Effect of different vegetation types on species abundance of soil fungal and bacterial at the phylum level (means ± se). (**a**) Fungal species *Ascomycota* and *Asidiomycota*, (**b**) fungal species *Mucoromycota*,

(**c**) bacterial species *Proteobacteria* and *Acidobacteria*, (**d**) bacterial species *Actinobacteria* and *Chloroflexi*, (**e**) bacterial species *Verrucomicrobia* and *Gemmatimonadetes*, (**f**) bacterial species *Bacteroidetes*. Different letters above columns in the same color indicate significantly different.

4. Discussion

4.1. Changes in Soil and Litter Chemical Characteristics

The transformation of vegetation-stand types has a certain impact on the physical and chemical properties of soil as well as litter and nutrient cycling [31]. Compared with Japanese cedar forests, the soil NH_4^+-N contents of mixed 2 and moso bamboo increased significantly, soil and litter organic matter carbon decreased significantly, and total nitrogen showed a trend of first increasing and then decreasing. The carbon–nitrogen ratio of Japanese cedar forest was significantly higher than that of moso bamboo, indicating that Japanese cedar forest has strong carbon-sequestration ability and strong carbon-assimilation ability. So, the expansion of moso bamboo in Japanese cedar forests has a certain impact on the soil carbon and nitrogen cycle. However, due to the characteristics of moso bamboo, the expansion of moso bamboo will lead to the death of other surrounding plants due to the failure of nutrient competition. Therefore, the underground biomass of plants will decrease, resulting in a decrease in soil organic carbon and total nitrogen content [10]. Conversely, the decreased soil organic carbon and total nitrogen may be related to the decomposition rate of the litter. The rate of decay of Japanese cedar litter is lower than that of moso bamboo litter. This may, therefore, accelerate the degradation of soil organic carbon and the total nitrogen absorption of moso bamboo forests [32,33]. The total nitrogen content of the soil under moso bamboo expansion decreased, indicating that the overall absorption of nitrogen during moso bamboo expansion was higher than that of Japanese cedar, which may be related to the increased available nitrogen in the plant. Moso bamboo expansion resulted in increased soil pH, which may have altered some ammonia-oxidizing bacteria and ammonia-oxidizing archaea or may also be due to changes in chemical properties such as cation exchange [34].

4.2. Changes in Fungal Community Structure

In this study, there were great differences in the soil fungal-community structure among the four stand structures. Changes in vegetation types will also alter soil fungal communities. The number of soil fungi OTUs of moso bamboo showed an increasing trend. The soil fungi diversity, as measured by the Chao1 and observed species indexes, indicated that *Basidiomycota* and *Mucoromycota* were significantly higher in mixed 2 and moso bamboo than in Japanese cedar. Therefore, the expansion of moso bamboo in Japanese cedar forests significantly increased soil fungal diversity, resulting in increased number of soil pathogens and symbiotic groups, altered soil fungal community structure and function, and enzyme activities [35]. The expansion of moso bamboo is also associated with differences in plant biomass inputs and outputs, thus affecting the composition of soil microbial communities [36,37].

4.3. Changes in Bacterial Community Structure

Soil bacteria are sensitive indicators of soil change. Increased plant diversity leads to increased soil microbial diversity. In this study, in terms of the simple index of bacterial diversity, *Acidobacteria*, *Chloroflexi*, and *Gemmatimonadetes* were significantly higher in Japanese cedar and mixed 1 than in moso bamboo. This is consistent with the findings of Lin et al. [24], whereby the expansion of moso bamboo reduced the plant diversity and the soil microbial bacterial-community structure and diversity. The allelopathy produced by the leaves of moso bamboo leads to a decrease in the diversity of the bacterial-community structure [13]. Additionally, plant diversity decreases with moso bamboo expansion, resulting in weakened competition between vegetation types, which, in turn, gradually increases the mutual benefit and encourages the frequent exchange of substances between the rhizosphere, and the resulting changes in rhizosphere exudates and decline in community productivity may lead to changes in the composition of the bacterial community [34].

4.4. Relationship between Soil Physical and Chemical Properties and Microbes of Moso Bamboo

The transformation of vegetation types plays a vital role in the structure of the soil microbial community [28]. This is mainly related to changes in the chemical composition of litter and root exudates and related soil carbon and nitrogen substrate concentrations, which alter the soil microbial activity and community structure [22,38]. Farmers picking bamboo shoots can cause disturbances to the soil. This process is similar to tillage and may reduce soil organic-matter content, which in turn alters bacterial communities [24]. Japanese cedar has high soil carbon and nitrogen content, has complex bacterial microbial diversity and community structure, and is associated with leaf density [26,39]. Understory plants and rhizosphere resources, such as root exudates and nutrient content, all affect the structure of the soil microbial community. Soil fungal communities are decomposers of soil organic matter [40]. Xiao et al. [41] found that soil organic carbon mineralization is directly affected by fungal diversity, and fungal diversity is also affected by pH. In this study, the expansion of soil organic carbon and total nitrogen in moso bamboo was significantly negatively correlated with fungal community diversity (Chao1 and observed species) and species abundance (*Basidiomycota* and *Mucoromycota*).

Compared with soil fungi, the soil bacterial-community structure is more sensitive to soil physical disturbances and changes in chemical properties [42]. In this study, moso bamboo increased the soil organic matter carbon and total nitrogen, and the simple bacterial-diversity index indicated that *Acidobacteria*, *Chloroflexi*, and *Gemmatimonadetes* all exhibited a downward trend. The number of soil bacterial OTUs was significantly higher than the number of fungal OTUs. There were certain differences in the soil microbial-community structure among the four forest-stand structures, demonstrating that the differences among different vegetation types have shaped different soil bacterial communities. The differing chemical compositions of the root exudates produced by different forest-stand structures will have a certain impact on the soil bacterial community structure.

5. Conclusions

With the expansion of moso bamboo in Japanese cedar, soil NH_4^+-N and pH increased, while soil organic carbon, total nitrogen, and litter carbon decreased. Moso bamboo expansion significantly increased soil fungal diversity, caused an increase in the number of soil pathogens and symbiotic groups, and altered the structure and function of the soil fungal communities. Moso bamboo expansion may promote changes in soil microbial-community structure by changing the basic phychemical properties of soil. Overall, with the continuous expansion of moso bamboo, the structure and diversity of soil bacterial communities decreased, and the structure and diversity of soil fungi have increased significantly. Therefore, the expansion of moso bamboo will cause significant changes in soil microbial activities and community structure, which should be considered in the management of moso bamboo expansion.

Author Contributions: Conceptualization, N.W., L.Z. and H.F.; methodology, L.Z., Y.L. and H.F.; software, H.F. and A.L.; validation, L.Z., M.L. and H.F.; formal analysis, H.F.; investigation, T.B., H.F., Q.Z. and Y.F.; resources, L.Z., W.D., M.L. and Y.L.; data curation, H.F., J.B., A.L. and J.Z.; writing—original draft preparation, H.F.; visualization, N.W., L.Z. and H.F.; supervision, X.L. and L.Z.; project administration, L.Z. All authors have read and agreed to the published version of the manuscript.

Funding: Study supported by National Natural Science Foundation of China (31770749, 31560203); Research Project of Lushan National Forest Ecosystem Research Station (9022206523); Jiangxi "Double Thousand Plan" Science and Technology Innovation High-end Talent Project (jxsq2019201078).

Institutional Review Board Statement: Not applicable.

Data Availability Statement: Data are contained within the article.

Acknowledgments: The authors acknowledge other colleagues from the Lushan National Nature Reserve for their assistance in the field work.

Conflicts of Interest: The authors declare no conflict of interest.

References

1. Ehrenfeld, J.G. Ecosystem Consequences of Biological Invasions. *Annu. Rev. Ecol. Evol. Syst.* **2010**, *41*, 59–80. [CrossRef]
2. Bradley, B.A.; Blumenthal, D.M.; Wilcove, D.S.; Ziska, L. Predicting plant invasions in an era of global change. *Trends Ecol. Evol.* **2010**, *25*, 310–318. [CrossRef] [PubMed]
3. Deng, B.L.; Liu, X.S.; Zheng, L.Y.; Liu, Q.; Guo, X.M.; Zhang, L. Effects of nitrogen deposition and UV-B radiation on seedling performance of Chinese tallow tree (*Triadica sebifera*): A photosynthesis perspective. *For. Ecol. Manag.* **2019**, *433*, 453–458. [CrossRef]
4. Xiao, S.; Callaway, R.M.; Graebner, R.; Hierro, J.; Montesinos, D. Modeling the relative importance of ecological factors in exotic invasion: The origin of competitors matters, but disturbance in the non-native range tips the balance. *Ecol. Model.* **2016**, *335*, 39–47. [CrossRef]
5. De Marco, A.; Arena, C.; Giordano, M.; Virzo De Santo, A. Impact of the invasive tree black locust on soil properties of Mediterranean stone pine-holm oak forests. *Plant Soil* **2013**, *372*, 473–486. [CrossRef]
6. Zou, J.W.; Rogers, W.E.; Siemann, E. Differences in Morphological and physiological traits between native and invasive populations of *Sapium sebiferum*. *Funct. Ecol.* **2007**, *21*, 721–730. [CrossRef]
7. De Marco, A.; Panico, S.C.; Memoli, V.; Santorufo, L.; Zarrelli, A.; Barile, R.; Maisto, G. Differences in soil carbon and nitrogen pools between afforested pine forests and natural shrublands in a Mediterranean area. *Appl. Soil Ecol.* **2022**, *170*, 104262. [CrossRef]
8. Wang, H.; Tian, G.; Chiu, C. Invasion of moso bamboo into a Japanese cedar plantation affects the chemical composition and humification of soil organic matter. *Sci. Rep.* **2016**, *6*, 32211. [CrossRef]
9. Liu, X.S.; Siemann, E.; Cui, C.; Liu, Y.Q.; Guo, X.M.; Zhang, L. Moso bamboo (*Phyllostachys edulis*) invasion effects on litter, soil and microbial PLFA characteristics depend on sites and invaded forests. *Plant Soil* **2019**, *438*, 85–99. [CrossRef]
10. Song, Q.N.; Ouyang, M.; Yang, Q.P.; Lu, H.; Yang, G.Y.; Chen, F.S.; Shi, J.M. Degradation of litter quality and decline of soil nitrogen mineralization after moso bamboo (*Phyllostachys pubscens*) expansion to neighboring broadleaved forest in subtropical China. *Plant Soil* **2016**, *404*, 113–124. [CrossRef]
11. Wu, J.; Jiang, P.; Chang, S.X. Dissolved soil organic carbon and nitrogen were affected by conversion of native forests to plantations in subtropical China. *Can. J. Soil Sci.* **2010**, *9*, 27–36. [CrossRef]
12. Li, Z.Z.; Zhang, L.; Deng, B.L.; Liu, Y.Q.; Kong, F.Q.; Huang, G.X.; Zou, Q.; Guo, X.M.; Fu, Y.Q.; Niu, D.K. Effects of moso bamboo (*Phyllostachys edulis*) invasions on soil nitrogen cycles depend on invasion stage and warming. *Environ. Sci. Pollut. Res.* **2017**, *24*, 24989–24999. [CrossRef]
13. Larpkern, P.; Moe, S.R.; Totland, Ø. Bamboo dominance reduces tree regeneration in a disturbed tropical forest. *Oecologia* **2011**, *165*, 161–168. [CrossRef]
14. Chang, E.; Chiu, C. Changes in soil microbial community structure and activity in a cedar plantation invaded by moso bamboo. *Appl. Soil Ecol.* **2015**, *91*, 1–7. [CrossRef]
15. Pan, J.; Liu, Y.Q.; Yuan, X.Y.; Xie, J.Y.; Niu, J.H.; Fang, H.F.; Wang, B.H.; Liu, W.; Deng, W.P.; Kong, F.Q.; et al. Root litter mixing with that of Japanese cedar altered CO_2 emissions from moso bamboo forest soil. *Forests* **2020**, *11*, 356. [CrossRef]
16. Song, X.Z.; Peng, C.H.; Ciais, P.; Li, Q.; Xiang, W.H.; Xiao, W.F.; Zhou, G.M.; Deng, L. Nitrogen addition increased CO_2 uptake more than non-CO_2 greenhouse gases emissions in a moso bamboo forest. *Sci. Adv.* **2020**, *6*, w5790. [CrossRef]
17. Tian, X.K.; Wang, M.Y.; Meng, P.; Zhang, J.S.; Zhou, B.Z.; Ge, X.G.; Yu, F.H.; Li, M.H. Native bamboo invasions into subtropical forests alter microbial communities in litter and soil. *Forests* **2020**, *11*, 314. [CrossRef]
18. De Marco, A.; Fioretto, A.; Giordano, M.; Innangi, M.; Menta, C.; Papa, S.; Virzo De Santo, A. C stocks in forest floor and mineral soil of two mediterranean Beech forests. *Forests* **2016**, *7*, 181. [CrossRef]
19. Ventorino, V.; De Marco, A.; Pepe, O.; Virzo De Santo, A.; Moschetti, G. Impact of innovative agricultural practices of carbon sequestration on soil microbial community. In *Carbon Sequestration in Agricultural Soils. A Multidisciplinary Approach to Innovative Methods*; Piccolo, A., Ed.; Springer: Berlin/Heidelberg, Germany, 2012; pp. 145–177.
20. Douterelo, I.; Goulder, R.; Lillie, M. Soil microbial community response to land-management and depth, related to the degradation of organic matter in English wetlands: Implications for the in situ preservation of archaeological remains. *Appl. Soil Ecol.* **2010**, *44*, 219–227. [CrossRef]
21. Fox, C.A.; Macdonald, K.B. Challenges related to soil biodiversity research in agroecosystems—Issues within the context of scale of observation. *Can. J. Soil Sci.* **2003**, *83*, 231–244. [CrossRef]
22. Rousk, J.; Brookes, P.C.; Bååth, E. Contrasting soil pH effects on fungal and bacterial growth suggest functional redundancy in carbon mineralization. *Appl. Environ. Microbiol.* **2009**, *75*, 1589–1596. [CrossRef]
23. Burton, J.; Chen, C.; Xu, Z.; Ghadiri, H. Soil microbial biomass, activity and community composition in adjacent native and plantation forests of subtropical Australia. *J. Soils Sediments* **2010**, *10*, 1267–1277. [CrossRef]

24. Lin, Y.T.; Tang, S.L.; Pai, C.W.; Whitman, W.B.; Coleman, D.C.; Chiu, C.Y. Changes in the Soil bacterial communities in a Cedar plantation invaded by moso bamboo. *Microb. Ecol.* **2014**, *67*, 421–429. [CrossRef]
25. Li, Z.; Siemann, E.; Deng, B.L.; Wang, S.L.; Gao, Y.; Liu, X.J.; Zhang, X.L.; Guo, X.M.; Zhang, L. Soil microbial community responses to soil chemistry modifications in alpine meadows following human trampling. *Catena* **2020**, *194*, 104717. [CrossRef]
26. Fang, H.F.; Gao, Y.; Zhang, Q.; Ma, L.L.; Wang, B.H.; Shad, N.; Deng, W.P.; Liu, X.J.; Liu, Y.Q.; Zhang, L. Moso bamboo and Japanese cedar seedlings differently affected soil N_2O emissions. *J. Plant Ecol.* **2021**, *15*, 277–285. [CrossRef]
27. Okutomi, K.; Shinoda, S.; Fukuda, H. Causal Analysis of the invasion of broad-leaved forest by bamboo in Japan. *J. Veg. Sci.* **1996**, *7*, 723–728. [CrossRef]
28. Grayston, S.J.; Campbell, C.D. Functional biodiversity of microbial communities in the rhizospheres of hybrid larch (*Larix eurolepis*) and Sitka spruce (*Picea sitchensis*). *Tree Physiol.* **1996**, *16*, 1031–1038. [CrossRef]
29. Merilä, P.; Malmivaara-Lämsä, M.; Spetz, P.; Stark, S.; Vierikko, K.; Derome, J.; Fritze, H. Soil organic matter quality as a link between microbial community structure and vegetation composition along a successional gradient in a boreal forest. *Appl. Soil Ecol.* **2010**, *46*, 259–267. [CrossRef]
30. Xie, J.Y.; Fang, H.F.; Zhang, Q.; Chen, M.Y.; Xu, X.T.; Pan, J.; Gao, Y.; Fang, X.M.; Guo, X.M.; Zhang, L. Understory plant functional types alter stoichiometry correlations between litter and soil in Chinese fir plantations with N and P addition. *Forests* **2019**, *10*, 742. [CrossRef]
31. Jiang, Y.M.; Chen, C.R.; Liu, Y.Q.; Xu, Z. Soil soluble organic carbon and nitrogen pools under mono- and mixed species forest ecosystems in subtropical China. *J. Soils Sediments* **2010**, *10*, 1071–1081. [CrossRef]
32. Chang, E.; Chen, T.; Tian, G.; Chiu, C. The effect of altitudinal gradient on soil microbial community activity and structure in moso bamboo plantations. *Appl. Soil Ecol.* **2016**, *98*, 213–220. [CrossRef]
33. Nakane, K. Soil carbon cycling in a Japanese cedar (*Cryptomeria japonica*) plantation. *For. Ecol. Manag.* **1995**, *72*, 185–197. [CrossRef]
34. Nicol, G.W.; Leininger, S.; Schleper, C.; Ranjard, L. The influence of soil pH on the diversity, abundance and transcriptional activity of ammonia oxidizing archaea and bacteria. *Environ. Microbiol.* **2008**, *10*, 2966–2978. [CrossRef] [PubMed]
35. Lejon, D.P.H.; Chaussod, R.; Ranger, J.; Ranjard, L. Microbial community structure and density under different tree species in an acid forest soil (Morvan, France). *Microb. Ecol.* **2005**, *50*, 614–625. [CrossRef] [PubMed]
36. Lal, R. Restoring Soil Quality to Mitigate Soil Degradation. *Sustainability* **2015**, *7*, 5875–5895. [CrossRef]
37. De Marco, A.; Esposito, F.; Berg, B.; Zarrelli, A.; Virzo De Santo, A. Litter Inhibitory Effects on Soil Microbial Biomass, Activity, and Catabolic Diversity in Two Paired Stands of Robinia pseudoacacia L. and Pinus nigra Arn. *Forests* **2018**, *9*, 766. [CrossRef]
38. Singh, B.K.; Millard, P.; Whiteley, A.S.; Murrell, J. Unravelling rhizosphere–microbial interactions: Opportunities and limitations. *Trends Microbiol.* **2004**, *12*, 386–393. [CrossRef]
39. Hortal, S.; Bastida, F.; Armas, C.; Lozano, Y.; Moreno, J.; García, C.; Pugnaire, F. Soil microbial community under a nurse-plant species changes in composition, biomass and activity as the nurse grows. *Soil Biol. Biochem.* **2013**, *64*, 139–146. [CrossRef]
40. Cantrell, S.A.; Molina, M.; Jean, L.; Rivera-Figueroa, F.; Ortiz-Hernández, M.; Marchetti, A.; Cyterski, M.; Perez-Jimenea, J. Effects of a simulated hurricane disturbance on forest floor microbial communities. *For. Ecol. Manag.* **2014**, *332*, 22–31. [CrossRef]
41. Xiao, D.; Huang, Y.; Feng, S.; Ge, Y.; Zhang, W.; He, X. Soil organic carbon mineralization with fresh organic substrate and inorganic carbon additions in a red soil is controlled by fungal diversity along a pH gradient. *Geoderma* **2018**, *321*, 79–89. [CrossRef]
42. Miya, R.K.; Firestone, M.K. Phenanthrene-degrader community dynamics in rhizosphere soil from a common annual grass. *J. Environ. Qual.* **2000**, *29*, 584–592. [CrossRef]

Article

Effects of Habitat Differences on Microbial Communities during Litter Decomposing in a Subtropical Forest

Hongrong Guo, Fuzhong Wu, Xiaoyue Zhang, Wentao Wei, Ling Zhu, Ruobing Wu and Dingyi Wang *

Key Laboratory for Humid Subtropical Eco-Geographical Processes of the Ministry of Education, School of Geographical Sciences, Fujian Normal University, Fuzhou 350007, China; qsx20201062@student.fjnu.edu.cn (H.G.); wufzchina@fjnu.edu.cn (F.W.); qbx20210127@student.fjnu.edu.cn (X.Z.); qsx20201041@student.fjnu.edu.cn (W.W.); qsx20201049@student.fjnu.edu.cn (L.Z.); qsx20201025@student.fjnu.edu.cn (R.W.)
* Correspondence: albertwdy@fjnu.edu.cn

Abstract: The differences between aquatic and terrestrial habitats could change microbial community composition and regulate litter decomposition in a subtropical forest, but the linkage remains uncertain. Using microbial phospholipid fatty acids (PLFAs), the litter decomposition associated with microbial organisms was monitored to characterize the differences of microbial communities in the forest floor, headwater stream, and intermittent stream. Habitat type did not significantly affect the concentrations of total PLFA. However, microbial community composition (fungi, G+ bacteria, and eukaryote) was significantly affected by the microenvironment among habitats. Compared with which in headwater stream, more individual PLFAs were identified in the natural forest floor and the intermittent stream during the whole decomposition period. The differences in individual PLFA concentrations were reflected in the forest floor and aquatic system in the early stage of litter decomposition, but they mainly reflected in the headwater stream and the intermittent stream in the later stage of litter decomposition. We linked the relationships between microbial community and litter decomposition and found that communities of decomposers drive differences in litter decomposition rate among habitats. Intriguingly, the microbial community showed the greatest correlation with the decomposition rate of litter in streams. These findings could contribute to the understanding of habitats difference on the microbial community during litter decomposition.

Keywords: leaf litter decomposition; habitat types; microbial community; phospholipid fatty acids (PLFAs); individual PLFA

Citation: Guo, H.; Wu, F.; Zhang, X.; Wei, W.; Zhu, L.; Wu, R.; Wang, D. Effects of Habitat Differences on Microbial Communities during Litter Decomposing in a Subtropical Forest. *Forests* **2022**, *13*, 919. https://doi.org/10.3390/f13060919

Academic Editor: Bartosz Adamczyk

Received: 10 April 2022
Accepted: 8 June 2022
Published: 13 June 2022

Publisher's Note: MDPI stays neutral with regard to jurisdictional claims in published maps and institutional affiliations.

1. Introduction

Litter decomposition, a crucial component of carbon and nutrient cycles, plays an indispensable role in regulating ecosystem structure and function in forests [1]. The litter produced by plants in terrestrial ecosystems most reaches the forest floor. Meanwhile, there are abundant headwater streams and intermittent streams in forest ecosystems. These aquatic ecosystems can receive substantial inputs of terrestrially-derived leaf litter. As a result, streams reconnect terrestrial and aquatic ecosystems by flows of organic matter that represent a great contribution to the global carbon cycle [2]. Some microbial decomposer communities, including fungi and bacteria, drives leaf litter decomposition which controls the flow of energy and nutrients in forest ecosystems [3,4]. Due to the different sensitivity of bacteria and fungi, they respond differently to relatively rapid shifts in microenvironmental conditions, such as temperature, precipitation, dissolved oxygen, or pH which affects their relative contribution to leaf decomposition [5,6]. After entering the ecosystems, habitats differ in properties such as nutrient availability, major environmental conditions, processes, and dynamics, which altogether influence the microbial abundance and community composition, and ultimately contribute to the decomposition process of litter [7,8]. Unfortunately, studies on microbial communities in litter decomposition have often been focused only on individual terrestrial or aquatic ecosystems, and only a handful of studies have dealt with the differences between them. This hinders our understanding of the impact of habitat differences on microbial community changes during litter decomposition.

The microenvironment in which microbial decomposers survive varies greatly between terrestrial and aquatic systems, due to the contrasting spatial and temporal scales at which litter is decomposed in different habitats [9]. These differences can control microbial decomposition by regulating microbial decomposers. Specifically, on the forest floor, the litter is in direct contact with the soil surface. At this time, the surface temperature and rainfall are important environmental factors to regulate the activities of microbial decomposers in terrestrial ecosystems [10]. Meanwhile, the microtopography, exposition, animal activities, or plant presence will also lead to the variations of temperature and humidity at regional scales of the forest floor [11] and then affect the activities of microbial decomposers. Compared with terrestrial ecosystems, the headwater stream is typified by relatively stable temperature, sufficient water source, low oxygen, and strong scour, which regulate microbial community composition [12,13]. When entering a stream, leaf litter will be promptly colonized by a diverse array of microbial decomposers, particularly aquatic fungi [14]. The hydrological regimes of intermittent streams are characterized by alternating flowing, nonflowing, and dry phases [13,15]. The microbial decomposer in litter decomposition in intermittence streams could be similar to headwater streams during flowing phases while it may be similar to the forest floor during dry phases. When it comes to the dry phases, the litter will be exposed to solar radiation, and its temperature and humidity will change dynamically [16,17]. Once the dry phase finishes, gradual rewetting promotes the onset of microbial decomposition [18]. As a result, microbial decomposition processes and rates dramatically change among these different habitat types but need further and unified understanding.

The phospholipid fatty acids (PLFAs) analysis method has been used for more than 42 years and is still popular as a means to characterize microbial communities in a diverse range of environmental matrices [19]. It is considered that degradation of PLFA biomarkers proceeds rapidly after cell death, providing robust information on the living microbial biomass and community structure [20]. Thus, the approach is advantageous there in that it can identify living microbial biomass and is more sensitive in detecting shifts in the microbial community compared to DNA-/RNA-based methods [21]. In addition, PLFA could provide more information on such a variety of microbial characteristics (both functional and structural) in a single analysis than other methods [19].

South China is rich in hydrothermal resources and breeds the world's largest humid subtropical mountain forest, which is typified by fast material circulation, rich biodiversity, and high productivity [22]. There are dense headwater streams hidden between mountains and gullies, and abundant intermittent streams on the forest surface in the rainy season [23]. Under the influence of water collection, a large number of forest litter gather in intermittent streams and headwater streams [24]. Microorganism plays a vital role in the biological decomposition of leaf litter in aquatic and terrestrial ecosystems. Due to its sensitivity to environmental conditions, microbial biomass and community composition may be different, and they ultimately influence the decomposition process of litter. Therefore, we hypothesized that the unique environmental conditions in different habitats will change the microbial community structure and influence the decomposition process of litter. To address this hypothesis, we collected leaf litter of the typical tree species *Castanopsis carlesii* (Hemsl.) Hayata. and use the litter bag decomposition method in the natural forest floor, headwater stream, and intermittent stream of the subtropical area. The intention of this study is not only to focus on the changes of microbial community in the process of litter decomposition in a single habitat but to focus on the main differences between habitats and its influence. This may help to improve our understanding of the abiotic drivers (microbial decomposers) on litter decomposition in forest ecosystems.

2. Materials and Methods

2.1. Study Site

This study was conducted at the Sanming Research Station of Forest Ecosystem, Fujian Province, China (26°19′ N, 117°36′ E). This area is dominated by low mountains with an average elevation of approximately 300 m and slopes of 25–45°. The soil is red soil with a thickness of more than 1 m. The climate is a maritime subtropical monsoon climate, with a mean annual temperature and precipitation of 19.3 °C and 1610 mm. Approximately 80% of rainfall occurs between March and August. The natural vegetation is a subtropical evergreen broadleaved forest. The nature *Castanopsis carlesii* is the dominant species, with *Schima superba* Gardn. et Champ., *Castanopsis fissa* (Champion ex Bentham) Rehder et E. H. Wilson, *Litsea elongate* (Wall. ex Nees) Benth. et Hook. f., and *Neolitsea aurata* (Hay.) Koidz. in the forest canopy and *Ormosia xylocarpa* Chun ex L. Chen, *Itea chinensis* Hook. et Arn., and *Ilex pubilimba* Merr. et Chun in the shrub layer.

2.2. Foliar Litter Decomposition Experiment

In January 2021, the fresh senescent foliar litter from *C. carlesii* was collected from the forest floor at the sample sites. The initial dry mass was obtained by measuring litter samples that were oven-drying (65 °C, 48 h) after being air-dried for more than 2 weeks at room temperature. Samples of air-dried foliar litter were placed inside nylon mesh bags (20 cm × 20 cm; 1.0 mm mesh size nylon mesh of bag; 10.00 g per bag) [25]. On 28 March 2021, litterbags were placed in three different habitats, including forest floor, headwater stream, and intermittent stream, respectively.

Three natural *C. carlesii* forest plots were selected for setting a 3 m × 3 m homogeneous quadrat in each site as three sample repeated plots. Before the formal experiments, we investigated the annual accumulation of litter per unit area one year and loaded required weight according to the area of the litter bag. Litter bags were placed on the forest floor after removing plants and litter from the soil surface of the quadrat. An interval of at least 2 cm was placed between each litterbag to avoid mutual disturbance upon collection. At a distance of about 200 m from the natural forest sample plot, three representative intermittent streams with basically the same environmental conditions, about 200 m apart and concentrated natural litter, were randomly selected as three repeated sample plots to place litter bags. The withered leaf bag is fixed to the root of the main branch of the shrub in the sample belt through the safety rope to prevent scouring and falling off. A stream about 100 m away from the natural forest sample was selected. Three points are randomly selected from the upstream to the downstream as repeated sample plots. Each sample plot is 1–2 km away from each other. The safety rope tied with litter bags is placed in the water body in the center of the stream, which is similar to the distribution of litter under natural conditions and can fluctuate with the flow of water. The safety rope is fixed at the main root of tall trees in the sample zone. As a result, a total of 98 litterbags (3 samples × 12 sampling dates (a whole year) × 3 replicates) were prepared in each habitat. All litter bags in terrestrial and aquatic ecosystems were carefully placed for avoiding overlap among samples.

After placing in the plots, three litter bags of each site were sampled and immediately brought back to the laboratory for refrigeration at the end of each month from April. However, due to the fast decomposition rate of the headwater stream and the intermittent stream, the litter is decomposed within 4 months. At the same time, we also chose to analyze the litter decomposition on the forest surface for four months. The specific sampling date is the 28th, 65th, 92nd, and 119th days after decomposition. Foreign materials, such as roots, soil debris, lichen, and bryophyte, were carefully removed from the litterbags. We freeze-dried (Wizard 2.0 hot plate vacuum freeze dryer VirTisUSA) all the collected samples until constant weight (about 3–4 days), which is used for the determination of residue rate. After that, three bags of litter in the same sample plots were mixed and ground and placed at −80 °C for the determination of PLFA (n = 3).

2.3. Monitoring Environmental Conditions

At the beginning of the experiment, rainfall buckets and thermohygrometers were set up in the sample plot to record the rainfall and atmospheric temperature every 1 h and 0.5 h respectively. The headwater stream water temperature was also monitored. Water samples from the headwater stream and intermittent stream were collected to determine the pH value. At the same time, according to rainfall data and water samples collected in intermittent streams in the same period, we found that the hydrological regimes of intermittent stream showed seven times alternating flowing during litter decomposition (during 31 March 2021–3 April 2021, 8–11 April 2021, 14 April 2021–7 July 2021, and 23–24 July 2021, there was water flow for 93 days in total; during 4–7 April 2021, 12–13 April 2021, and 8–23 July 2021, there was no water flow for 21 days in total).

2.4. PLFA Analysis

Total lipids were extracted from 1.0 g freeze-dried litter soaked in a mixture of chloroform: methanol: phosphate buffer (1:2:0.8, $v/v/v$) according to the method of Zelles and Otaki [26,27]. Phospholipids were isolated on silica columns and hydrolyzed and methylated using a methanolic KOH solution. Fatty acid methyl esters (FAME) were separated into saturated, polyunsaturated and monounsaturated fatty acids using amino propyl-modified and silver-impregnated SPE columns. The PLFA module of the American MIDI Sherlock microbial identification system platform is used for sample analysis, with hydrogen as the carrier gas, Agilent 6890N meteorological chromatograph and FID detector as the hardware platform. The chromatographic column is Agilent 19091B-102 (25 m $\times$ 200 μm $\times$ 0.33 μm). The injection volume of each sample is 1 μL. According to the MIDI platform specification, all tests are calibrated with standards. PLFAs were identified and analyzed based on the MIDI map recognition software Sherlock system 6.2.

2.5. Data Analysis

We examined how habitats impacted the concentrations of microbial PLFAs and the ratios of biomarkers by using repeated measures MANOVA with habitat type and decomposition time as factors. We used one-way ANOVA in the case of normal distribution of data and nonparametric test in the case of non-normal distribution of data to compare the differences of microbial PLFAs concentrations among habitats in the same decomposition period. Microbial community structure was characterized by performing principal components analysis (PCA) on the $\log_{10}$-transformed mole of various microbial groups PLFA biomarkers. Subsequently, we used linear regression to analyze the influence of habitat type on the first two principal component scores obtained after PCA of various microbial groups PLFA biomarkers. Meanwhile, we selected individual PLFA that simultaneously acted on litter decomposition during the whole decomposition period in three habitats. We selected individual PLFA that acts on three habitats in the same decomposition period to compare the number of individual PLFA acting on litter decomposition in different habitats and different decomposition periods. Then, we used one-way ANOVA in the case of normal distribution of data and nonparametric test in the case of non-normal distribution of data to compare the differences of individual PLFA concentrations in different decomposition periods of the same habitat and between different habitats in the same decomposition period. We used standard linear regressions to investigate correlations between the concentrations of microbial PLFAs and the ratios of biomarkers and leaf litter mass loss.

3. Results

3.1. Microbial Total Biomass and Community Composition

Total PLFA concentrations, a proxy for microbial biomass, increased significantly ($p < 0.05$) in the nature forest floor, the intermittent stream and the headwater stream from the 28th day of decomposition to the 119th day. The total PLFA concentrations in the headwater stream were significantly higher than those in the intermittent stream and the natural forest floor during the 62nd day of decomposition. However, the total PLFA concentrations decreased greatly on the 95th day of decomposition in both the intermittent stream and the headwater stream, which made habitat type don't significantly affect the concentrations of total PLFA at the end of litter decomposition (Figure 1a).

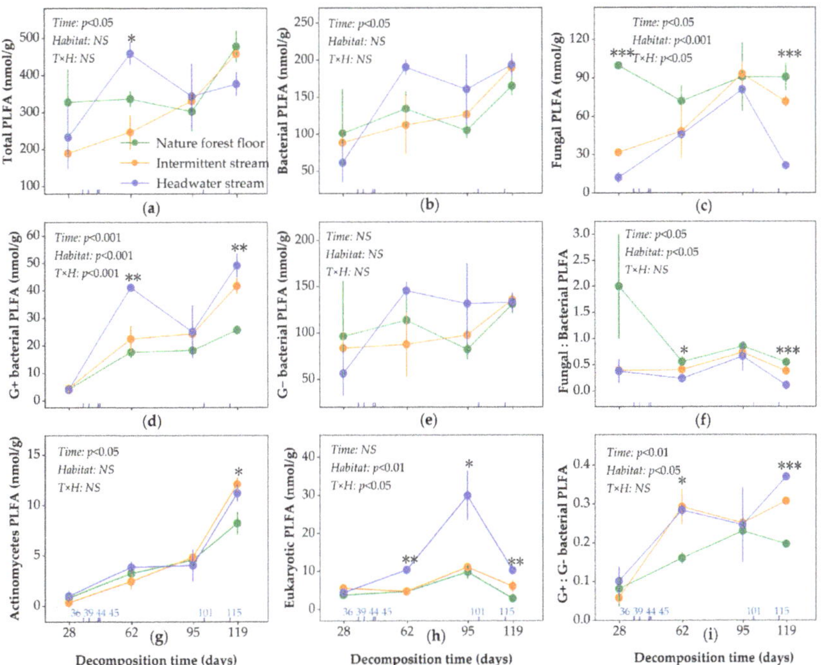

Figure 1. The concentrations of microbial total PLFA (**a**) and PLFA of microbial groups (**b–e,g,h**) in the natural forest floor, intermittent stream, and headwater stream leaf litter and the ratios of biomarkers (**f,i**). The blue font indicates the number of days the intermittent stream was dry. The *p* values showed the results from repeated measures ANOVA testing for the effect of habitat type over time. The expression data were shown as mean ± SE. *NS*: not significant. Asterisks denote significant differences among habitat types: * $p < 0.05$, ** $p < 0.01$, *** $p < 0.001$.

Some microbial community compositions varied over time, and this variation showed different patterns among habitat types. In three habitats, the G+ bacterial and actinomycetes PLFA and the ratios of G+:G− bacteria PLFA (G+/G−) peaked on the 119th day and its concentrations and ratios in the stream and the intermittent stream were significantly higher than that in the nature forest floor. Meanwhile the G+ bacterial PLFA concentrations and the ratios of G+/G− peaked on the 62nd day in the stream (Figure 1d,g,i)). At this point, the G+ bacterial PLFA concentrations in the headwater stream were significantly higher than those in the other two habitats. The ratios of G+/G− in the headwater stream and the intermittent stream were significantly higher than that in the nature forest floor. In the nature forest floor, the fungal PLFA concentrations were significantly higher than those in the other two habitats during the early and late stages of decomposition (Figure 1c). The ratios of fungal: bacterial PLFA (F/B) had a similar trend as the fungal PLFA (Figure 1f). In the headwater stream, the eukaryotic PLFA concentrations were significantly higher than those in the other two habitats during the late stages of decomposition (Figure 1h).

3.2. Distribution Patterns of PLFA

We used the PCA to illustrate the differences among the three habitats for selected PLFAs related to litter decomposition (Figure 2 and Table 1). The PCA performed on the composition of PLFAs showed that 61.0% and 13.2% of variations were explained by the first and second axes respectively. The total biomass, bacteria, G+ bacteria, G− bacteria, and actinomycete had positive loadings on the PCA-1 in the three habitats. However, fungal and eukaryotic PLFA was unspecific in three habitats. The fungal PLFA had positive loadings on PCA-1 in the intermittent stream and headwater stream and had positive loadings on PCA-2 in the nature forest floor. The eukaryotic PLFA had positive loadings on PCA-1 in the headwater stream and had positive loadings on PCA-2 in both intermittent stream and headwater streams.

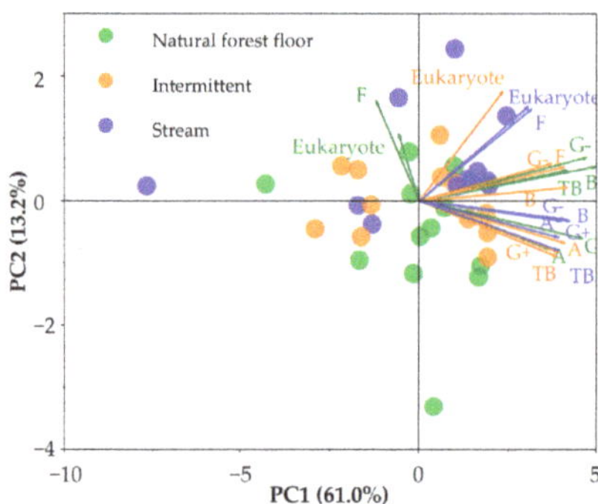

Figure 2. Principal components analyses of PLFA profiles ($\log_{10}$ transformed mol) on litter at three habitats. PLFAs in green, orange and blue font represent the natural forest floor, the intermittent stream and stream samples, respectively. TB: Total Biomass, B: Bacteria, F: Fungi, G+: G+ bacteria, G−: G− bacteria, A: Actinomycete.

Table 1. Linear regression for three habitats between the first and second principal components (PCA-1, PCA-2) after the PCA and the concentrations of microbial PLFA. Asterisks denote significant influences of the concentrations of microbial PLFA on the data variation represented by PCA-1 and PCA-2: * $p < 0.05$, ** $p < 0.01$, *** $p < 0.001$.

Index	PCA-1			PCA-2		
	Natural Forest Floor	Intermittent Stream	Headwater Stream	Natural Forest Floor	Intermittent Stream	Headwater Stream
Total Biomass	0.467 **	0.703 ***	0.816 ***	0.030	−0.030	−0.027
Bacteria	0.820 ***	0.803 ***	0.943 ***	0.039	−0.011	−0.089
Fungi	0.002	0.851 ***	0.337 *	0.035	0.018	0.520 **
G+ bacteria	0.716 ***	0.727 ***	0.882 ***	0.141	−0.069	−0.100
G− bacteria	0.723 ***	0.612 **	0.866 ***	0.143	−0.029	−0.085
Actinomycete	0.366 *	0.816 ***	0.851 ***	0.378	−0.084	−0.079
Eukaryote	−0.086	0.232	0.376 *	−0.098	0.322 *	0.383 *

3.3. Individual PLFA Analysis

The comparison of individual PLFAs in terms of the whole decomposition period considered revealed a major change in the composition of the microbial communities in the three habitats. We identified more individual PLFAs in the natural forest floor and the intermittent stream over the decomposition of the litter (Figure 3). Among them, the concentrations of individual PLFA indicative of G+ bacteria (i15:0, a15:0) increased steadily with decomposition time in the forest floor and the intermittent stream. The concentrations of PLFA indicative of G− bacteria (18:1 w5c) fluctuated greatly with decomposition time (Figure 3a,b). In contrast, there were only two PLFAs (n19:0, 18:2 w6c) in the headwater stream (Figure 3c).

Under the same decomposition period, we identified more individual PLFAs in the later stage of decomposition (Figure 4). The individual PLFA differed more markedly among the three habitats with the increase of decomposition time. In the early stage of decomposition, the difference of individual PLFA was mainly reflected between the forest floor and the aquatic system (Figure 4a,b). There was insignificant difference between the headwater stream and the intermittent stream. On the contrary, in the later stage of decomposition, the difference of individual PLFA was mainly reflected in the headwater stream and the intermittent stream (Figure 4c,d).

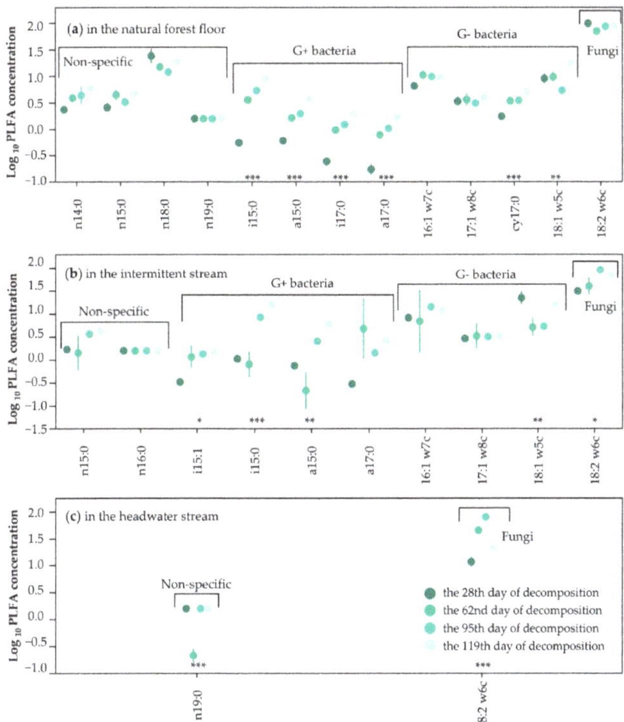

Figure 3. Concentrations of individual PLFA in (**a**) the natural forest floor and (**b**) the intermittent stream and (**c**) the headwater stream leaf litter. Asterisks indicate significant differences among decomposition time: * $p < 0.05$, ** $p < 0.01$, *** $p < 0.001$. Dots indicate the mean and whiskers the standard error.

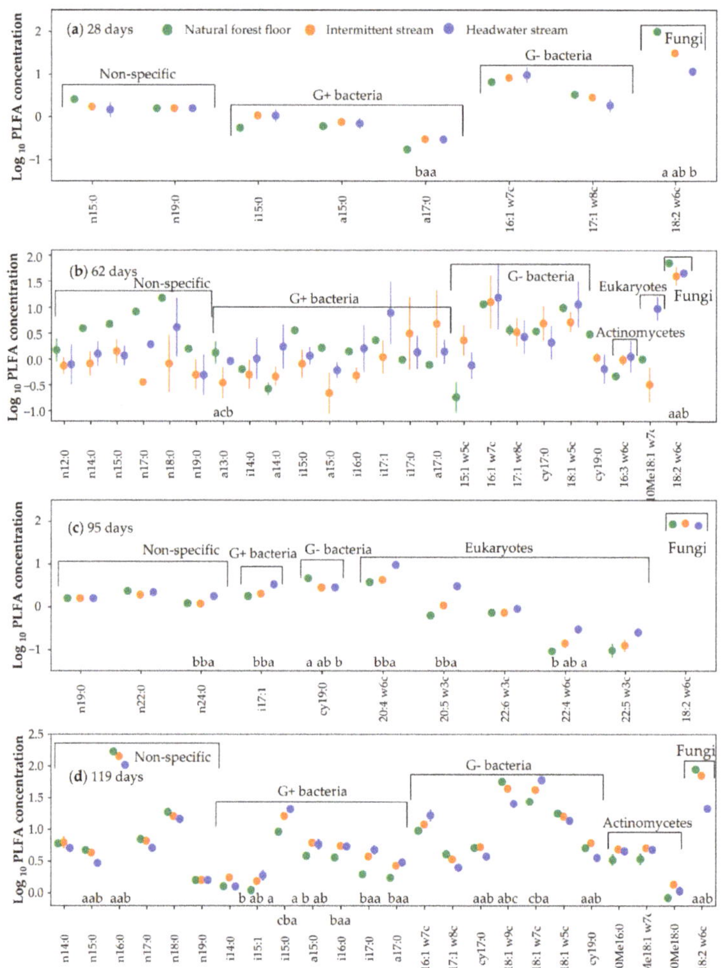

Figure 4. Concentrations of individual PLFA on (**a**) the 28th day of decomposition and (**b**) the 62nd day of decomposition and (**c**) the 95th day of decomposition and (**d**) the 119th day of decomposition leaf litter. Letters indicate significant differences among transect regions ($p < 0.05$). Dots indicate the mean and whiskers the standard error.

3.4. PLFA Biomass and Litter Decomposition

Total PLFA did not correlate with mass loss in three habitats ($p > 0.05$) (Figure 5a). However, the communities of microbial decomposers showed a strong correlation with litter mass loss in three habitats. Among them, the PLFA concentrations of microbial groups and the ratios of biomarkers showed the highest correlation with mass loss in the headwater stream (Figure 5). Specifically, the concentrations of bacterial PLFA, fungal PLFA, G+ bacterial PLFA, G− bacterial PLFA, eukaryotic PLFA, and the G+/G− ratios were related to the loss of litter quality ($p < 0.05$) (Figure 5b,c,e–g,i)). However, only the concentrations of G+ bacterial PLFA, actinomycetes PLFA, and the G+/G− ratios correlated with mass loss in the natural forest floor ($p < 0.001$) (Figure 5d,e,i)). Only the G+/G− ratios correlated with mass loss in the intermittent stream ($p < 0.05$) (Figure 5i).

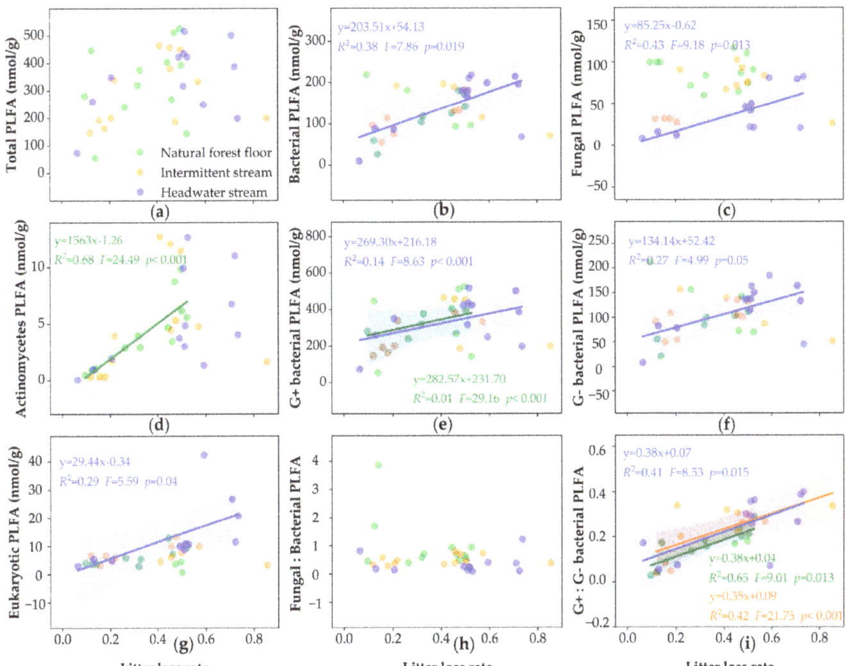

Figure 5. Linear regression for three habitats between litter mass loss and the concentrations of microbial total PLFA (**a**) and PLFA of microbial groups (**b–g**), and the ratios of biomarkers (**h,i**).

4. Discussions

We analyzed how the microbial community changes during litter decomposition in the natural forest floor, intermittent stream, and headwater stream by PLFA analyses. We found that habitats didn't affect total PLFA concentrations but strongly influenced microbial community composition during the litter decomposition. Meanwhile, microbial total PLFA increased significantly with time among habitat types. Some specific microbial groups showed different trends with decomposition time in three habitats. Meanwhile, linear regression analysis between PLFA concentration and different environmental variables in the three habitats showed that some PLFA were significantly affected by environmental factors (Table 2). For example, bacteria were significantly affected by temperature and pH while fungi didn't show the same pattern. In general, fungi are assumed to be less sensitive to changes in temperature and moisture than bacteria due to chitinous cell walls [3]. In addition, microorganisms ultimately led to direct litter mass loss. Some communities of microbial decomposers such as bacteria (G+ and G− bacteria), fungi, and eukaryote showed a strong correlation with litter mass loss in the headwater stream, which verifies our hypothesis. This indicated that the environmental differences among habitats will affect the development of microbial communities, and ultimately alter litter decomposition [28,29].

Table 2. Linear regression for three habitats between environmental factors (Temperature, Rainfall, pH, Flow rate) and the concentrations of microbial PLFAs.

Microbial Composition	Environmental Factor								
	Temperature			Rainfall			pH		
Natural forest floor	R^2	F	p	R^2	F	p	R^2	F	p
Total Biomass	0.16	3.11	>0.05	0.01	1.16	>0.05	-	-	-
Bacteria	0.03	1.29	>0.05	−0.09	0.12	>0.05	-	-	-
Fungi	−0.09	0.11	>0.05	0.10	2.18	>0.05	-	-	-
G+	0.85	64.8	<0.001	0.01	1.05	>0.05	-	-	-
G−	−0.07	0.27	>0.05	−0.10	0.01	>0.05	-	-	-
Actinomycetes	0.84	56.7	<0.001	0.25	4.66	>0.05	-	-	-
Eukaryote	−0.09	0.12	>0.05	−0.10	0.01	>0.05	-	-	-
Intermittent stream	R^2	F	p	R^2	F	p	R^2	F	p
Total Biomass	0.71	27.3	<0.001	0.26	4.71	>0.05	0.44	9.48	<0.05
Bacteria	0.46	10.5	<0.01	0.17	3.18	>0.05	0.14	2.75	>0.05
Fungi	0.39	7.98	<0.05	0.09	2.13	>0.05	−0.10	0.01	>0.05
G+	0.87	73.1	<0.001	0.10	2.58	>0.05	0.57	15.47	<0.01
G−	0.20	3.68	>0.05	0.11	2.34	>0.05	−0.02	0.83	>0.05
Actinomycetes	0.83	53.6	<0.001	0.41	8.75	<0.05	0.28	5.28	<0.05
Eukaryote	0.01	1.03	>0.05	0.06	1.67	>0.05	0.03	1.32	>0.05
Headwater stream	R^2	F	p	R^2	F	p	R^2	F	p
Total Biomass	0.08	0.84	>0.05	0.01	1.02	>0.05	−0.01	0.93	>0.05
Bacteria	0.33	0.51	<0.05	−0.10	0.01	>0.05	−0.13	0.10	>0.05
Fungi	0.06	1.64	>0.05	−0.09	0.12	>0.05	−0.01	1.10	>0.05
G+	0.43	9.43	<0.05	−0.10	0.02	>0.05	0.46	7.69	<0.05
G−	0.18	3.43	>0.05	−0.09	0.08	>0.05	−0.13	0.10	>0.05
Actinomycetes	0.62	19.1	<0.001	0.22	4.08	>0.05	0.32	4.77	>0.05
Eukaryote	0.15	2.93	>0.05	−0.06	0.43	>0.05	−0.04	0.73	>0.05

The sensitivity of microorganisms to the environment is different, which is greatly reflected in our results. First, the total PLFA concentrations in the intermittent stream and the natural forest floor were significantly lower than that in the headwater stream during the 62nd day of decomposition. This may be that we observed two transient droughts in intermittent streams between days 28th and 62nd of decomposition. At this time, the litter was exposed to solar radiation. The disappearance of water means great physiological stress to microbial decomposers and results in marked decline of its biomass [30]. Moreover, some differences of microbial communities in the three habitats are mainly reflected in fungi, G+ bacteria, and eukaryotes. First, the fungal PLFA concentrations in the natural forest floor were significantly higher than those in other habitats. However, fungal PLFA concentrations were little difference between the headwater stream and the intermittent stream during the whole decomposition period. It is generally accepted that fungi are the main decomposers of refractory substances such as lignin [31]. Once entrained in the ecosystems, litter is rapidly colonized by microorganisms [14]. However, under the conditions of headwater streams and intermittent streams during flow, leaf litter was more vulnerable to leaching [32], which may reduce the colonization of fungi acting on lignin. Some research results also showed that water flow is related to the growth and activity of fungi [17]. Hypoxia conditions in streams may limit the respiration of fungal organisms and inhibit mycelial growth [30]. Meanwhile, the research showed that intermittent stream should be dominated by fungal species with traits of higher desiccation resistance [33]. This makes the fungal concentrations of the intermittent stream similar to that of the headwater stream even under drought. Second, the difference in bacterial concentrations among habitats is mainly reflected in G+ bacteria. Some studies associate G+ bacterial PLFA with stress conditions found that G+ bacteria may be dominant in the condition that the available C is relatively less [34]. Thus, the higher G+ bacterial PLFA in the headwater stream and the intermittent stream suggested lower concentrations of available C received by aquatic ecosystems from land [35]. Meanwhile, the difference in ratios of biomarkers among habitats also showed the different flow of energy and nutrients through the microbial community in different ecosystems [36,37].

Total PLFA concentrations increased significantly with time among habitat types. This may be that some microbial groups (including bacteria, G+ bacteria and actinomycete) also increased significantly with decomposition time in the three habitats. Although the PLFA concentrations of fungi and eukaryotes decreased significantly at the end of decomposition, their concentrations also increased on the whole. It is generally accepted that fungal biomass increases exponentially, and it is later stabilized or even decreases [38]. However, bacteria play a predominant role in the later stages of litter decomposition [39]. Our results also verify this statement. In addition, Otaki et al. [40] also found similar results in a three-year litter decomposition study. Their study found that the bacterial biomass increased steadily until the end of the experiment while fungal biomass reached its peak in the first year. In a four-month study (from the late summer to the early winter periods), Wilkinson et al. [41] found that bacterial and fungal biomass increased through time on spruce litter in Germany, which is also consistent with our results.

The analysis of individual PLFAs in terms of the whole decomposition period considered revealed a major change in the composition of the microbial communities in the three habitats. It showed that compared with stream conditions, similar strains acting on litter decomposition were more concentrated on the forest floor and intermittent stream. It has been reported that the decomposition environment of stream litter is being influenced by more complex factors such as water chemistry, temperature, velocity, and turbulence, which may result in less individual PLFAs in the whole decomposition period [42]. Under the same decomposition period, we identified more individual PLFAs in the later stage of decomposition. And in the early stage of decomposition, the number of individual PLFA with significant differences among the three habitats was less. This may be due to the large environmental differences among habitats, which makes different microorganisms act on litter decomposition. What is more, their difference was mainly reflected between the forest floor and the aquatic system. The possible reason was that the environmental difference between the two ecosystems may be mainly reflected in the temperature. Temperature is the main driving factor of microbial activity [42,43]. With the increase of temperature, the water temperature is relatively stable, while the atmospheric temperature fluctuates greatly. This led to the difference in individual PLFA between forest floor and water environment habitat. In contrast, in the later stage of decomposition, the difference in individual PLFA is gradually reflected among the three habitats. It was mainly reflected in the headwater stream and the intermittent stream. During this period, the intermittent stream has been in a dry state for a long time, and the litter is exposed to the soil surface. Temporary stress events, such as high temperatures or desiccation, result in dynamic changes in microbial communities over time [44]. This may explain our results.

Litter decomposition is often thought to be regulated by abiotic factors such as microclimate and chemical quality [45]. However, this view changed considerably with the development of molecular tools. Recent studies showed that soil microbial communities differ substantially over the litter decomposition course and space [46,47]. Spatial variation of a rather basic microbial parameter, such as microbial biomass, can be an important determinant of decomposition [48]. Though we did not find direct significant correlations between total PLFA and litter mass loss in three habitats. However, some communities of microbial decomposers showed a strong correlation with litter mass loss. This suggests that, in addition to abiotic factors and potential nutrient transfer among habitat types, the development of some microbial decomposers also drives the decomposition of litter, which further lead to different litter decomposition rates [49]. Among them, the PLFA concentrations of microbial groups showed the highest correlation with mass loss in the headwater stream. Meanwhile, we found fewer common individual PLFAs in the headwater stream throughout the decomposition. This indicated that the diversity of microorganisms is more abundant in the stream environment. A tremendous diversity of microbes contributes to litter decomposition and interacts in a cross-kingdom functional succession of communities [46]. What is more, linking decomposer identity to decomposition could provide a better understanding of the relationship between microbial community structure and leaf litter decomposition.

5. Conclusions

Our findings suggest that habitat do not affect the total biomass of microorganisms, but microbial community composition is significantly influenced by the habitat micro-environment, which might alter the decomposition rate of litter. Results here implied that the number of individual PLFA in the forest floor and intermittent stream was significantly higher than that in the stream in the whole decomposition period. Compared with the late stage of litter decomposition, the difference of individual PLFA concentrations among the three habitats was reflected in the forest floor and aquatic system in the early stage of litter decomposition, but it was mainly reflected in the headwater stream and the intermittent stream in the late stage of litter decomposition. Interestingly, we linked the relationship between microbial community structure and litter decomposition and found that the microbial community showed the greatest correlation with the litter decomposition rate in streams. In conclusion, our results further clarify the dynamics of microbial communities during the litter decomposition in different habitats.

Author Contributions: Conceptualization, H.G., F.W. and X.Z.; data curation, H.G.; formal analysis, H.G.; funding acquisition, F.W. and D.W.; investigation, H.G., X.Z., W.W., L.Z. and R.W.; visualization, H.G.; writing-original draft, H.G.; writing-review and editing, H.G., F.W. and D.W. All authors have read and agreed to the published version of the manuscript.

Funding: The National Natural Science Foundation of China, grant number 32001965, 32171641 and 32101509 which funded this study.

References

1. Hatton, P.J.; Castanha, C.; Torn, M.S.; Bird, J.A. Litter type control on soil C and N stabilization dynamics in a temperate forest. *Glob. Change Biol.* **2015**, *21*, 1358–1367. [CrossRef]
2. Battin, T.J.; Luyssaert, S.; Kaplan, L.A.; Aufdenkampe, A.K.; Richter, A.; Tranvik, L.J. The boundless carbon cycle. *Nat. Geosci.* **2009**, *2*, 598–600. [CrossRef]
3. Zhao, B.; Xing, P.; Wu, Q.L. Interactions between bacteria and fungi in macrophyte leaf litter decomposition. *Environ. Microbiol.* **2021**, *23*, 1130–1144. [CrossRef]
4. Ramirez, K.S.; Craine, J.M.; Fierer, N. Consistent effects of nitrogen amendments on soil microbial communities and processes across biomes. *Glob. Change Biol.* **2012**, *18*, 1918–1927. [CrossRef]
5. Parton, W.; Silver, W.L.; Burke, I.C.; Grassens, L.; Harmon, M.E.; Currie, W.S.; King, J.Y.; Adair, E.C.; Brandt, L.A.; Hart, S.C.; et al. Global-scale similarities in nitrogen release patterns during long-term decomposition. *Science* **2007**, *315*, 361–364. [CrossRef] [PubMed]
6. Garcia-Palacios, P.; Shaw, E.A.; Wall, D.H.; Hattenschwiler, S. Temporal dynamics of biotic and abiotic drivers of litter decomposition. *Ecol. Lett.* **2016**, *19*, 554–563. [CrossRef] [PubMed]
7. Graca, M.A.S.; Ferreira, V.; Canhoto, C.; Encalada, A.C.; Guerrero-Bolano, F.; Wantzen, K.M.; Boyero, L. A conceptual model of litter breakdown in low order streams. *Int. Rev. Hydrobiol.* **2015**, *100*, 1–12. [CrossRef]
8. Bradford, M.A.; Berg, B.; Maynard, D.S.; Wieder, W.R.; Wood, S.A. Understanding the dominant controls on litter decomposition. *J. Ecol.* **2016**, *104*, 229–238. [CrossRef]
9. Yue, K.; Garcia-Palacios, P.; Parsons, S.A.; Yang, W.; Peng, Y.; Tan, B.; Huang, C.; Wu, F. Assessing the temporal dynamics of aquatic and terrestrial litter decomposition in an alpine forest. *Funct. Ecol.* **2018**, *32*, 2464–2475. [CrossRef]
10. Joly, F.X.; Kurupas, K.L.; Throop, H.L. Pulse frequency and soil-litter mixing alter the control of cumulative precipitation over litter decomposition. *Ecology* **2017**, *98*, 2255–2260. [CrossRef]
11. Bradford, M.A.; Warren, R.J.; Baldrian, P.; Crowther, T.W.; Maynard, D.S.; Oldfield, E.E.; Wieder, W.R.; Wood, S.A.; King, J.R. Climate fails to predict wood decomposition at regional scales. *Nat. Clim. Change* **2014**, *4*, 625–630. [CrossRef]
12. Shumilova, O.; Zak, D.; Datry, T.; von Schiller, D.; Corti, R.; Foulquier, A.; Obrador, B.; Tockner, K.; Allan, D.C.; Altermatt, F.; et al. Simulating rewetting events in intermittent rivers and ephemeral streams: A global analysis of leached nutrients and organic matter. *Glob. Change Biol.* **2019**, *25*, 1591–1611. [CrossRef] [PubMed]
13. Fenoy, E.; Pradhan, A.; Pascoal, C.; Rubio-Rios, J.; Batista, D.; Moyano-Lopez, F.J.; Cassio, F.; Casas, J.J. Elevated temperature may reduce functional but not taxonomic diversity of fungal assemblages on decomposing leaf litter in streams. *Glob. Change Biol.* **2022**, *28*, 115–127. [CrossRef] [PubMed]
14. Kuehn, K.A. Lentic and lotic habitats as templets for fungal communities: Traits, adaptations, and their significance to litter decomposition within freshwater ecosystems. *Fungal Ecol.* **2016**, *19*, 135–154. [CrossRef]
15. Larned, S.T.; Datry, T.; Arscott, D.B.; Tockner, K. Emerging concepts in temporary-river ecology. *Freshw. Biol.* **2010**, *55*, 717–738. [CrossRef]

16. Febria, C.M.; Hosen, J.D.; Crump, B.C.; Palmer, M.A.; Williams, D.D. Microbial responses to changes in flow status in temporary headwater streams: A cross-system comparison. *Front. Microbiol.* **2015**, *6*, 522. [CrossRef] [PubMed]

17. Mora-Gomez, J.; Duarte, S.; Cassio, F.; Pascoal, C.; Romani, A.M. Microbial decomposition is highly sensitive to leaf litter emersion in a permanent temperate stream. *Sci. Total Environ.* **2018**, *621*, 486–496. [CrossRef]

18. Schlief, J.; Mutz, M. Leaf Decay Processes during and after a Supra-Seasonal Hydrological Drought in a Temperate Lowland Stream. *Int. Rev. Hydrobiol.* **2011**, *96*, 633–655. [CrossRef]

19. Willers, C.; van Rensburg, P.J.J.; Claassens, S. Phospholipid fatty acid profiling of microbial communities-a review of interpretations and recent applications. *J. Appl. Microbiol.* **2015**, *119*, 1207–1218. [CrossRef]

20. Zelles, L. Phospholipid fatty acid profiles in selected members of soil microbial communities. *Chemosphere* **1997**, *35*, 275–294. [CrossRef]

21. Drenovsky, R.E.; Elliott, G.N.; Graham, K.J.; Scow, K.M. Comparison of phospholipid fatty acid (PLFA) and total soil fatty acid methyl esters (TSFAME) for characterizing soil microbial communities. *Soil Biol. Biochem.* **2004**, *36*, 1793–1800. [CrossRef]

22. Yu, G.R.; Chen, Z.; Piao, S.L.; Peng, C.H.; Ciais, P.; Wang, Q.F.; Li, X.R.; Zhu, X.J. High carbon dioxide uptake by subtropical forest ecosystems in the East Asian monsoon region. *Proc. Natl. Acad. Sci. USA* **2014**, *111*, 4910–4915. [CrossRef] [PubMed]

23. Bardgett, R.D.; Walker, L.R. Impact of coloniser plant species on the development of decomposer microbial communities following deglaciation. *Soil Biol. Biochem.* **2004**, *36*, 555–559. [CrossRef]

24. Tonin, A.M.; Goncalves, J.F., Jr.; Bambi, P.; Couceiro, S.R.M.; Feitoza, L.A.M.; Fontana, L.E.; Hamada, N.; Hepp, L.U.; Lezan-Kowalczuk, V.G.; Leite, G.F.M.; et al. Plant litter dynamics in the forest-stream interface: Precipitation is a major control across tropical biomes. *Sci. Rep.* **2017**, *7*, 10799. [CrossRef]

25. Zhao, Y.; Wu, F.; Yang, W.; Tan, B.; He, W. Variations in bacterial communities during foliar litter decomposition in the winter and growing seasons in an alpine forest of the eastern Tibetan Plateau. *Can. J. Microbiol.* **2016**, *62*, 35–48. [CrossRef]

26. Zelles, L.; Bai, Q.Y. Fractionation of fatty acids derived from soil lipids by solid phase extraction and their quantitative analysis by GC-MS. *Soil Biol. Biochem.* **1993**, *25*, 495–507. [CrossRef]

27. Otaki, M.; Takeuchi, F.; Tsuyuzaki, S. Changes in microbial community composition in the leaf litter of successional communities after volcanic eruptions of Mount Usu, northern Japan. *J. Mt. Sci.* **2016**, *13*, 1652–1662. [CrossRef]

28. Gessner, M.O.; Swan, C.M.; Dang, C.K.; McKie, B.G.; Bardgett, R.D.; Wall, D.H.; Hattenschwiler, S. Diversity meets decomposition. *Trends Ecol. Evol.* **2010**, *25*, 372–380. [CrossRef]

29. Kardol, P.; Cregger, M.A.; Campany, C.E.; Classen, A.T. Soil ecosystem functioning under climate change: Plant species and community effects. *Ecology* **2010**, *91*, 767–781. [CrossRef]

30. Foulquier, A.; Artigas, J.; Pesce, S.; Datry, T. Drying responses of microbial litter decomposition and associated fungal and bacterial communities are not affected by emersion frequency. *Freshw. Sci.* **2015**, *34*, 1233–1244. [CrossRef]

31. Arnstadt, T.; Hoppe, B.; Kahl, T.; Kellner, H.; Krüger, D.; Bauhus, J.; Hofrichter, M. Dynamics of fungal community composition, decomposition and resulting deadwood properties in logs of Fagus sylvatica, Picea abies and Pinus sylvestris. *For. Ecol. Manag.* **2016**, *382*, 129–142. [CrossRef]

32. Marks, J.C. Revisiting the Fates of Dead Leaves That Fall into Streams. In *Annual Review of Ecology, Evolution, and Systematics*; Futuyma, D.J., Ed.; Annual Reviews: Palo Alto, CA, USA, 2019; Volume 50, pp. 547–568.

33. Shearer, C.A.; Descals, E.; Kohlmeyer, B.; Kohlmeyer, J.; Marvanova, L.; Padgett, D.; Porter, D.; Raja, H.A.; Schmit, J.P.; Thorton, H.A.; et al. Fungal biodiversity in aquatic habitats. *Biodivers. Conserv.* **2007**, *16*, 49–67. [CrossRef]

34. Bird, J.A.; Herman, D.J.; Firestone, M.K. Rhizosphere priming of soil organic matter by bacterial groups in a grassland soil. *Soil Biol. Biochem.* **2011**, *43*, 718–725. [CrossRef]

35. Cole, J.J.; Prairie, Y.T.; Caraco, N.F.; McDowell, W.H.; Tranvik, L.J.; Striegl, R.G.; Duarte, C.M.; Kortelainen, P.; Downing, J.A.; Middelburg, J.J.; et al. Plumbing the global carbon cycle: Integrating inland waters into the terrestrial carbon budget. *Ecosystems* **2007**, *10*, 171–184. [CrossRef]

36. Hogberg, M.N.; Hogbom, L.; Kleja, D.B. Soil microbial community indices as predictors of soil solution chemistry and N leaching in *Picea abies* (L.) Karst. forests in S. Sweden. *Plant Soil* **2013**, *372*, 507–522. [CrossRef]

37. Bragazza, L.; Bardgett, R.D.; Mitchell, E.; Buttler, A. Linking soil microbial communities to vascular plant abundance along a climate gradient. *New Phytol.* **2015**, *205*, 1175–1182. [CrossRef]

38. Artigas, J.; Gaudes, A.; Munoz, I.; Romani, A.M.; Sabater, S. Fungal and Bacterial Colonization of Submerged Leaf Litter in a Mediterranean Stream. *Int. Rev. Hydrobiol.* **2011**, *96*, 221–234. [CrossRef]

39. Duarte, S.; Pascoal, C.; Alves, A.; Correia, A.; Cassio, F. Assessing the dynamic of microbial communities during leaf decomposition in a low-order stream by microscopic and molecular techniques. *Microbiol. Res.* **2010**, *165*, 351–362. [CrossRef]

40. Otaki, M.; Tsuyuzaki, S. Succession of litter-decomposing microbial organisms in deciduous birch and oak forests, northern Japan. *Acta Oecologica* **2019**, *101*, 103485. [CrossRef]

41. Wilkinson, S.C.; Anderson, J.M.; Scardelis, S.P.; Tisiafouli, M.; Taylor, A.; Wolters, V. PLFA profiles of microbial communities in decomposing conifer litters subject to moisture stress. *Soil Biol. Biochem.* **2002**, *34*, 189–200. [CrossRef]

42. Brown, J.H.; Gillooly, J.F.; Allen, A.P.; Savage, V.M.; West, G.B. Toward a metabolic theory of ecology. *Ecology* **2004**, *85*, 1771–1789. [CrossRef]

43. Chen, Y.M.; Gao, J.T.; Xiong, D.C.; Yuan, P.; Zeng, X.M.; Xie, J.S.; Yang, Y.S. Effects of soil warming on soil microbial community structure and soil available nitrogen in subtropical young Chinese fir plantation. *J. Subtropic. Resour. Environ.* **2016**, *11*, 1–8.

44. Penuelas, J.; Rico, L.; Ogaya, R.; Jump, A.S.; Terradas, J. Summer season and long-term drought increase the richness of bacteria and fungi in the foliar phyllosphere of Quercus ilex in a mixed Mediterranean forest. *Plant Biol.* **2012**, *14*, 565–575. [CrossRef] [PubMed]
45. Huang, J.X.; Huang, L.M.; Lin, Z.C.; Cheng, G.S. Controlling factors of litter decomposition rate in China's forests. *J. Subtropic. Resour. Environ.* **2010**, *5*, 56–63.
46. Herzog, C.; Hartmann, M.; Frey, B.; Stierli, B.; Rumpel, C.; Buchmann, N.; Brunner, I. Microbial succession on decomposing root litter in a drought-prone Scots pine forest. *ISME J.* **2019**, *13*, 2346–2362. [CrossRef]
47. Baldrian, P. Forest microbiome: Diversity, complexity and dynamics. *Fems Microbiol. Rev.* **2017**, *41*, 109–130. [CrossRef]
48. Bradford, M.A.; Veen, G.F.; Bonis, A.; Bradford, E.M.; Classen, A.T.; Cornelissen, J.H.C.; Crowther, T.W.; De Long, J.R.; Freschet, G.T.; Kardol, P.; et al. A test of the hierarchical model of litter decomposition. *Nat. Ecol. Evol.* **2017**, *1*, 1836. [CrossRef]
49. Zhang, X.; Wang, L.; Zhou, W.; Hu, W.; Hu, J.; Hu, M. Changes in litter traits induced by vegetation restoration accelerate litter decomposition in Robinia pseudoacacia plantations. *Land Degrad. Dev.* **2022**, *33*, 179–192. [CrossRef]

Article

Alpine Litter Humification and Its Response to Reduced Snow Cover: Can More Carbon Be Sequestered in Soils?

Dingyi Wang, Xiangyin Ni, Hongrong Guo and Wenyuan Dai *

Key Laboratory for Humid Subtropical Eco-Geographical Processes of the Ministry of Education, School of Geographical Sciences, Fujian Normal University, Fuzhou 350007, China; albertwdy@fjnu.edu.cn (D.W.); nixy@fjnu.edu.cn (X.N.); QSX20201062@student.fjnu.edu.cn (H.G.)
* Correspondence: dwy_geo@163.com

Abstract: While carbon loss from plant litter is well understood, the mechanisms by which this carbon is sequestered in the decomposing litter substrate remains unclear. Here we assessed humus accumulations in five foliar litters during four years of decomposition and their responses to reduced snow cover in an alpine forest. In contrast to the traditional understanding (i.e., the three-stage model), we found that fresh litter had a high humus content (8–13% across species), which consistently increased during litter decomposition and such an increase primarily depended on the accumulation of humic acid. Further, reduced snow cover decreased humus accumulation at early stages but increased it at late stages. These results suggested that humification simultaneously occurred with decomposition during early litter decay, but this process was more sensitive to the changing climate in seasonally snow-covered ecosystems, as previously expected.

Keywords: litter humification; humus accumulation; reduced snow cover; alpine forest

Citation: Wang, D.; Ni, X.; Guo, H.; Dai, W. Alpine Litter Humification and Its Response to Reduced Snow Cover: Can More Carbon Be Sequestered in Soils?. *Forests* **2022**, *13*, 897. https://doi.org/10.3390/f13060897

Academic Editor: Bartosz Adamczyk

Received: 24 April 2022
Accepted: 6 June 2022
Published: 9 June 2022

Publisher's Note: MDPI stays neutral with regard to jurisdictional claims in published maps and institutional affiliations.

1. Introduction

Decomposition and humification are two continual processes that determine the fate of plant litter [1] but remain largely uncoupled in current studies [2]. In recent decades, litter decomposition has become increasingly well understood [3,4]. However, the humified and not the decomposed components are more important for the formation and stabilization of soil organic matter [5], but the humification process is poorly understood [6]. A guiding paradigm suggests that litter decomposes to a "limit value" and recalcitrant residues that cannot be decomposed are polymerized into humus. Thus, humus is considered to accumulate at a very late stage of litter decomposition [4]. However, recent evidence challenges this traditional understanding [7] and shows that considerable litter carbon is input into mineral soil during early litter decay [8,9]. The humic acid (HA) and fulvic acid (FA) fractions from the alkali extraction method suggested by the International Humic Substances Society still provide historical comparative data and accurately represent natural organic matter across multiple environments, source materials and research objectives [10]. These isotope-labeling studies have greatly advanced our knowledge of carbon flux from plant litter to soils [11,12]; however, the mechanisms by which humus accumulates in the decomposing litter substrate remain unknown. Clearly, the potential mechanisms that govern the decomposition and humification of plant litter and control SOM formation must be determined [13], and the feedback of these fundamental processes must be incorporated into ongoing research on climate change.

Snow covers half of the land area in the Northern Hemisphere [14] and acts as a thermal insulator that modulates key ecological processes in many cold ecosystems, including Asian and American Arctic tundra ecosystems [15,16]. However, seasonal snow cover has decreased by 7% and is projected to decrease by 25% by the end of this century [17]. Such a decline in snow cover was demonstrated to slow the carbon flux from aboveground plants to soil via litter decomposition [18,19]. In fact, the carbon turnover in cold biomes covered by seasonal snow is inherently slow due to the limitations of low temperature [20,21]. Thus, reduced snow cover may exert a profound influence on litter humification, which sequesters carbon in the decomposing litter, rather than on litter decomposition, which allows litter to decay in cold forest ecosystems, as shown by our preliminary results [22]. Recent evidence has reinforced this hypothesis. For example, the release of labile components, which have been reported as major precursors for SOM formation [13], is decreased by reduced snow cover. This decreased energy supply from labile carbon [23] and increased soil freezing [24] induced by a reduction in snow cover may limit the contribution of soil microorganisms [25,26] to SOM formation and stabilization, as suggested by an increasing number of theories [9,11,27]. This effect of reduction in winter snow cover may impact long-term "recalcitrant" pools of soil carbon (e.g., SOM) in addition to short-term "labile" pools of soil carbon [28] (e.g., decomposing plant litter) in these cold biomes; however, our current understanding is unclear.

Here, we address two questions: (1) Is humus formed at a very late stage of litter decomposition? (2) Will reduced snow cover alter this carbon sequestration process in high-altitudinal ecosystems? Our overall objective was to explore how plant litter is decomposed and/or humified after leaf fall. As a preliminary test, we conducted an in situ litterbag experiment in an alpine forest on the eastern Tibetan Plateau in 2012 to explore the chemical changes [29], elemental dynamics [30], humus accumulation [31] and microbial community [32] during the decomposition and humification of six foliar litters and one twig litter, which were used to decrease the uncertainty among litter species and types. Seasonal snow cover and associated soil freezing and thawing modulate certain key biogeochemical cycles [33]; therefore, we also evaluated the effects of the reduction in snow cover. Winter litter decomposition is a significant ecological process in this alpine forest [34]; thus, samplings were scheduled at the end of snow formation, coverage and melt stages during winter and at the end of the growing season (Table 1), and the focus was on wintertime to explore the temporal effects of the reduction in snow cover [22]. In this paper, we present data on the humus content (alkali-extractable substances, including humic acid and fulvic acid) in three tree and two shrub foliar litters with different initial qualities (Table 2) over four years (from 2012 to 2016) and its response to the reduction in snow cover. Our results should provide a possible consequence of long-term SOM formation with reduced winter snow cover under a changing climate.

Table 1. Sampling compared with the timings and lengths of winters during the four years of decomposition. Hourly temperature with a 50% probability (continual 12 h per day) of freezing (below 0 °C) is defined as the beginning and end of winter in this study (modified from Ladwig et al. [35]). This definition is not conservative because solar radiation is very strong in this alpine forest and increases in temperature were recorded using a data logger. Sampling dates were scheduled approximately in late April and October within one-week intervals referring to the timing of the 2012/2013 winter.

	Timing of Winter		Length of Winter (Days)	Sampling Date	
Winter	Beginning	End		End of Growing Season	End of Snowmelt
2012/2013 winter	30 October 2012	24 April 2013	176	Experiment began	24 April 2013
2013/2014 winter	31 October 2013	30 April 2014	181	30 October 2013	24 April 2014
2014/2015 winter	29 October 2014	26 April 2015	179	29 October 2014	23 April 2015
2015/2016 winter	30 October 2015	26 April 2016	179	2 November 2015	24 April 2016

Table 2. Initial chemical compounds in the five foliar litters used in our long-term decomposition experiment. Data from Ni et al. [31,36]. Values are means and standard deviations in parentheses ($n = 3$). Different superscript letters in the same column represent significant ($p < 0.05$) differences between litter species. C: carbon, N: nitrogen, P: phosphorus, WSS: water-soluble substances, OSS: organic-soluble substances, ASS: acid-soluble substances, AUR: acid-unhydrolyzable residues.

Litter	C (%)	N (%)	P (%)	WSS (%)	OSS (%)	ASS (%)	AUR (%)	C/N Ratio	AUR/N Ratio
Cypress	51.64 (1.77) [a]	0.88 (0.010) [c]	0.12 (0.006) [ab]	35.74 (0.69) [c]	33.16 (3.43) [a]	32.43 (1.29) [a]	20.60 (3.41) [b]	58.86 (2.21) [b]	23.48 (3.89) [b]
Larch	54.35 (0.63) [a]	0.86 (0.041) [c]	0.13 (0.002) [a]	40.08 (1.08) [b]	19.11 (0.68) [c]	29.24 (0.87) [ab]	21.46 (0.94) [b]	63.32 (3.49) [b]	25.01 (2.04) [b]
Birch	49.69 (1.45) [ab]	1.33 (0.022) [a]	0.09 (0.004) [c]	25.06 (1.96) [d]	11.43 (0.75) [d]	27.74 (0.94) [b]	50.96 (0.96) [a]	37.24 (1.35) [c]	38.19 (1.01) [a]
Willow	45.23 (1.65) [b]	1.15 (0.028) [b]	0.11 (0.002) [b]	41.71 (0.32) [ab]	18.48 (1.57) [c]	28.56 (1.88) [b]	26.15 (3.29) [b]	39.49 (2.18) [c]	22.79 (2.45) [b]
Azalea	50.29 (1.60) [a]	0.67 (0.020) [d]	0.11 (0.009) [bc]	43.14 (1.16) [a]	25.84 (2.29) [b]	27.00 (0.59) [b]	21.84 (3.42) [b]	75.54 (4.47) [a]	32.90 (6.13) [ab]

2. Materials and Methods

2.1. Experimental Site

This study was conducted at the Long-Term Research Station of Alpine Forest Ecosystems ($31°14'$ N, $102°53'$ E, approximately 3600 m a.s.l.). Our experimental sites are located in an alpine forest on the eastern Tibetan Plateau [22]. The annual mean temperature and precipitation are 2.7 °C and 850 mm, respectively. Seasonal snow cover develops in late October and melts in the following April, with a maximum snow depth of approximately 50 cm and a soil freezing time of approximately 120 days [34]. This alpine forest is dominated by coniferous cypress (*Sabina saltuaria* (Rehder & E.H.Wilson) W.C.Cheng & W.T.Wang) and deciduous larch (*Larix mastersiana* Rehd. et Wils.) and birch (*Betula albosinensis* Burkill). The main shrubs are dwarf willow (*Salix paraplesia* Schneid.) and azalea (*Rhododendron lapponicum* (L.) Wahl.) and there are also some sedge (*Carex* sp.) and moss at the site. Canopy opening induced by natural tree fall and other climatic extremes covers 13–23% of the landscape. Decomposition of plant detritus is limited by the low temperature; thus, soils are classified as dark brown soil. The depth of the forest floor (Oi + Oe + Oa) is 7 cm with a carbon stock of 1.6 t C ha^{-1}. Carbon, nitrogen, phosphorus and humus contents (the sum of alkali extracted organic matter) in the organic horizon are 16, 0.58, 0.17 and 6.1%, respectively [36].

2.2. Snow Manipulation

Three long-term sites (as three replicates; 3579–3582 m a.s.l., 500–2000 m apart) were established in this alpine forest in 2009 with similar topographies and canopy covers so that similar snowfall and litter fall were received. Each site consisted of four plots: two plots located in an open canopy to simulate full snow cover (control) and two plots located in a closed canopy to simulate reduced snow cover [22] ($n = 6$). This experimental design was used to decrease the risk of needle litter (fir, cypress and larch) escaping from litterbags, which must have sufficiently large mesh sizes to permit soil fauna access [23]. Each plot had five subplots (3 m × 3 m in size and 3–4 m apart) for incubating the five dominant foliar litters with different initial qualities (Table 2). The snow depth was manually measured in triplicate in each control and reduced snow cover plot ($n = 18$) at each sampling date. The temperatures at the litter surface were recorded using data loggers (iButton DS1923-F5, Sunnyvale, CA, USA), which were placed in marked litterbags in each plot ($n = 6$). A freeze–thaw cycle was defined as a transition above or below 0 °C for at least 3 h and then a transition back according to hourly temperature data [37].

2.3. Litterbag Experiment

Five dominant foliar fresh plant litters in this alpine forest (cypress and larch needles and birch, willow and azalea leaves) were studied in our long-term decomposition experiment using an in situ litterbag technique [38]. In the fall of 2012, senesced foliage from each litter fall species was collected on tarpaulins from twenty or more trees or shrubs by shaking their limbs at the experimental sites. Green or partly decomposed needles (leaves), twigs and bark were removed, and only newly shed foliar litter was air dried at room temperature for two weeks. The litterfall samples (10 ± 0.05 g) were placed in nylon litterbags (20 cm × 25 cm in size, with mesh sizes of 1.0 mm on the top and 0.5 mm on the bottom), and a total of approximately 8600 litterbags (1720 samples per specie) were carefully transferred to the corresponding subplots on 15 and 16 November 2012, for long-term litter decomposition/humification monitoring in this alpine forest. Litterbags in the same subplots were strung together and placed 2–5 cm apart. After all the litterbags were established on the soil surface, five litterbags of each litter species were randomly collected and returned to the laboratory to determine the losses during establishment and the water content of the air-dried foliar litter.

Samplings were scheduled at the snow formation stage (15 November 2012, 30 October 2013, 29 October 2014, 2 November 2015 and 30 October 2016), snow cover stage (26 December 2012 and 23 December 2013) and snow melt stage (26 December 2012 and 23 December 2013) during winter, as well as during the growing season (24 April 2013, 24 April 2014, 23 April 2015 and 24 April 2016), during decomposition, with a focus on wintertime [34] based on experimental procedures in other cold biomes [19,32] and our temperature data in this alpine forest (Figure 1). The beginning and end of winter were defined as hourly temperatures with a 50% probability (continual over 12 h per day) of freezing [35] (below 0 °C; Table 1). Based on the snow dynamics and changes in temperature, winters were further divided into three stages: snow formation, coverage and melt stages (Figure 1b). In this study, we sampled at the end of snow formation, coverage and melt stages in winter and at the end of the growing season (beginning of snowfall) from 2012 to 2016 to explore the temporal effect of snow reduction on litter humification. We sampled only twice per year (at the end of the snow melt stage and growing season) beginning in winter 2014–2015. All sampling dates are presented in Figure 1. At each sampling date above, two litterbags per subplot were randomly collected carefully. The 120 litter bags of each species were carefully removed from snow, leaves and twigs on the surface of the bags then placed in separate plastic bags and transported to the laboratory.

2.4. Chemical Analysis

Once the samples of five species were transferred to the laboratory, all litter samples were carefully removed from the bags, in which visible roots, mosses and soils were completely removed from the litter samples. One cleaned subsample was oven dried at 105 °C (for more than 48 h) to measure the dry mass and water content. The remaining mass was evaluated based on the dry mass as presented elsewhere. Another subsample was used to determine the total humus, humic acid and fulvic acid contents in the foliar litter.

We used the alkali method to extract the humus in the foliar litter. Although this method was recently questioned [7], alkali extraction provides a quantitative measurement of the magnitude of humus accumulation in decomposing litter. Specifically, a 1.00 g air-dried subsample was placed in a 150 mL bottle and extracted with a 100 mL mixed solution of 0.1 M NaOH and $Na_4P_2O_7$ at 80 °C for 1 h after shaking for 30 min. The dissolved solution was filtered and stored under anaerobic conditions as the total extractive humus.

Humic acid and fulvic acid were isolated using 0.05 M HCl to pH 2.0 and kept at 80 °C for 30 min. Humic acid (not soluble in acid but soluble in alkali) was separated out as floccules and filtered and re-dissolved with 0.05 M NaOH. The isolated humic acid and total extractive humus were filtered through 0.45 μm meshes, and 200 μL solutions were determined using a TOC analyzer (multi N/C 2100, Analytik Jena, Thüringen, Germany).

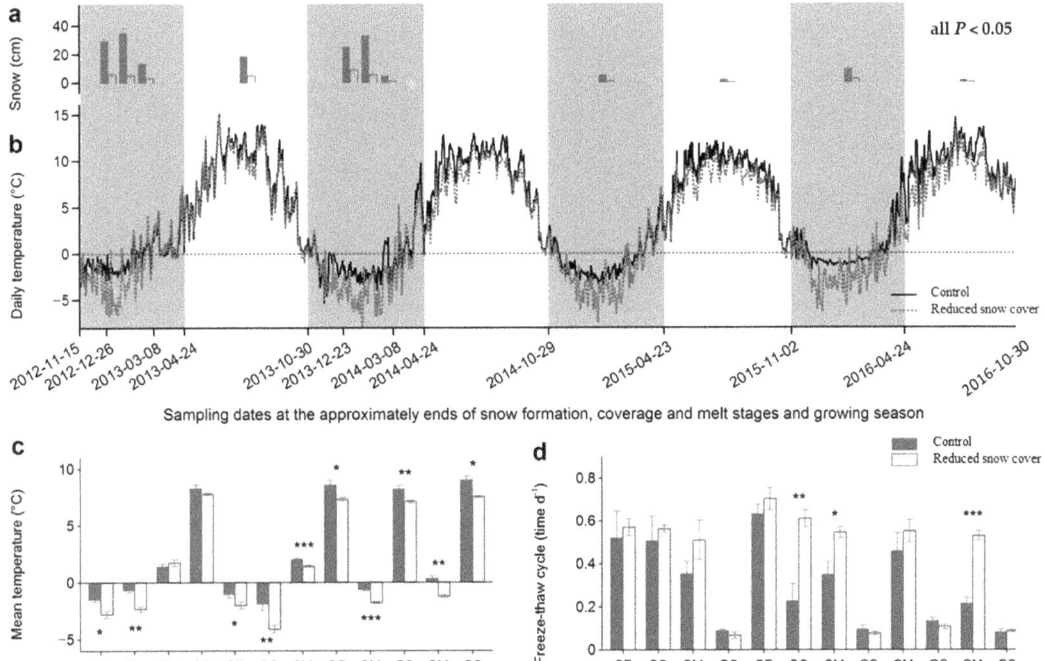

Figure 1. Temperatures and freeze–thaw cycles in the control and reduced snow cover plots. (**a**) Snow depths ($\pm$SE, $n = 18$) on each sampling date. Differences between control and reduced snow cover plots were significant ($p < 0.05$) on all sampling dates. (**b**) Daily temperatures ($n = 6$) at the litter surface during the four years of decomposition. Sampling dates were scheduled at approximately the ends of the snow formation, coverage and melt stages, and the growing season. Winter times, with the snow formation, coverage and melt stages, are shaded. The dashed line is drawn at a daily temperature of zero. (**c**) Mean temperatures ($\pm$SE, $n = 6$) at individual stages. Times between two adjacent sampling dates were defined as separate stages to better understand the effect of snow reduction at the snow formation, coverage and melt stages. SF: snow form stage, SC: snow cover stage, SM: snowmelt stage, GS: growing season. (**d**) Freeze–thaw cycles ($\pm$SE, $n = 6$) at the individual stages. Values were calculated using the quotients of the total number of freeze–thaw cycles and the days of certain stages. * $p < 0.05$, ** $p < 0.01$, *** $p < 0.001$.

The humus content (including humic acid and fulvic acid) was defined as alkali-extractable substances and calculated as a percentage of the litter mass. The fulvic acid content (FA, % of litter mass) represented the difference between the total extractive humus and humic acid.

$$FA\ (\%) = Hu - HA \tag{1}$$

where Hu and HA are the contents of the total extractive humus and humic acid, respectively.

A 0.5 g air-dried subsample was oven dried at 105 °C for 48 h to determine the litter moisture, which was used to adjust the final value of the humus content.

2.5. Data Analysis

We first examined the effect of snow reduction and its variation over time among the litter species using a three-way analysis of variance (ANOVA). Differences between the control and reduced snow cover plots were assessed via paired t-tests or Wilcoxon signed-rank tests in MATLAB R2012a (MathWorks Inc., Natick, MA, USA) if the sample sizes were unequal. Then, we compared the relationships between the remaining litter mass and the contents of the total humus, humic acid and fulvic acid. We also monitored certain approximate chemical compounds and mineral elements in this long-term experiment and used a partial least squares analysis [39] in SIMCA 14.0 (MKS, Umeå, Sweden) to distinguish the importance of the factors that may impact the humus content. The relative influence of each independent explanatory variable (chemical compounds and mineral elements) were estimated using the variable importance of projection (VIP) and values greater than 1 were considered relevant and significant for explaining the humus content [40].

3. Results

3.1. Humus

The initial humus content (alkali-extractable substances) was 8–13% in the foliar litters and increased to 16–21% across all species in the control plots after four years of humification (Figure 2). The humus content was stable but greatly increased beginning in the second growing season. Snow manipulation did not significantly impact the humus content ($F = 0.02$, $p = 0.90$; Table 3), although its interactions with both sampling time ($F = 2.2$, $p = 0.015$) and litter species ($F = 2.9$, $p = 0.022$) were significant. During the first three years of humification, lower humus contents ($p < 0.05$) were observed in the reduced snow cover plots for larch, birch, willow and azalea litters, except at the second snow cover stage for willow litter (Figure 2d). In the fourth year, higher humus contents ($p < 0.05$) were observed in the reduced snow cover plots, except at the fourth snowmelt stage for larch litter (Figure 2b). At the end of this experiment, higher values ($p < 0.05$) were observed in the reduced snow cover plots for larch, birch and willow litter (Figure 2b–d). When all litter species were combined, the humus content significantly increased ($R^2 = 0.40$, $p < 0.001$; Figure 3a) with decreases in the remaining litter mass. Manganese (Mn) was the most significant factor that impacted the humus content, and the effect of acid-unhydrolyzable residue (AUR) was greater than that of labile carbon, although neither was significant and presented VIP values below 1 (Figure 4).

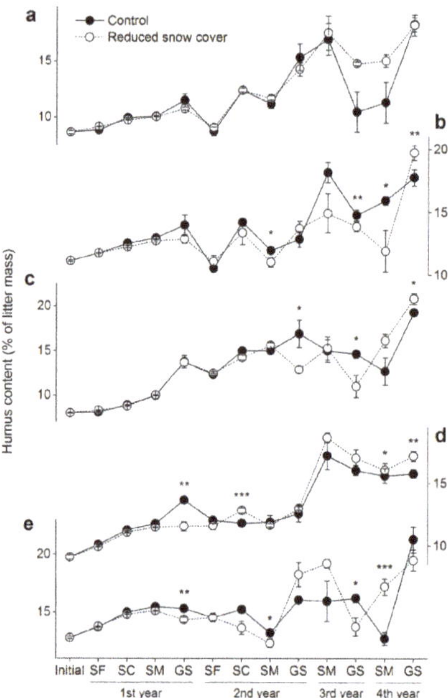

Figure 2. Humus contents ($\pm$SE, $n = 6$) in the five foliar litters. (**a**) Cypress, (**b**) larch, (**c**) birch, (**d**) willow and (**e**) azalea. SF: snow formation stage, SC: snow cover stage, SM: snowmelt stage, GS: growing season. * $p < 0.05$, ** $p < 0.01$, *** $p < 0.001$.

Table 3. Results of three-way ANOVA testing for the effects of time, litter species and snow manipulation. Bold values are significant ($p < 0.05$).

Source of Variation	df	Humus		Humic Acid		Fulvic Acid	
		F Value	p-Value	F Value	p-Value	F Value	p-Value
Time	11	112.0	<0.001	300.7	<0.001	63.1	<0.001
Litter	4	59.6	<0.001	50.0	<0.001	22.6	<0.001
Snow	1	0.02	0.90	6.0	0.015	2.1	0.14
Time × Litter	44	7.5	<0.001	15.9	<0.001	6.4	<0.001
Time × Snow	11	2.2	0.015	5.8	<0.001	3.1	<0.001
Litter × Snow	4	2.9	0.022	2.0	0.097	5.3	<0.001

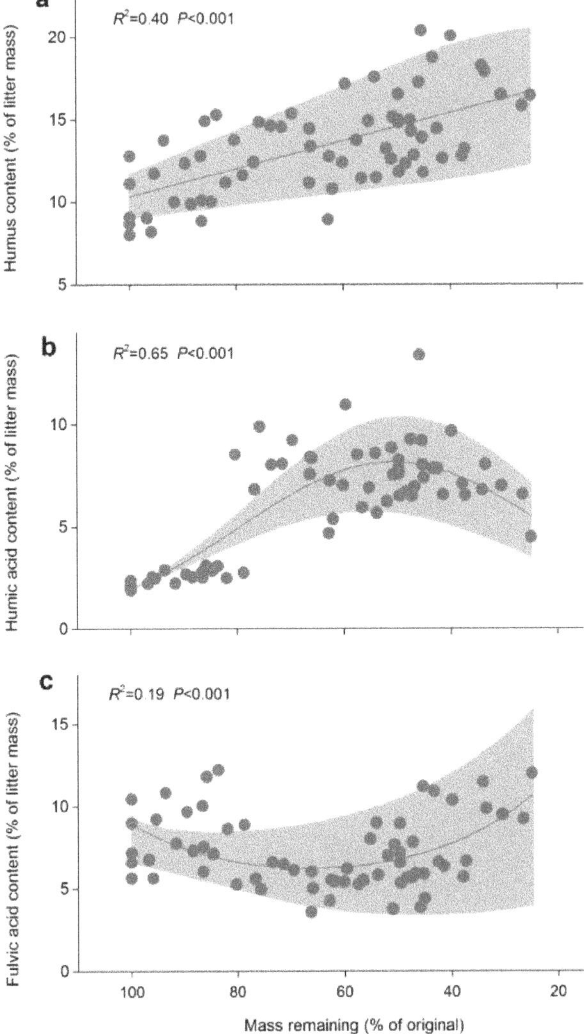

Figure 3. Relationships between the remaining mass and humus contents (**a**), humuc acid contents (**b**) and fulvic acid contents (**c**). Shaded areas are 95% confidence intervals. Adjusted R^2 and p-values from linear or exponential regressions are shown in each panel (all $n = 78$).

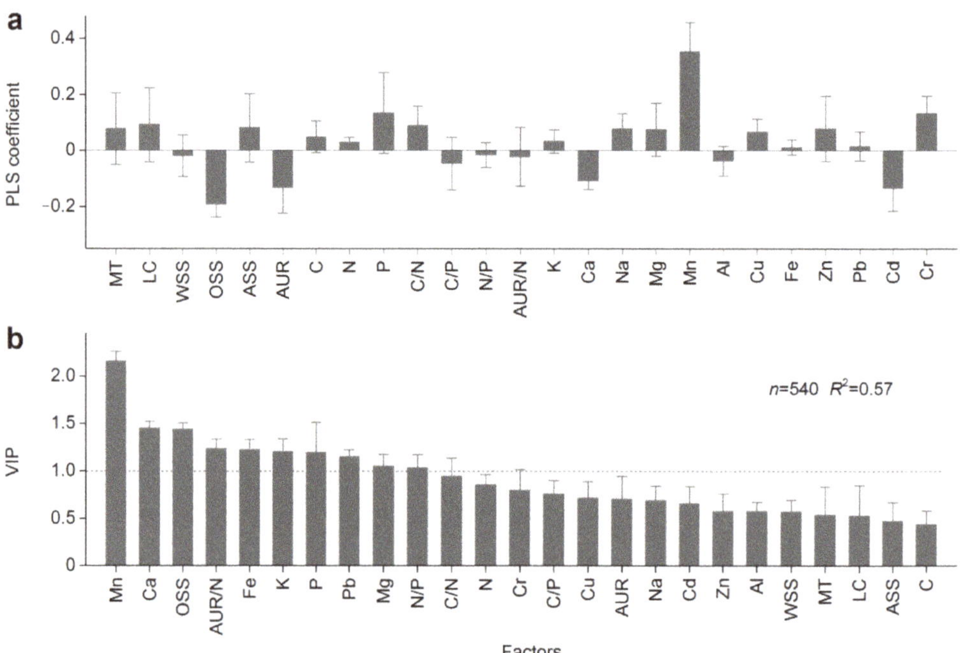

Figure 4. Partial least squares (PLS) analysis results for testing factors that impacted the humus content. (**a**) PLS coefficient. (**b**) Variable importance of projection (VIP). Error bars represent standard errors. VIP values greater than 1 (above the dashed line) were significant. MT: mean temperature, LC: labile carbon, WSS: water-soluble substances, OSS: organic-soluble substances, ASS: acid-soluble substances, AUR: acid-unhydrolyzable residue, C: carbon: N: nitrogen, P: phosphorus, K: potassium, Ca: calcium, Na: sodium, Mg: magnesium, Mn: manganese, Al: aluminum, Cu: copper, Fe: iron, Zn: zinc, Pb: lead, Cd: cadmium, Cr: chromium. All these factors were monitored in our long-term experiment.

3.2. Humic Acid

The initial humus content (alkali-extractable substances) was 1.9–2.4% in the foliar litter and increased to 4.3–8.9% across all species in the control plots after four years of humification (Figure 5). The humic acid content was stable but greatly increased in the first growing season. Snow manipulation significantly ($F = 6.0$, $p = 0.015$; Table 3) impacted the humic acid content. During four years of humification, lower humic acid contents ($p < 0.05$) were observed in the reduced snow cover plots for cypress (Figure 5a), willow (Figure 5d) and azalea (Figure 5e) litter, as well as in the control plots for larch (Figure 5b) and birch (Figure 5c) litter. At the end of this experiment, higher humic acid content was observed in the reduced snow cover plots for both larch and birch litters (both $p < 0.01$). When all litter species were combined, the humic acid content greatly increased as the remaining mass decreased until it reached 50–60% ($R^2 = 0.65$, $p < 0.001$; Figure 3b).

3.3. Fulvic Acid

The initial fulvic acid contents were 5.6–10.4% in the foliar litter and increased to 10.6–12.4% across all species in the control plots after four years of humification (Figure 6). The fulvic acid content was stable during the first winter but greatly decreased during the first growing season and then increased in the third year. Snow manipulation did not significantly impact the fulvic acid content ($F = 2.1$, $p = 0.14$; Table 3), although its interactions with both sampling time ($F = 3.1$, $p < 0.001$) and litter species ($F = 5.3$, $p < 0.001$)

were significant. At the end of this experiment, higher fulvic acid contents were observed in the reduced snow cover plot for willow litter ($p < 0.01$; Figure 6d), whereas lower values were observed in the reduced snow cover plot for azalea litter ($p < 0.05$; Figure 6e). When all litter species were combined, the fulvic acid content decreased when the remaining litter mass was approximately >70% but then greatly increased ($R^2 = 0.19$, $p < 0.001$; Figure 3c).

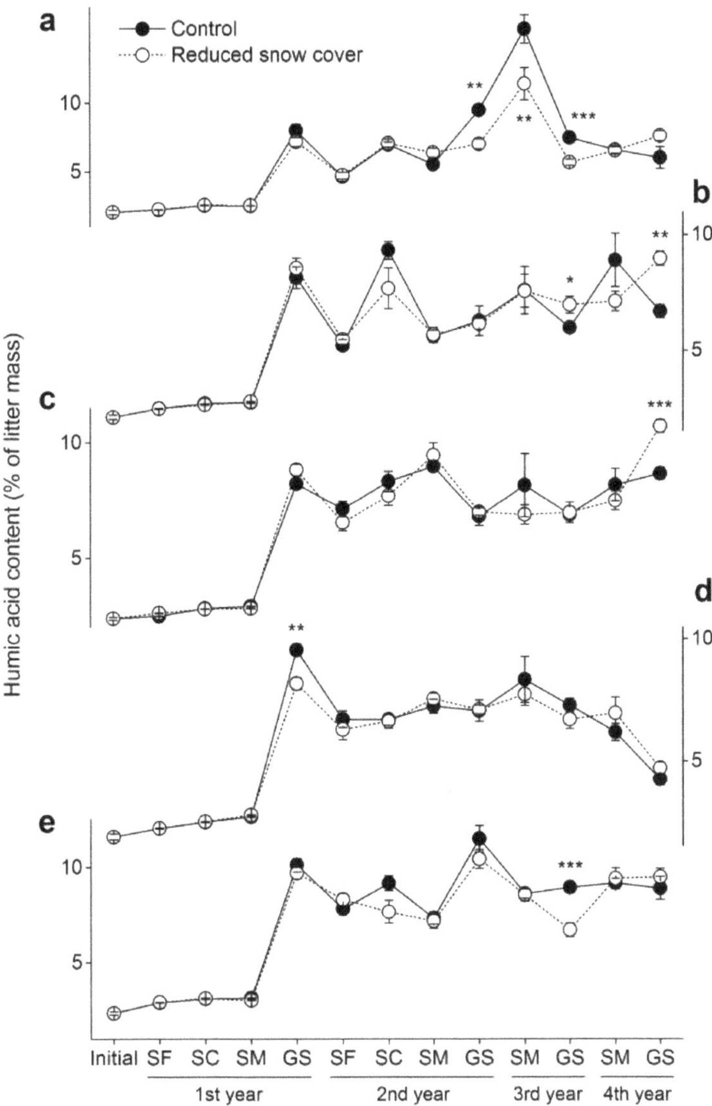

Figure 5. Humic acid contents ($\pm$SE, $n = 6$) in the five foliar litters. (**a**) Cypress, (**b**) larch, (**c**) birch, (**d**) willow and (**e**) azalea. SF: snow formation stage, SC: snow cover stage, SM: snowmelt stage, GS: growing season. * $p < 0.05$, ** $p < 0.01$, *** $p < 0.001$.

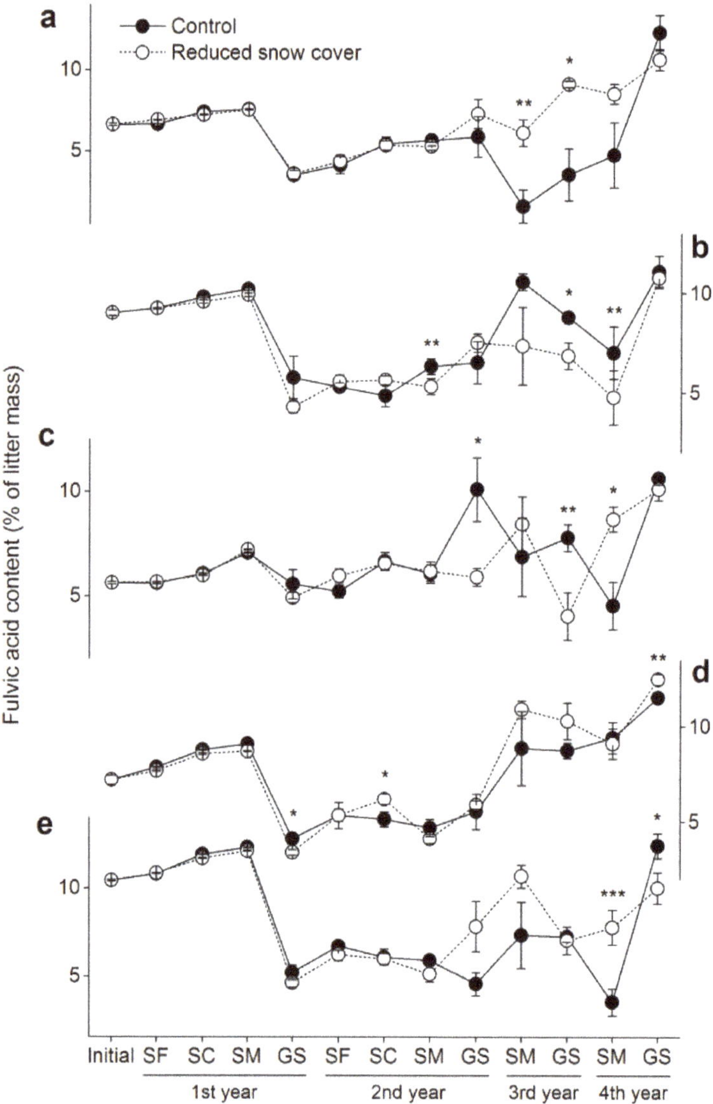

Figure 6. Fulvic acid contents (±SE, $n = 6$) in the five foliar litters. (**a**) Cypress, (**b**) larch, (**c**) birch, (**d**) willow and (**e**) azalea. SF: snow formation stage, SC: snow cover stage, SM: snowmelt stage, GS: growing season. * $p < 0.05$, ** $p < 0.01$, *** $p < 0.001$.

4. Discussion

SOM sequesters twice as much carbon as the total carbon contained in the atmosphere and terrestrial plants [41], and forecasting its feedback to climate change requires a clear understanding, not only of SOM decomposition [12] but also of SOM formation. Recent studies have changed our traditional understanding of the mechanisms underlying SOM formation [7,13]. A considerable amount of carbon is known to be transferred from plant litter to mineral soils at the early stages of litter decomposition [8]. However, the mechanisms underlying the accumulation of humus in the decomposing litter substrate remain unclear.

4.1. Is Humus Really Formed at a Very Late Stage of Litter Decomposition?

The conventional view suggests that plant litter decomposes to a "limit value", beyond which, litter cannot be decomposed and recalcitrant residues are humified and retained in stable SOM; therefore, humus is considered to form at a very late stage of litter decay [1]. In this study, we found that the initial humus content accounted for 8–13% of the litter mass (Figure 2), suggesting a high degree of humification in freshly senesced foliage. Furthermore, the humus content consistently increased as decomposition proceeded ($R^2 = 0.40$, $p < 0.001$; Figure 3a) and eventually reached 16–21% after four years of humification. Our results demonstrated that considerable humus accumulation occurred, and it sequestered carbon in the remaining litter substrate and did not allow for the net release of carbon from litter during the early stages of decomposition (four years) in this high-altitude forest.

Increasing evidence suggests that recalcitrant compounds (e.g., AUR) are not selectively preserved during litter decomposition [42]. In contrast, labile components increase microbial substrate use efficiency and promote SOM formation and persistence [9,13]. In this experiment, we monitored certain chemical compounds and mineral elements during litter decomposition and performed a partial least squares analysis to distinguish the importance of these factors that may impact the humus content. We found that the effect of AUR was greater than that of labile carbon, although neither was significant (Figure 4b). However, this result from our litterbag experiment was insufficient to support the microbial efficiency-matrix stabilization theory [13]. Our findings indicated that the litter Mn concentration was the dominant factor (VIP = 2.2) and positively related to the humus content (the coefficient was 0.35; Figure 4a), which suggested that Mn not only stimulated litter decomposition [43] but also promoted humus accumulation, thus supporting the hypothesis proposed by Berg et al. [44]. Our results also implied that bonding between humus and mineral elements (e.g., calcium and iron) may be more important than the approximate chemical compounds (e.g., water-soluble substances and AUR) in maintaining the organo-mineral stabilization of newly formed humus during the early stages of litter decomposition.

Fulvic acid (5.6–10.4%; Figure 6) formed earlier than humic acid (1.9–2.4%; Figure 5) in the decomposing foliar litter. However, the newly formed fulvic acid was not stable and degraded at earlier stages of humification (remaining mass >70%; Figure 3c), whereas humic acid greatly increased before the remaining mass reached 50–60%. It is not surprising that FA and HA were most present in the alkali-extractable substances since individual, lower molecular weight organic substances are extracted from fresh plant sediment. A study by Qualls et al. [45] also found that humic acid increased from 2.1% to 15.1% and that fulvic acid decreased from 7.5% to 6.1% in 3-year-old pine (*Pinus strobus* L.) litter compared with the fresh sample. These results show that humus accumulation during early litter decay was primarily dependent on the increase in humic acid. Our findings emphasized that more long-term ecological research should be performed to address how carbon is sequestered in soils instead of how carbon is lost from plant litter in a changing climate.

4.2. Will Reduced Snow Cover Alter This Carbon Sequestration Process in High-Altitudinal Ecosystems?

Winter snow cover has decreased in cold biomes [17], and numerous studies assessed its influence on the decomposition of plant litter; however, the results indicated high uncertainty. For example, experiments found that reduced snow cover decreases the litter decomposition rates by 5–47% in forests [18,46] and tundra [47], and a surprising threefold lower decomposition rate was reported under reduced snow cover in a subalpine meadow [19]. However, other experiments suggested that reduced snow cover has only a minor [28,48] or even null effect [49,50] on litter decomposition in subarctic and alpine ecosystems. Indeed, in our mid-latitudinal alpine forest, we also found that reduced snow cover decreased the loss of litter mass but only during the first two years and not during the subsequent periods.

In fact, the humified and not the decomposed components in the decomposing litter are more important for SOM formation and stabilization. Sequestering more carbon into the SOM with a longer mean resistance time [51] rather than allowing it to decompose is more effective under a changing climate [5]. Unfortunately, the mechanisms involved in SOM formation and its feedback on climate change are largely unknown. In this study, we found that reduced snow cover decreased the humus content in decomposing litter during the first three years but increased it in the fourth year (Figure 2). This result suggested that short-term assessments may erroneously estimate the real influence of the reduction in winter snow cover, which has a more complex effect on soil carbon sequestration, as indicated by previous snow manipulation studies. Thus, long-term ecological studies should be performed to decrease the uncertainty and accurately evaluate the SOM–climate feedback.

5. Conclusions

Our in situ litterbag experiment quantified 16–21% of the remaining plant litter sequestered as stable humic substances in the substrate material after four years of humification. Although this proportion was small compared with the decomposed amount (57–76%), carbon was retained to a degree in decomposing litter during its early decay periods, when an organo-mineral association formed via the interactions of mineral elements with newly accumulated humus. Clearly, the early carbon sequestration process from plant litter will be limited under a reduction in winter snow cover, although the effect is more complex than previously expected, and more long-term assessments should address how carbon is sequestrated and not decomposed from plant litter under a changing climate in seasonal snow-covered ecosystems.

Author Contributions: D.W. and W.D. designed the experiments; X.N. collected and exampled the samples; H.G. analyzed the data and drew the figures; D.W. drafted the manuscript; W.D. revised the manuscript. All authors have read and agreed to the published version of the manuscript.

Funding: This research was funded by the National Natural Science Foundation of China, grant numbers 32001965 and 41971261.

Institutional Review Board Statement: Not applicable.

Informed Consent Statement: Not applicable.

Data Availability Statement: The data are included in the manuscript submitted for peer review and will be archived online as supplementary materials of the manuscript is accepted for publication.

Conflicts of Interest: The authors declare no conflict of interest.

References

1. Berg, B.; McClaugherty, C. *Plant Litter: Decomposition, Humus Formation, Carbon Sequestration*, 4th ed.; Springer: New York, NY, USA, 2020; pp. 13–37.
2. Bailey, V.L.; Pries, C.H.; Lajtha, K. What do we know about soil carbon destabilization? *Environ. Res. Lett.* **2019**, *14*, 083004. [CrossRef]
3. Tie, L.; Hu, J.; Penuelas, J.; Sardans, J.; Wei, S.; Liu, X.; Zhou, S.; Huang, C. The amounts and ratio of nitrogen and phosphorus addition drive the rate of litter decomposition in a subtropical forest. *Sci. Total Environ.* **2022**, *833*, 155163. [CrossRef]
4. Moore, C.J.; Mihaylov, D.P.; Lasenby, A.; Gilmore, G. Astrometric search method for individually resolvable gravitational wave sources with Gaia. *Phys. Rev. Lett.* **2017**, *119*, 261102. [CrossRef]
5. Prescott, C.E. Litter decomposition: What controls it and how can we alter it to sequester more carbon in forest soils? *Biogeochemistry* **2010**, *101*, 133–149. [CrossRef]
6. Schmidt, M.W.I.; Torn, M.S.; Abiven, S.; Dittmar, T.; Guggenberger, G.; Janssens, I.A.; Kleber, M.; Kogel-Knabner, I.; Lehmann, J.; Manning, D.A.C.; et al. Persistence of soil organic matter as an ecosystem property. *Nature* **2011**, *478*, 49–56. [CrossRef]
7. Lehmann, J.; Kleber, M. The contentious nature of soil organic matter. *Nature* **2015**, *528*, 60–68. [CrossRef]
8. Rubino, M.; Dungait, J.A.J.; Evershed, R.P.; Bertolini, T.; De Angelis, P.; D'Onofrio, A.; Lagomarsino, A.; Lubritto, C.; Merola, A.; Terrasi, F.; et al. Carbon input belowground is the major C flux contributing to leaf litter mass loss: Evidences from a C-13 labelled-leaf litter experiment. *Soil Biol. Biochem.* **2010**, *42*, 1009–1016. [CrossRef]
9. Haddix, M.L.; Paul, E.A.; Cotrufo, M.F. Dual, differential isotope labeling shows the preferential movement of labile plant constituents into mineral-bonded soil organic matter. *Glob. Chang. Biol.* **2016**, *22*, 2301–2312. [CrossRef]

Olk, D.C.; Bloom, P.R.; Perdue, E.M.; McKnight, D.M.; Chen, Y.; Farenhorst, A.; Senesi, N.; Chin, Y.-P.; Schmitt-Kopplin, P.; Hertkorn, N.; et al. Environmental and agricultural relevance of humic fractions extracted by alkali from soils and natural waters. *J. Environ. Qual.* **2019**, *48*, 217–232. [CrossRef]

11. Cotrufo, M.F.; Soong, J.L.; Horton, A.J.; Campbell, E.E.; Haddix, M.L.; Wall, D.H.; Parton, A.J. Formation of soil organic matter via biochemical and physical pathways of litter mass loss. *Nat. Geosci.* **2015**, *8*, 776. [CrossRef]

12. Pries, C.E.H.; Bird, J.A.; Castanha, C.; Hatton, P.J.; Torn, M.S. Long term decomposition: The influence of litter type and soil horizon on retention of plant carbon and nitrogen in soils. *Biogeochemistry* **2017**, *134*, 5–16. [CrossRef]

13. Cotrufo, M.F.; Wallenstein, M.D.; Boot, C.M.; Denef, K.; Paul, E. The Microbial Efficiency-Matrix Stabilization (MEMS) framework integrates plant litter decomposition with soil organic matter stabilization: Do labile plant inputs form stable soil organic matter? *Glob. Chang. Biol.* **2013**, *19*, 988–995. [CrossRef] [PubMed]

14. Luojus, K.; Cohen, J.; Ikonen, J.; Pulliainen, J.; Takala, M.; Veijola, K.; Lemmetyinen, J.; Nagler, T.; Derksen, C.; Brown, R.; et al. Assessment of seasonal snow cover mass in northern hemisohere during the satellite-era. In Proceedings of the 38th IEEE International Geoscience and Remote Sensing Symposium (IGARSS), Valencia, Spain, 22–27 July 2018.

15. Li, Z.; Yang, W.; Yue, K.; Justine, M.F.; He, R.; Yang, K.; Zhuang, L.; Wu, F.; Tan, B.; Zhang, L.; et al. Effects of snow absence on winter soil nitrogen dynamics in a subalpine spruce forest of southwestern China. *Geoderma* **2017**, *307*, 107–113. [CrossRef]

16. Cooper, E.J. Warmer Shorter Winters Disrupt Arctic Terrestrial Ecosystems. In *Annual Review of Ecology, Evolution, and Systematics*; Futuyma, D.J., Ed.; Annual Reviews: Palo Alto, CA, USA, 2014; Volume 45, p. 271.

17. Marco-Barba, J.; Holmes, J.A.; Mesquita-Joanes, F.; Miracle, M.R. The influence of climate and sea-level change on the Holocene evolution of a Mediterranean coastal lagoon: Evidence from ostracod palaeoecology and geochemistry. *Geobios* **2013**, *46*, 409–421. [CrossRef]

18. Kreyling, J.; Haei, M.; Laudon, H. Erratum to: Absence of snow cover reduces understory plant cover and alters plant community composition in boreal forests. *Oecologia* **2013**, *173*, 1157. [CrossRef]

19. Saccone, P.; Morin, S.; Baptist, F.; Bonneville, J.M.; Colace, M.P.; Domine, F.; Faure, M.; Geremia, R.; Lochet, J.; Poly, F.; et al. The effects of snowpack properties and plant strategies on litter decomposition during winter in subalpine meadows. *Plant Soil* **2013**, *363*, 215–229. [CrossRef]

20. Bastida, F.; Eldridge, D.J.; Garcia, C.; Kenny Png, G.; Bardgett, R.D.; Delgado-Baquerizo, M. Soil microbial diversity-biomass relationships are driven by soil carbon content across global biomes. *Isme J.* **2021**, *15*, 2081–2091. [CrossRef]

21. Gavazov, K.S. Dynamics of alpine plant litter decomposition in a changing climate. *Plant Soil* **2010**, *337*, 19–32. [CrossRef]

22. Ni, X.Y.; Yang, W.Q.; Li, H.; Xu, L.Y.; He, J.; Tan, B.; Wu, F.Z. The responses of early foliar litter humification to reduced snow cover during winter in an alpine forest. *Can. J. Soil Sci.* **2014**, *94*, 453–461. [CrossRef]

23. Bradford, M.A.; Keiser, A.D.; Davies, C.A.; Mersmann, C.A.; Strickland, M.S. Empirical evidence that soil carbon formation from plant inputs is positively related to microbial growth. *Biogeochemistry* **2013**, *113*, 271–281. [CrossRef]

24. Xiang, B.; Liu, E.; Yang, L. Influences of freezing-thawing actions on mechanical properties of soils and stress and deformation of soil slope in cold regions. *Sci. Rep.* **2022**, *12*, 5387. [CrossRef] [PubMed]

25. Aanderud, Z.T.; Jones, S.E.; Schoolmaster, D.R., Jr.; Fierer, N.; Lennon, J.T. Sensitivity of soil respiration and microbial communities to altered snowfall. *Soil Biol. Biochem.* **2013**, *57*, 217–227. [CrossRef]

26. Hui, R.; Liu, L.; Xie, M.; Yang, H. Variation in snow cover drives differences in soil properties and microbial biomass of BSCs in the Gurbantunggut Desert-3 years of snow manipulations. *Ecohydrology* **2019**, *12*, e2118. [CrossRef]

27. Castellano, M.J.; Mueller, K.E.; Olk, D.C.; Sawyer, J.E.; Six, J. Integrating plant litter quality, soil organic matter stabilization, and the carbon saturation concept. *Glob. Chang. Biol.* **2015**, *21*, 3200–3209. [CrossRef]

28. Aerts, R.; Callaghan, T.V.; Dorrepaal, E.; van Logtestijn, R.S.P.; Cornelissen, J.H.C. Seasonal climate manipulations have only minor effects on litter decomposition rates and N dynamics but strong effects on litter P dynamics of sub-arctic bog species. *Oecologia* **2012**, *170*, 809–819. [CrossRef]

29. Li, H.; Wu, F.Z.; Yang, W.Q.; Xu, L.Y.; Ni, X.Y.; He, J.; Tan, B.; Hu, Y. Effects of forest gaps on litter lignin and cellulose dynamics vary seasonally in an alpine forest. *Forests* **2016**, *7*, 27. [CrossRef]

30. He, J.; Yang, W.Q.; Xu, L.Y.; Ni, X.Y.; Li, H.; Wu, F.Z. Copper and zinc dynamics in foliar litter during decomposition from gap center to closed canopy in an alpine forest. *Scand. J. Forest Res.* **2016**, *31*, 355–367. [CrossRef]

31. Ni, X.Y.; Yang, W.Q.; Tan, B.; He, J.; Xu, L.Y.; Li, H.; Wu, F.Z. Accelerated foliar litter humification in forest gaps: Dual feedbacks of carbon sequestration during winter and the growing season in an alpine forest. *Geoderma* **2015**, *241*, 136–144. [CrossRef]

32. Zhao, Y.; Wu, F.; Yang, W.; He, W.; Tan, B.; Xu, Z. Bacterial community changes during fir needle litter decomposition in an alpine forest in eastern Tibetan Plateau. *Russ. J. Ecol.* **2016**, *47*, 145–157. [CrossRef]

33. Wu, F.; Peng, C.; Zhu, J.; Zhang, J.; Tan, B.; Yang, W. Impacts of freezing and thawing dynamics on foliar litter carbon release in alpine/subalpine forests along an altitudinal gradient in the eastern Tibetan Plateau. *Biogeosciences* **2014**, *11*, 6871. [CrossRef]

34. Wu, F.Z.; Yang, W.Q.; Zhang, J.; Deng, R.J. Litter decomposition in two subalpine forests during the freeze-thaw season. *Acta Oecol.* **2010**, *36*, 135–140. [CrossRef]

35. Ladwig, L.M.; Ratajczak, Z.R.; Ocheltree, T.W.; Hafich, K.A.; Churchill, A.C.; Frey, S.J.K.; Fuss, C.B.; Kazanski, C.E.; Munoz, J.D.; Petrie, M.D.; et al. Beyond arctic and alpine: The influence of winter climate on temperate ecosystems. *Ecology* **2016**, *97*, 372–382. [CrossRef] [PubMed]

36. Ni, X.Y.; Yang, W.Q.; Tan, B.; Li, H.; He, J.; Xu, L.Y.; Wu, F.Z. Forest gaps slow the sequestration of soil organic matter: A humification experiment with six foliar litters in an alpine forest. *Sci. Rep.* **2016**, *6*, 19744. [CrossRef] [PubMed]

37. Konestabo, H.S.; Michelsen, A.; Holmstrup, M. Responses of springtail and mite populations to prolonged periods of soil freeze-thaw cycles in a sub-arctic ecosystem. *Appl. Soil Ecol.* **2007**, *36*, 136–146. [CrossRef]

38. Lan, L.Y.; Yang, W.Q.; Wu, F.Z.; Liu, Y.W.; Yang, F.; Guo, C.H.; Chen, Y.; Tan, B. Effects of soil fauna on microbial community during litter decomposition of Populus simonii and Fargesia spathacea in the subalpine forest of western Sichuan, China. *J. Appl. Ecol.* **2019**, *30*, 2983–2991.

39. Mehmood, T.; Liland, K.H.; Snipen, L.; Saebo, S. A review of variable selection methods in Partial Least Squares Regression. *Chemometr. Intell. Lab.* **2012**, *118*, 62–69. [CrossRef]

40. Trap, J.; Akpa-Vinceslas, M.; Margerie, P.; Boudsocq, S.; Richard, F.; Decaens, T.; Aubert, M. Slow decomposition of leaf litter from mature Fagus sylvatica trees promotes offspring nitrogen acquisition by interacting with ectomycorrhizal fungi. *J. Ecol.* **2017**, *105*, 528–539. [CrossRef]

41. Oldfield, E.E.; Crowther, T.W.; Bradford, M.A. Substrate identity and amount overwhelm temperature effects on soil carbon formation. *Soil Biol. Biochem.* **2018**, *124*, 218–226. [CrossRef]

42. Hall, S.J.; Huang, W.; Timokhin, V.I.; Hammel, K.E. Lignin lags, leads, or limits the decomposition of litter and soil organic carbon. *Ecology* **2020**, *101*, e03113. [CrossRef]

43. Berg, B.; Sun, T.; Johansson, M.B.; Sanborn, P.; Ni, X.Y.; Akerblom, S.; Lonn, M. Magnesium dynamics in decomposing foliar litter-A synthesis. *Geoderma* **2021**, *382*, 114756. [CrossRef]

44. Berg, B.; Erhagen, B.; Johansson, M.B.; Nilsson, M.; Stendahl, J.; Trum, F.; Vesterdal, L. Manganese in the litter fall-forest floor continuum of boreal and temperate pine and spruce forest ecosystems-A review. *Forest Ecol. Manag.* **2015**, *358*, 248–260. [CrossRef]

45. Qualls, R.G.; Takiyama, A.; Wershaw, R.L. Formation and loss of humic substances during decomposition in a pine forest floor. *Soil Sci. Soc. Am. J.* **2003**, *67*, 899–909. [CrossRef]

46. Christenson, L.M.; Mitchell, M.J.; Groffman, P.M.; Lovett, G.M. Winter climate change implications for decomposition in northeastern forests: Comparisons of sugar maple litter with herbivore fecal inputs. *Glob. Chang. Biol.* **2010**, *16*, 2589–2601. [CrossRef]

47. Blok, D.; Elberling, B.; Michelsen, A. Initial stages of tundra shrub litter decomposition may be accelerated by deeper winter snow but slowed down by spring warming. *Ecosystems* **2016**, *19*, 155–169. [CrossRef]

48. Baptist, F.; Yoccoz, N.G.; Choler, P. Direct and indirect control by snow cover over decomposition in alpine tundra along a snowmelt gradient. *Plant Soil* **2010**, *328*, 397–410. [CrossRef]

49. Bokhorst, S.; Bjerke, J.W.; Melillo, J.; Callaghan, T.V.; Phoenix, G.K. Impacts of extreme winter warming events on litter decomposition in a sub-Arctic heathland. *Soil Biol. Biochem.* **2010**, *42*, 611–617. [CrossRef]

50. Bokhorst, S.; Metcalfe, D.B.; Wardle, D.A. Reduction in snow depth negatively affects decomposers but impact on decomposition rates is substrate dependent. *Soil Biol. Biochem.* **2013**, *62*, 157–164. [CrossRef]

51. Dungait, J.A.J.; Hopkins, D.W.; Gregory, A.S.; Whitmore, A.P. Soil organic matter turnover is governed by accessibility not recalcitrance. *Glob. Chang. Biol.* **2012**, *18*, 1781–1796. [CrossRef]

Article

Acid Hydrolysable Components Released from Four Decomposing Litter in an Alpine Forest in Sichuan, China

Shu Liao, Kai Yue, Xiangyin Ni and Fuzhong Wu *

Key Laboratory for Humid Subtropical Eco-Geographical Processes of the Ministry of Education, School of Geographical Sciences, Fujian Normal University, Fuzhou 350007, China; liaos@fjnu.edu.cn (S.L.); kkyue@fjnu.edu.cn (K.Y.); nixy@fjnu.edu.cn (X.N.)
* Correspondence: wufzchina@fjnu.edu.cn

Abstract: Acid hydrolysable components have been thought to release from plant litter at early periods of decomposition and to be sensitive to hydrological change. Variations in snow depth and timing may alter the release of acid hydrolysable components from decomposing litter in seasonally snow-covered ecosystems. Here, we measured the release of acid hydrolyzable components from four foliar litters (fir, cypress, larch and birch) in deep and shallow snow plots during winter (snow formation, snow coverage and snowmelt stages) and growing seasons in an alpine forest from 2012 to 2016. We found that the content of acid hydrolysable components was 16–21% in fresh litter across species, and only 4–5% of these components remained in the litter after four years of decomposition when 53–66% of litter mass was lost. The content of acid hydrolysable components greatly decreased within 41 days and during the growing seasons of the fourth year of decomposition, suggesting that acid hydrolysable components in plant litter are not only released at early periods but also at a very late period during litter decay. However, the content of acid hydrolysable components increased significantly at snowmelt stages. Reduced snow cover increased the content and remaining level of acid hydrolysable components during the four years of decomposition by altering leaching, microbial biomass and stoichiometry. We propose that more effective partitioning of chemical fractions should be incorporated to distinguish the carbon and nutrient release during litter decomposition within a complex context of the changing environment.

Keywords: acid hydrolysable components; litter decomposition; snow cover; winter

Citation: Liao, S.; Yue, K.; Ni, X.; Wu, F. Acid Hydrolysable Components Released from Four Decomposing Litter in an Alpine Forest in Sichuan, China. *Forests* **2022**, *13*, 876. https://doi.org/10.3390/f13060876

Academic Editor: Sadanandan Nambiar

Received: 6 April 2022
Accepted: 1 June 2022
Published: 3 June 2022

Publisher's Note: MDPI stays neutral with regard to jurisdictional claims in published maps and institutional affiliations.

1. Introduction

The release of labile components from plant litter is one of the fastest pathways for carbon (C) turnover in forest ecosystems due to its high decomposability [1]. These labile fractions, such as water soluble and acid hydrolysable components, are primarily leached during early decomposition [2] and are rapidly consumed by soil decomposers [3]. However, "labile" is defined relative to "recalcitrant", and our current knowledge of the entire fraction of "labile" components is a nebulous concept in litter decomposition studies [1]. Most studies have focused on dissolved organic carbon (DOC), but other forms of labile fractions, such as acid hydrolysable components, are also released from plant litter during early stages of decomposition and regulate some important ecological processes [4]. These acid hydrolysable components decomposed from fresh litter supply the C that is available for soil microorganisms, which in turn contribute to litter breakdown and C and nutrient cycling [5].

Acid hydrolysable components in plant litter are consumed or released in different stages of decomposition [2]. In the early stage, acid hydrolysable components are released rapidly from senesced litter because fresh litter contains large amounts of soluble compounds, including some polysaccharides, hemicellulose, cellulose and some soluble lignin [6]. Additionally, the snowmelt stage may be another peak of acid hydrolysable components release because higher microbial activity under a warmer environment facilitates the degradation of hemicellulose and cellulose, and this decomposed detritus could be leached by snowmelt water [7]. Moreover, repeated soil freezing and thawing during winter causes previously stabilized organic matter to be available as labile C and nutrients [8]. These previous findings suggest that shifts in the ambient environment during different decomposition periods strongly control the release of acid hydrolysable components from plant litter.

Snow cover in the Northern Hemisphere has decreased by 7% and is projected to decrease by 25% by the end of this century [9]. This reduction has a profound influence on litter decomposition and nutrient cycling in many cold ecosystems [10,11]. Notably, a reduction in snow depth directly decreases water availability due to the decreased snowmelt water supply [12]. Furthermore, some acid hydrolysable components in plant litter are water soluble and primarily released by leaching [13]; thereby, a decline in winter snow depth may directly decrease the release of acid hydrolysable components from decomposing litter materials [14]. Additionally, reduced snow cover has been demonstrated to reduce soil microbial biomass and extracellular enzymatic activity, and this effect could decrease the depolymerization of acid hydrolysable components during litter decomposition [15]. However, reduced snow cover also leads to more fluctuations of soil freezing and thawing [16], which can physically impact litter breakdown and therefore promote the release of acid hydrolysable components [17].

Here, we hypothesize that (1) the decline in snow depth decreases the release of acid hydrolysable components from plant litter, and (2) the formation, coverage and melting of winter snow cover alters the release of acid hydrolysable components. We conducted a litterbag experiment in a high-elevation alpine forest using four dominant tree litter species and measured the contents and release patterns of acid hydrolysable components during four years of decomposition from 2012 to 2016. Our objectives were as follows: (1) to follow the long-term release pattern of acid hydrolysable components, which have been thought to release rapidly during early periods of litter decomposition; and (2) to assess the potential influence of variations in snow depth and timing on the release of acid hydrolysable components in decomposing litter in this alpine forest.

2. Materials and Methods

2.1. Experimental Site

This study was conducted at the Long-term Research Station of Alpine Forest Ecosystems, Sichuan, China (31°14′ N, 102°53′ E, 3579–3582 m). The experimental sites are located in a primeval alpine forest on the eastern Tibetan Plateau and have been extensively described elsewhere [18]. The mean annual temperature and precipitation are 2.7 °C and 850 mm, respectively. Snow cover develops in late October and melts in April of the following year with a maximum snow depth of approximately 50 cm and a soil freezing time of approximately 120 days [19]. This alpine forest is dominated by coniferous fir (*Abies faxoniana* Rehd. et Wils.), cypress (*Sabina saltuaria* Rehder & E.H.Wilson), deciduous larch (*Larix mastersiana* Rehd. et Wils.) and birch (*Betula albosinensis* Burk.). Canopy gaps induced by natural tree fall and other climatic extremes cover 13–23% of the forest landscape. Soils are classified as Cambisols (WRB soil taxonomy) [20] and the litter depth (Oi/Oe) is 7 cm with a C stock of 1.6 t ha^{-1} in autumn (the peak of litter fall). The contents of C, nitrogen (N), phosphorus (P) and humic substances in the organic soils (Oa/A) are 16%, 0.58%, 0.17% and 6.1%, respectively [21].

2.2. Experimental Design

We used a 2 (snow depth) × 4 (litter species) factorial design in this study. Litter decay at different snow depths was facilitated by incubating litter samples in the canopy gap (deep snow plot) and under the canopy (shallow snow plot). The alternative to this strategy was to perform repeated manual shoveling to reduce snow cover, but this strategy increases the risk of losing needle litter (fir, cypress and larch) from litterbags, which must have sufficiently large mesh sizes to permit access by soil fauna [22]. To decrease the uncertainty among litter species, four dominant tree species (fir, cypress, larch and birch) with different initial qualities (Table 1) were studied. Specifically, three sites (as replicates, 500–2000 m apart) with similar topographies and canopy covers that receive similar snowfall and litter fall were set up in this alpine forest in 2009. Each site had two pairs of deep/shallow snow plots ($n = 6$), and each plot had four subplots (3 × 3 m in size and 3–4 m apart) for incubating the four foliar litters (a total of 24 subplots).

Table 1. Initial chemical compounds in the four foliar litters.

Litter	C (%)	N (%)	P (%)	WSS (%)	OSS (%)	ASS (%)	AUR (%)	C/N Ratio	N/P Ratio	AUR/N Ratio
Fir	50.56 (2.96) [a]	0.88 (0.003) [b]	0.11 (0.010) [b]	23.76 (0.62) [b]	27.62 (2.28) [a]	27.36 (1.33) [b]	23.92 (2.54) [b]	57.77 (3.53) [a]	7.7 (0.62) [c]	27.33 (2.92) [b]
Cypress	51.64 (1.77) [a]	0.88 (0.010) [b]	0.12 (0.006) [ab]	13.81 (1.21) [c]	33.16 (3.43) [a]	32.43 (1.29) [a]	20.60 (3.41) [b]	58.86 (2.21) [a]	7.1 (0.27) [bc]	23.48 (3.89) [b]
Larch	54.35 (0.63) [a]	0.86 (0.041) [b]	0.13 (0.002) [a]	30.19 (0.43) [a]	19.11 (0.68) [b]	29.24 (0.87) [ab]	21.46 (0.94) [b]	63.32 (3.49) [a]	6.5 (0.25) [c]	25.01 (2.04) [b]
Birch	49.69 (1.45) [a]	1.33 (0.022) [a]	0.09 (0.004) [c]	9.87 (0.95) [c]	11.43 (0.75) [c]	27.74 (0.94) [b]	50.96 (0.96) [a]	37.24 (1.35) [b]	14.6 (0.50) [a]	38.19 (1.01) [a]

The values are means with standard deviations shown in parentheses ($n = 3$). The values in the same columns with different superscript letters are significantly ($p < 0.05$) different among litter species. C: carbon, N: nitrogen, P: phosphorus, WSS: water-soluble substances, OSS: organic-soluble substances, ASS: acid-soluble substances, and AUR: acid-unhydrolyzable residues.

To assess the effect of variation in winter snow timing on litter decomposition, samplings were scheduled at the ends of the snow formation, snow coverage and snowmelt stages during winter from 2012 to 2014. From 2014 to 2016, we collected samples only once each winter (at the end of the snowmelt stage). The beginning and end of winter were defined as having hourly temperatures with a 50% probability (continuous over 12 h per day) of freezing (below 0 °C; Table 2) [23]. The winter was further divided into snow formation, snow coverage and snowmelt stages based on the snow dynamics and changes in temperature according to our previous meteorological monitoring in this alpine forest (Figure 1) [18] and the experiments that were conducted in other cold biomes [24]. We also performed sampling at the end of the growing season (beginning of the snow formation stage) of each year.

Table 2. Sampling comparison of the timing and length of winters during the four years of decomposition.

Winter	Timing of Winter [a]		Length of Winter (Days)	Sampling Date [b]	
	Beginning	End		End of Growing Season	End of Snowmelt
2012/2013 winter	30 October 2012	24 April 2013	176	Experiment begins	24 April 2013
2013/2014 winter	31 October 2013	30 April 2014	181	30 October 2013	24 April 2014
2014/2015 winter	29 October 2014	26 April 2015	179	29 October 2014	23 April 2015
2015/2016 winter	30 October 2015	26 April 2016	179	2 November 2015	24 April 2016

[a] Hourly temperature with 50% probability (continual 12 h per day) of freezing (below 0 °C) is defined as the beginning and end of winter in this study. This definition is not conservative because solar radiation is very strong in this alpine forest, which increases the temperature recorded by the data logger. [b] Sampling dates are scheduled approximately in late April and October within one week of intervals referring to the timing of the 2012/2013 winter.

2.3. Litterbag Experiment

Foliar litters of the four dominant trees species in this alpine forest were studied using an in situ litterbag method [22]. In the fall of 2012, senesced needles or leaves of each tree species were collected on tarpaulins from 20 or more trees (15–20 cm in diameter) by shaking their limbs. Green or partly decomposed needles (leaves), twigs and bark were removed, and only newly shed foliar litter was air-dried at room temperature for two weeks. Litter samples (10 ± 0.05 g) were placed into nylon litterbags (20 × 25 cm in size with mesh sizes of 1.0 mm on the top and 0.5 mm on the bottom). A total of 1172 litterbags (2 litterbags × 6 replicates × 4 litter species × 2 snow plots × 12 sampling times + 5 litterbags × 4 litter species) were incubated in the deep and shallow snow plots on November 15 and 16, 2012. Litterbags in the same subplots were strung together and kept 2–5 cm apart. Five litterbags of each litter species were randomly collected to determine the losses during sample establishment and the gravimetric water contents of these air-dried litter samples before field incubation.

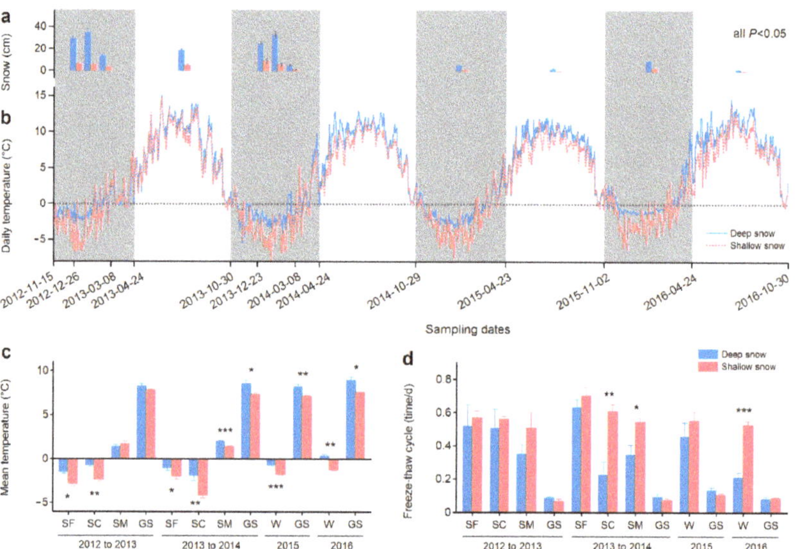

Figure 1. Snow depth, temperature and freeze–thaw cycle. (**a**) Snow depths (±SE, *n* = 18) in deep and shallow snow plots on each sampling date. The samplings were scheduled at the ends of the snow formation, snow coverage and snowmelt stages during winter from 2012 to 2014 (only the snowmelt stages from 2014 to 2016) and at the end of the growing season (beginning of the snow formation stage). The differences between deep and shallow snow plots are significant (*p* < 0.05) on all sampling dates. (**b**) Daily temperatures (*n* = 6) at the litter surface during the four years of decomposition. Winter times including the snow formation, snow coverage and snowmelt stages are shaded. The dashed line is drawn at the zero value. (**c**) Mean temperatures (±SE, *n* = 6) at individual stages. The times between two adjacent sampling dates were defined as separate stages. (**d**) Freeze–thaw cycles (±SE, *n* = 6) at individual stages. The values were calculated by the quotients of the total number of freeze–thaw cycles and the number of days of certain stages. For panel c and d, the winters in 2015 and 2016 were not divided into snow formation, snow coverage and snowmelt stages. SF: snow formation stage, SC: snow coverage stage, SM: snowmelt stage, W: winter, and GS: growing season. * *p* < 0.05, ** *p* < 0.01 and *** *p* < 0.001.

2.4. Sampling and Analysis

On each sampling date, snow depths were manually measured in triplicate in each deep and shallow snow plot (*n* = 18). Temperatures at the litter surface were recorded using data loggers (iButton DS1923-F5, Sunnyvale, CA, USA) placed in marked litterbags in deep and shallow snow plots (both *n* = 6). Based on hourly temperature data, a freeze–thaw cycle was defined as a transition above or below 0 °C for at least 3 h and then a transition back [25].

Two litterbags per subplot were randomly collected, carefully placed in separate plastic bags and returned to the laboratory. Roots, mosses and soils were carefully removed from the litter samples. One cleaned subsample was oven-dried at 105 °C for 48 h to measure the dry mass remaining and gravimetric water content in the decomposing litter. The remaining mass of decomposing litter was evaluated on a dry matter basis. Another subsample was air-dried, milled and used to extract the acid hydrolysable components using an acid hydrolysis method [26]. In our long-term study in this alpine forest, we also measured proximate fractions, elements, humic substances and microbial activities during the decomposition of these foliar litters. In this study, we focused on the release pattern of acid hydrolysable components.

Here, the acid hydrolysable components were defined as the fraction that can be hydrolyzed by 2.5 mol/L H_2SO_4 [26]. These components were different with acid-soluble substances hydrolyzed by 72% H_2SO_4 as well as with dissolved organic carbon or hot-water extractable carbon, which are considered to be more labile with higher decomposability than acid hydrolysable components. In this study, the acid hydrolysable components include some polysaccharide, hemicellulose, cellulose and soluble lignin or phenols [6].

Specifically, a 0.10 g subsample was hydrolyzed by 2.5 mol/L H_2SO_4 at 105 °C for 30 min. The extracted acid hydrolysable components were diluted with deionized water, filtered through a 0.45 μm mesh and measured using a TOC analyzer (multi N/C 2100, Analytic Jena, Thüringen, Germany). Litter C and N contents were determined by elemental analyzer (Vario Max CN, Elementar, Germany) and were used to calculate litter C/N stoichiometry. Microbial biomass C content was measured using the chloroform fumigation-incubation method [27]. At the same time, a 0.5 g subsample was oven-dried at 105 °C for 48 h to determine the gravimetric water content in the air-dried decomposing litter. The content of acid hydrolysable components was calculated based on the dry mass of decomposing litter.

2.5. Data Analysis

The content of acid hydrolysable components (%) was presented as % of litter dry mass, and the remaining acid hydrolysable components (AHC_r, % of original; Equation (1)) were calculated as follows:

$$AHC_r(\%) = (AHC_t \times M_t)/(AHC_0 \times M_0) \times 100 \tag{1}$$

where $AHCC_0$ and AHC_t are the initial content of acid hyrolysable components and their contents at time t ($t = 1, 2, \ldots, 12$), respectively; and M_0 and M_t are the initial litter dry mass and the masses at time t ($t = 1, 2, \ldots, 12$).

We first examined the overall effects of variations in snow depth and litter species over time via variance analysis (ANOVA). Differences between deep and shallow snow plots were assessed by paired t-tests, or by Wilcoxon signed-rank tests if sample sizes were unequal in MATLAB R2012a (MathWorks Inc., Natick, MA, USA). A path analysis was used to distinguish how variation in snow depth affects acid hydrolysable components during litter decomposition by altering the environment (litter temperature and gravimetric water content), litter quality (e.g., C/N ratio) and soil microbes (e.g., microbial biomass C; MBC) in AMOS 22.0 (IBM SPSS, Chicago, IL, USA). We also compared the relationships between litter mass remaining and the remaining acid hydrolysable components for each litter species.

3. Results

3.1. Snow Depth, Temperature and Freeze–Thaw Cycle

The snow depth in the shallow snow plot was 79% lower than that in the deep snow plot (Figure 1a). The snow depths were higher at the ends of the snow formation and snow coverage stages relative to those at the ends of the snowmelt stages in 2013 and 2014. The daily temperature was higher in the deep snow plot than that in the shallow snow plot, particularly during winter (Figure 1b). The mean temperatures at snow formation and snow coverage stages were below 0, but those at the snowmelt stages were above 0 during the winters in 2013 and 2014. The mean temperatures were significantly ($p < 0.05$) higher in the deep snow plots at all stages during the four years of decomposition, except those at the snowmelt stage and growing season in 2013 (Figure 1c).

The average freeze–thaw cycles in winters during the four years of decomposition were 0.41 and 0.57 times per day for the deep and shallow snow plots, respectively. For individual stages, the freeze–thaw cycles were significantly ($p < 0.05$) lower in the deep snow plots than those in the shallow snow plots at the snow coverage and snowmelt stages in 2014 as well as during the winter in 2016 (Figure 1d).

3.2. Mass Remaining

Litter mass remaining varied greatly among litter species ($F = 35.9$, $p < 0.001$) but did not differ between deep and shallow snow plots ($F = 1.8$, $p = 0.25$, Table 3) over time. After four years of decomposition, 53%, 66%, 58% and 61% of litter mass were lost at the deep snow plots, respectively, and 55%, 65%, 55% and 60% of litter mass were lost at the shallow snow plots for fir, cypress, larch and birch litter, respectively (Figure 2).

Table 3. Results of repeated measures ANOVA testing for the effects of litter species and variation in snow depth over time.

Source of Variation	Mass Remaining		Gravimetric Water Content		AHC Content		AHC Remaining	
	F Value	*p* Value	*F* Value	*p* Value	*F* Value	*p* Value	*F* Value	*p* Value
Time	5060.8	**<0.001**	775.7	**<0.001**	274.1	**<0.001**	333.7	**<0.001**
Litter	35.9	**<0.001**	438.3	**<0.001**	1.5	0.23	114.2	**<0.001**
Snow	1.8	0.25	13.7	**<0.001**	3.0	0.083	6.8	**0.0094**
Time × Litter	45.4	**<0.001**	94.7	**<0.001**	5.7	**<0.001**	8.5	**<0.001**
Time × Snow	2.3	0.073	11.0	**<0.001**	1.8	0.055	2.2	**0.015**
Litter × Snow	1.5	0.23	1.4	0.25	1.3	0.27	2.1	0.10

Bold *p* values are significant ($p < 0.05$). AHC: acid hydrolysable components.

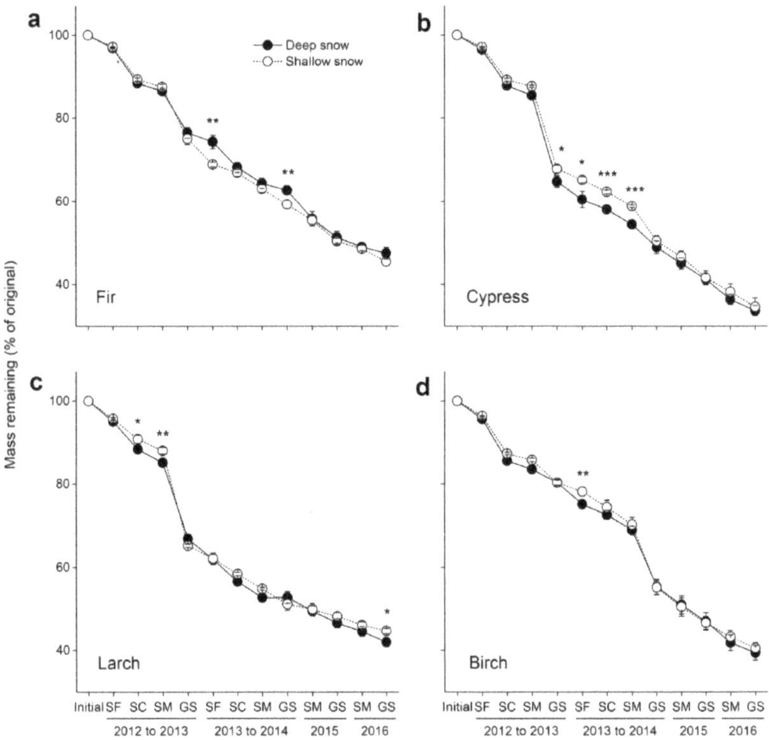

Figure 2. Mass remaining in decomposing litter in the deep and shallow snow plots. (**a**) Fir, (**b**) cypress, (**c**) larch, and (**d**) birch. SF: snow formation stage, SC: snow coverage stage, SM: snowmelt stage, and GS: growing season. * $p < 0.05$, ** $p < 0.01$ and *** $p < 0.001$.

3.3. Gravimetric Water Content

The initial gravimetric water contents in the fir, cypress, larch and birch litters were 8%, 9%, 10% and 13%, respectively, but varied greatly over time ($F = 775.7$, $p < 0.001$; Table 3). The gravimetric water content in all the foliar litters showed an increasing tendency in 2013 but greatly decreased at the snow formation and snow coverage stages in 2014 (Figure 3). For all the litter species, the gravimetric water content notably increased in 2016 to 40%, 30%, 23% and 24%, respectively, in the deep snow plots after four years of decomposition. The variation in snow depth significantly ($F = 13.7$, $p < 0.001$; Table 3) changed the gravimetric water content in the decomposing litter, and higher ($p < 0.05$) gravimetric water contents were observed in deep snow plots for all litter species (Figure 3a–d).

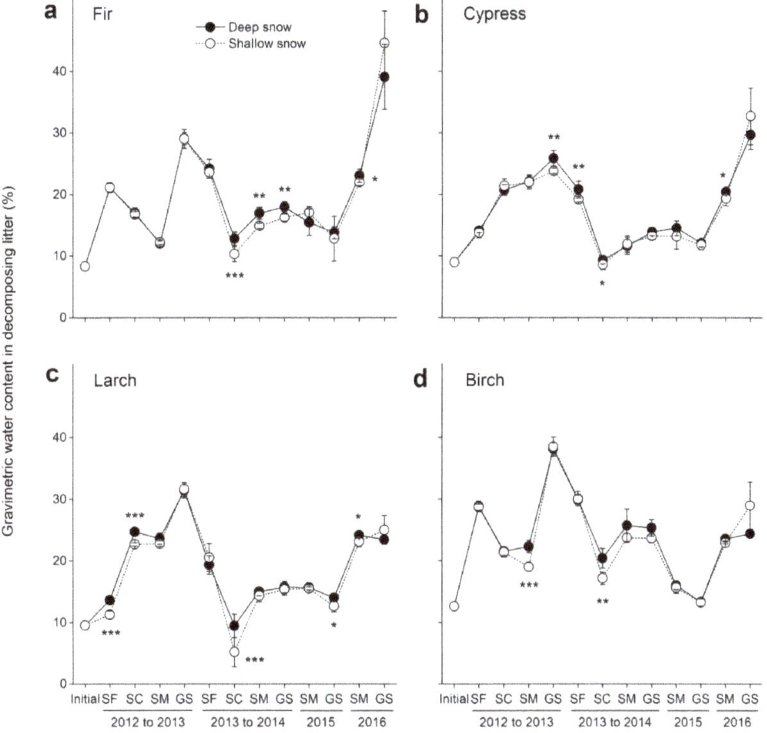

Figure 3. Gravimetric water contents ($\pm$SE, $n = 6$) in decomposing litter in the deep and shallow snow plots. (**a**) Fir, (**b**) cypress, (**c**) larch, and (**d**) birch. SF: snow formation stage, SC: snow coverage stage, SM: snowmelt stage, and GS: growing season. * $p < 0.05$, ** $p < 0.01$ and *** $p < 0.001$.

3.4. Content of Acid Hydrolysable Components

The initial contents of the acid hydrolysable components were 19%, 19%, 21% and 16% for the fir, cypress, larch and birch litters, respectively (Figure 4a–d). The content of acid hydrolysable components significantly ($F = 274.1$, $p < 0.001$; Table 3) changed over time, but this change did not differ among the litter species ($F = 1.5$, $p = 0.23$). During the four years of decomposition, the content of the acid hydrolysable components greatly decreased at the snow formation stage in 2012 and the growing season in 2016. The contents of the acid hydrolysable components also decreased during the growing season in 2013 to the snow coverage stage in 2014, as well as at the snowmelt stage and growing season in 2015. However, the content of acid hydrolysable components increased at both the snow coverage and snowmelt stages in 2013, while only at the snowmelt stages in 2014 and 2016 for all litter species. After four years of decomposition, the content of the acid hydrolysable components was only 1.8–2.7% in the deep snow plots across the litter species.

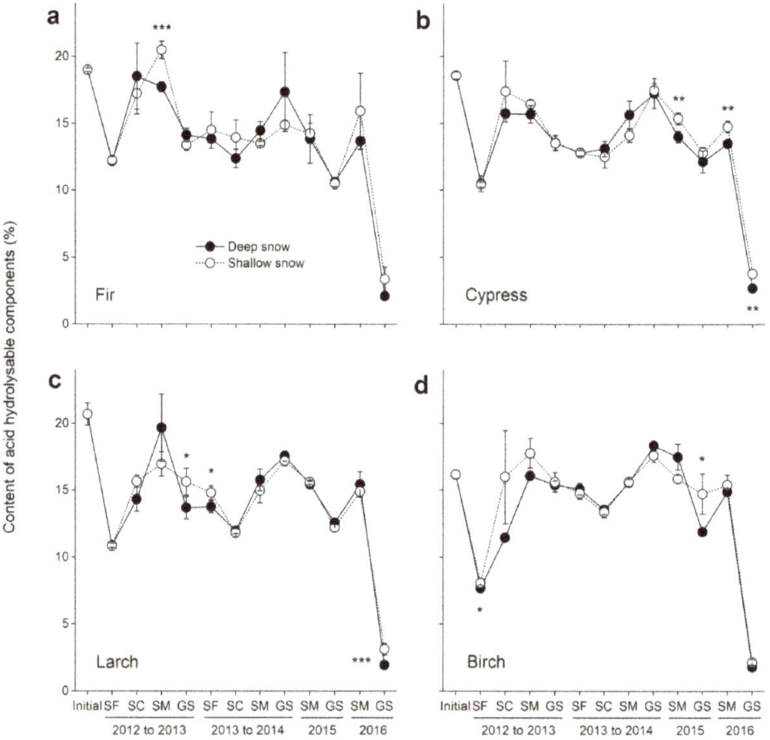

Figure 4. Contents of acid hydrolysable components ($\pm$SE, $n = 6$) in decomposing litter in the deep and shallow snow plots. (**a**) Fir, (**b**) cypress, (**c**) larch, and (**d**) birch. SF: snow formation stage, SC: snow coverage stage, SM: snowmelt stage, and GS: growing season. * $p < 0.05$, ** $p < 0.01$ and *** $p < 0.001$.

Overall, the variation in snow depth did not significantly ($F = 3.0$, $p = 0.083$; Table 3) change the content of the acid hydrolysable components for the decomposing litter across species. However, for individual litter species, the effect of variation in snow depth on the content of acid hydrolysable components was greater for birch litter ($F = 6.0$, $p = 0.016$) compared with those for other foliar litters ($p > 0.05$; Table 4). During the four years of decomposition, the contents of the acid hydrolysable components were significantly higher ($p < 0.05$) in shallow snow plots than those in deep snow plots for all litter species at certain stages. At the end of the experiment, higher ($p < 0.01$) contents of acid hydrolysable components were also observed in the shallow snow plots for cypress and larch litters.

The path analysis results showed that snow depth was positively ($p < 0.001$; Figure 5) correlated with litter temperature ($r = 0.38$), gravimetric water content ($r = 0.18$) and MBC ($r = 0.43$) but negatively ($r = -0.21$, $p < 0.001$) correlated with the litter C/N ratio. The litter temperature and gravimetric water content did not have significant effects on either the C/N ratio or MBC (all $p > 0.05$), but both were negatively correlated with the content of acid hydrolysable components ($r = -0.11$, $p < 0.01$ for litter temperature and $r = -0.25$, $p < 0.001$ for gravimetric water content). The litter C/N ratio and MBC had, respectively, negative ($r = -0.34$) and positive ($r = 0.15$; both $p < 0.001$) effects on the content of the acid hydrolysable components.

Table 4. Results of repeated measures ANOVA testing for the effects of variation in snow depth over time for each litter species.

Litter	Time		Snow		Time × Snow	
	F Value	*p* Value [a]	*F* Value	*p* Value	*F* Value	*p* Value
	Content of acid hydrolysable components					
Fir	45.0	**<0.001**	0.61	0.43	1.5	0.13
Cypress	105.8	**<0.001**	0.82	0.37	1.9	0.051
Larch	116.2	**<0.001**	0.4	0.53	2.6	**0.0055**
Birch	98.4	**<0.001**	6.0	**0.016**	3.4	**<0.001**
	Remaining acid hydrolysable components					
Fir	77.7	**<0.001**	0	0.96	1.6	0.12
Cypress	142.5	**<0.001**	6.6	**0.011**	0.81	0.63
Larch	133.0	**<0.001**	0.17	0.68	1.7	0.081
Birch	74.0	**<0.001**	6.4	**0.013**	2.5	**0.0078**

[a] Bold *p* values are significant ($p < 0.05$).

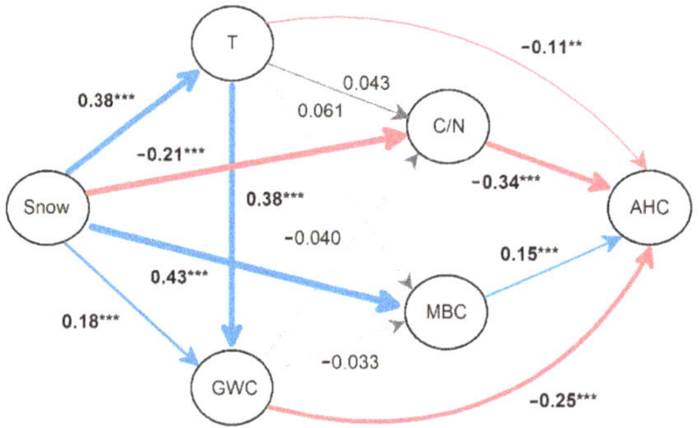

Figure 5. Path analysis shows direct and indirect effects of the variation in snow depth on the release of acid hydrolysable components. The color (blue: positive, and red: negative) and gray arrows denote significant and non-significant effects, respectively. The widths of the arrows and associated numbers (standard regression weights) represent the strength of the path coefficients, and the bold coefficients with asterisks are significant (** $p < 0.01$ and *** $p < 0.001$). T: temperature, GWC: gravimetric water content, C/N: carbon to nitrogen ratio, MBC: microbial biomass carbon, and AHC: acid hydrolysable components. $n = 576$, $R^2 = 0.531$, $p < 0.001$ for the model.

3.5. Remaining Acid Hydrolysable Components

Overall, the remaining acid hydrolysable components consistently decreased over time ($F = 333.7$, $p < 0.001$; Table 3), and marked declines were found at the snow formation stage in 2012 and during the growing seasons in 2012 and 2016 for all litter species. However, the remaining acid hydrolysable components increased during the snow coverage and snowmelt stages in 2013 as well as the snowmelt stages in 2014 and 2016 (Figure 6a–d). After four years of decomposition, the remaining acid hydrolysable components were only 4–5% across the litter species.

The variation in snow depth had a significant ($F = 6.8$, $p = 0.0094$; Table 3) effect on the remaining acid hydrolysable components across litter species, and these effects were greater for cypress ($F = 6.6$, $p = 0.011$; Table 4) and birch litters ($F = 6.4$, $p = 0.013$) compared with fir and larch litters (both $p > 0.05$). During the four years of decomposition, the remaining acid hydrolysable components were significantly higher ($p < 0.05$) in shallow snow plots than those in deep snow plots for all litter species at certain stages. At the end of the experiment, higher ($p < 0.001$) remaining acid hydrolysable components were also observed in the shallow snow plot for larch litter (Figure 5c). The remaining acid hydrolysable components in the decomposing litter were significantly correlated with the remaining litter mass for all litter species (all $p < 0.001$; Figure 7).

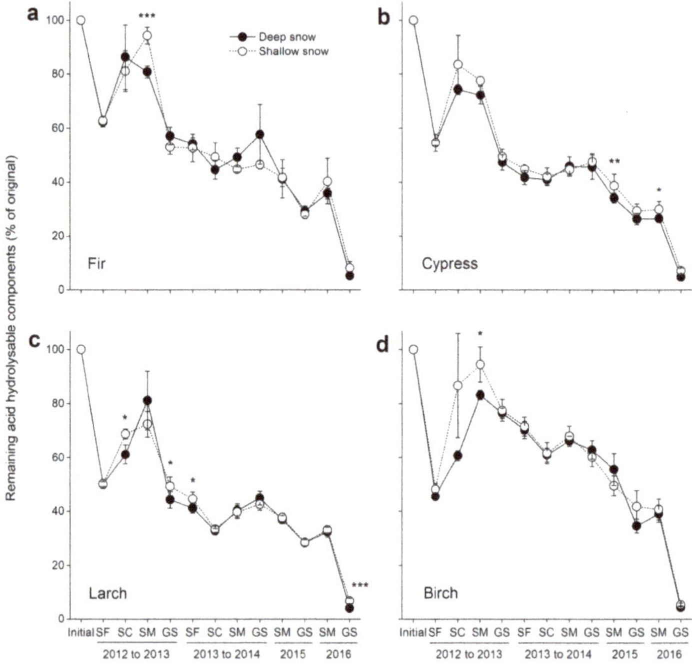

Figure 6. Remaining acid hydrolysable components ($\pm$SE, $n = 6$) in decomposing litter in the deep and shallow snow plots. (**a**) Fir, (**b**) cypress, (**c**) larch, and (**d**) birch. SF: snow formation stage, SC: snow coverage stage, SM: snowmelt stage, and GS: growing season. $*$ $p < 0.05$, $**$ $p < 0.01$ and $***$ $p < 0.001$.

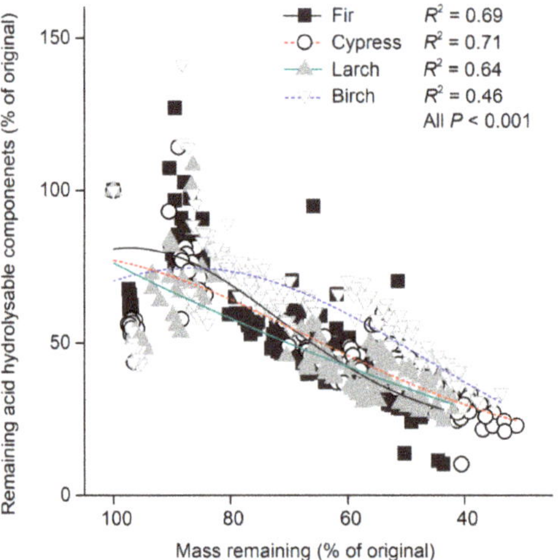

Figure 7. Remaining litter mass versus acid hydrolysable components for each litter species. Adjusted R^2 values from Gaussian functions are shown for each litter species, and all $p < 0.001$ ($n = 144$ for each litter species).

4. Discussion

Different components with hierarchical decomposability have been widely used to assess chemical changes during litter decomposition [6]. Labile components are highly decomposable, but a clear boundary is absent in soil organic matter studies and Earth system models [1]. Dissolved organic C has been considered as a labile component and is rich in fresh plant litter [4]. In this study, the initial contents of water-soluble substances (extracted at 80 °C for 1 h) were 24%, 14%, 30% and 10%, respectively, for fresh fir, cypress, larch and birch litters (Table 1). However, the initial contents of acid hydrolysable components were only 19%, 19%, 21% and 16% for fresh fir, cypress, larch and birch litters, respectively, which were similar with the content of water-soluble substances but lower than the contents of acid-soluble substances (extracted with 72% H_2SO_4) (Table 1). These results suggest that the acid hydrolysable components under the conditions of this study (extracted by 2.5 mol/L H_2SO_4) are just parts of the soluble components that can be dissolved in hot water.

Acid hydrolysable components have been thought to release during the early or intermediate periods of litter decomposition in both field and laboratory experiments [4,28], and this hierarchical pattern of decomposability is similar with dissolved organic C [1]. For example, in a ^{13}C tracer experiment, Kammer et al. [3] found that approximately 80% of the annual dissolved organic C was leached from the litter layer during the first winter of decomposition after litter fall. Our results showed that 38%–55% of acid hydrolysable components were released during the first 41 days, suggesting that the acid hydrolysable components are highly soluble and release rapidly. However, another peak for the release of acid hydrolysable components was observed in the fourth year of decomposition, possibly due to the large increase in litter water content (Figure 3). These results suggest that the acid hydrolysable components are different with dissolved organic C and could be released at a later period than previously expected. In addition, a significant release of acid hydrolysable components occurred during the growing season of the first year of decomposition. These releases were in accordance with the increase in gravimetric water content in the decomposing litter, implying that water content may be the dominant driver for the release of acid hydrolysable components. The path analysis results also indicated that gravimetric water content had a strong effect on the release of acid hydrolysable components (Figure 5), suggesting that hydrological leaching strongly controls the release of these labile components during litter decomposition.

However, we also found that a high level of acid hydrolysable components was maintained over a long term after a rapid release and even increased during snowmelt (Figure 6). Several reasons can explain the high level of acid hydrolysable components during litter decomposition, particularly during snowmelt. First, some acid hydrolysable are intermediate decomposable carbohydrates [2], and these components were decomposed to smaller carbohydrates [6], increasing the proportion of acid hydrolysable components measured under the conditions of this study. Our previous studies also found that cellulose content greatly decreased during the snow coverage and snowmelt stages in the first year of decomposition at these sites [29]. Moreover, we found that repeated soil freezing and thawing during winter strongly contributes to C decay during litter decomposition in this alpine forest [19]. In addition, earlier studies found that cold-tolerant microbes maintain high activities at temperatures above −4 °C [30] but die during snowmelt due to C starvation over winter and their inability to adapt to an increase in ambient temperature [31]. Our previous research conducted in this alpine forest also found that the microbial biomass was higher during snow cover than that during snowmelt [32], suggesting that dead microbial necromass (i.e., glucosamine and muramic acid) may contribute to the increase in acid hydrolysable components during snowmelt. Our results suggest that acid hydrolysable components do not consistently decrease as decomposition proceeds, and winter snow exerts an important influence on these labile components in plant litter during snow formation, snow coverage and snowmelt in this alpine forest.

Reduced snow cover increases soil freezing (Figure 1d) [16] and alters soil microbial processes in cold areas [33,34]. In this study, a reduction in snow depth decreased the gravimetric water content, which had a negative effect ($r = -0.25$, $p < 0.001$) on the content of acid hydrolysable components, and this effect was greater than that from the change in temperature driven by the shift in snow depth ($r = -0.11$, $p < 0.01$). These results indicated that a decline in winter snow cover may primarily reduce leaching and, thus, decrease the release of acid hydrolysable components from plant litter [4]. A reduction in snow cover has also been demonstrated to strongly decrease the microbial biomass [35], which was positively correlated with the content of acid hydrolysable components. This result suggests that microorganisms can either degrade the intermediate decomposable compounds from plant-derived carbohydrates [36] or contribute the dead biomass to microbial-derived fractions [37] that can be measured by acid hydrolysis.

5. Conclusions

The acid hydrolysable components released rapidly at the initial stages of decomposition as well as at later periods. However, these labile components greatly increased during snowmelt. Only 4–5% of the acid hydrolysable components remained in litter after four years of decomposition in this alpine forest. A reduction in snow depth decreased the release of acid hydrolysable components from plant litter by altering leaching, microbial biomass and stoichiometry.

Author Contributions: X.N. and F.W. conceived the study; S.L., X.N. and K.Y. performed the field experiment and chemical analysis; S.L. wrote the first draft. All authors contributed critically to the drafts and revisions. All authors have read and agreed to the published version of the manuscript.

Funding: The National Natural Science Foundation of China (32101509, 32022056 and 32171641).

Institutional Review Board Statement: Not applicable.

Informed Consent Statement: Not applicable.

Data Availability Statement: The data are included in the manuscript submitted for peer review and will be archived online as supplementary materials of the manuscript is accepted for publication.

Conflicts of Interest: The authors declare no conflict of interest.

References

1. Adair, E.C.; Parton, W.J.; Grosso, S.J.D.; Silver, W.L.; Harmon, M.E.; Hall, S.A.; Burke, I.C.; Hart, S.C. Simple three-pool model accurately describes patterns of long-term litter decomposition in diverse climates. *Glob. Chang. Biol.* **2008**, *14*, 2636–2660. [CrossRef]
2. Berg, B. Decomposition patterns for foliar litter—A theory for influencing factors. *Soil Biol. Biochem.* **2014**, *78*, 222–232. [CrossRef]
3. Kammer, A.; Schmidt, M.W.I.; Hagedorn, F. Decomposition pathways of ^{13}C-depleted leaf litter in forest soils of the Swiss Jura. *Biogeochemistry* **2012**, *108*, 395–411. [CrossRef]
4. Soong, J.L.; Parton, W.J.; Calderon, F.; Campbell, E.E.; Cotrufo, M.F. A new conceptual model on the fate and controls of fresh and pyrolized plant litter decomposition. *Biogeochemistry* **2015**, *124*, 27–44. [CrossRef]
5. Schimel, J.P.; Schaeffer, S.M. Microbial control over carbon cycling in soil. *Front. Microbiol.* **2012**, *3*, 348. [CrossRef]
6. Preston, C.M.; Nault, J.R.; Trofymow, J.A.; Smyth, C.; CIDET Working Group. Chemical changes during 6 years of decomposition of 11 litters in some Canadian forest sites. part 1. elemental composition, tannins, phenolics, and proximate fractions. *Ecosystems* **2009**, *12*, 1053–1077. [CrossRef]
7. Campbell, E.E.; Parton, W.J.; Soong, J.L.; Paustian, K.; Hobbs, N.T.; Cotrufo, M.F. Using litter chemistry controls on microbial processes to partition litter carbon fluxes with the Litter Decomposition and Leaching (LIDEL) model. *Soil Biol. Biochem.* **2016**, *100*, 160–174. [CrossRef]
8. Schimel, J.; Balser, T.C.; Wallenstein, M. Microbial stress-response physiology and its implications for ecosystem function. *Ecology* **2007**, *88*, 1386–1394. [CrossRef]
9. IPCC. Summary for Policymakers. In *Climate Change 2020: The Physical Science Basis*; Cambridge University Press: New York, NY, USA, 2021.
10. Baptist, F.; Yoccoz, N.G.; Choler, P. Direct and indirect control by snow cover over decomposition in alpine tundra along a snowmelt gradient. *Plant Soil.* **2010**, *328*, 397–410. [CrossRef]
11. Aerts, R.; Callaghan, T.V.; Dorrepaal, E.; Van Logtestijn, R.S.P.; Cornelissen, J.H.C. Seasonal climate manipulations have only minor effects on litter decomposition rates and N dynamics but strong effects on litter P dynamics of sub-arctic bog species. *Oecologia* **2012**, *170*, 809–819. [CrossRef]

12. Brooks, P.D.; Grogan, P.; Templer, P.H.; Groffman, P.; Öquist, M.G.; Schimel, J. Carbon and nitrogen cycling in snow-covered environments. *Geogr. Comp.* **2011**, *5*, 682–699. [CrossRef]
13. Neff, J.C.; Asner, G.P. Dissolved organic carbon in terrestrial ecosystems: Synthesis and a model. *Ecosystems* **2001**, *4*, 29–48. [CrossRef]
14. Edwards, A.C.; Scalenghe, R.; Freppaz, M. Changes in the seasonal snow cover of alpine regions and its effect on soil processes: A review. *Quatern. Int.* **2007**, *162*, 172–181. [CrossRef]
15. Kreyling, J.; Haei, M.; Laudon, H. Snow removal reduces annual cellulose decomposition in a riparian boreal forest. *Can. J. Soil Sci.* **2013**, *93*, 427–433. [CrossRef]
16. Groffman, P.M.; Driscoll, C.T.; Fahey, T.J.; Hardy, J.P.; Fitzhugh, R.; Tierney, G.L. Colder soils in a warmer world: A snow manipulation study in a northern hardwood forest ecosystem. *Biogeochemistry* **2001**, *56*, 135–150. [CrossRef]
17. Matzner, E.; Borken, W. Do freeze-thaw events enhance C and N losses from soils of different ecosystems? A review. *Eur. J. Soil Sci.* **2008**, *59*, 274–284. [CrossRef]
18. Ni, X.; Yang, W.; Li, H.; Xu, L.; He, J.; Tan, B.; Wu, F. The responses of early foliar litter humification to reduced snow cover during winter in an alpine forest. *Can. J. Soil Sci.* **2014**, *94*, 453–461. [CrossRef]
19. Wu, F.; Peng, C.; Zhu, J.; Zhang, J.; Tan, B.; Yang, W. Impacts of freezing and thawing dynamics on foliar litter carbon release in alpine/subalpine forests along an altitudinal gradient in the eastern Tibetan Plateau. *Biogeosciences* **2014**, *11*, 6471–6481.
20. IUSS Working Group WRB. *World Reference Base for Soil Resources 2014*; FAO: Rome, Italy, 2015.
21. Ni, X.; Yang, W.; Tan, B.; He, J.; Xu, L.; Li, H.; Wu, F. Accelerated foliar litter humification in forest gaps: Dual feedbacks of carbon sequestration during winter and the growing season in an alpine forest. *Geoderma* **2015**, *241*, 136–144. [CrossRef]
22. Bradford, M.A.; Tordoff, G.M.; Eggers, T.; Jones, T.H.; Newington, J.E. Microbiota, fauna, and mesh size interactions in litter decomposition. *Oikos* **2002**, *99*, 317–323. [CrossRef]
23. Ladwig, L.M.; Ratajczak, Z.R.; Ocheltree, T.W.; Hafich, K.A.; Churchill, A.C.; Frey, S.J.K.; Fuss, C.B.; Kazanski, C.E.; Muñoz, J.D.; Petrie, M.D.; et al. Beyond arctic and alpine: The influence of winter climate on temperate ecosystems. *Ecology.* **2016**, *97*, 272–283. [CrossRef] [PubMed]
24. Olsson, P.Q.; Sturm, M.; Racine, C.H.; Romanovsky, V.; Liston, G.E. Five stages of the Alaskan arctic cold season with ecosystem implications. *Arc. Ant. Alp. Res.* **2003**, *35*, 74–81. [CrossRef]
25. Konestabo, H.S.; Michelsen, A.; Holmstrup, M. Responses of springtail and mite populations to prolonged periods of soil freeze-thaw cycles in a sub-arctic ecosystem. *Appl. Soil Ecol.* **2007**, *36*, 136–146. [CrossRef]
26. Rovira, P.; Vallejo, V.R. Labile and recalcitrant pools of carbon and nitrogen in organic matter decomposition at different depths in soil: An acid hydrolysis approach. *Geoderma* **2002**, *107*, 109–141. [CrossRef]
27. Jenkinson, D.S.; Powlson, D.S. The effects of biocidal treatments on metabolism in soil V. A method for measuring soil biomass. *Soil Biol. Biochem.* **1976**, *8*, 209–213. [CrossRef]
28. Cotrufo, M.F.; Soong, J.L.; Horton, A.J.; Campbell, E.E.; Haddix, M.L.; Wall, D.H.; Parton, W.J. Formation of soil organic matter via biochemical and physical pathways of litter mass loss. *Nat. Geosci.* **2015**, *8*, 776–779. [CrossRef]
29. Li, H.; Wu, F.; Yang, W.; Xu, L.; Ni, X.; He, J.; Tan, B.; Hu, Y. Effects of forest gaps on litter lignin and cellulose dynamics vary seasonally in an alpine forest. *Forests.* **2016**, *7*, 27. [CrossRef]
30. Schmidt, S.K.; Lipson, D.A. Microbial growth under the snow: Implications for nutrient and allelochemical availability in temperate soils. *Plant Soil* **2004**, *259*, 1–7. [CrossRef]
31. Lipson, D.A.; Schmidt, S.K.; Monson, R.K. Carbon availability and temperature control the post-snowmelt decline in alpine soil microbial biomass. *Soil Biol. Biochem.* **2004**, *32*, 441–448. [CrossRef]
32. Yang, Y.; Wu, F.; He, Z.; Xu, Z.; Liu, Y.; Yang, W.; Tan, B. Effects of snow pack removal on soil microbial biomass carbon and nitrogen and the number of soil culturable microorganisms during wintertime in alpine *Abies faxoniana* forest of western Sichuan, Southwest China. *Chin. J. Appl. Ecol.* **2012**, *23*, 1809–1816, (In Chinese with English abstract).
33. Monson, R.K.; Lipson, D.L.; Burns, S.P.; Turnipseed, A.A.; Delany, A.C.; Williams, M.W.; Schmidt, S.K. Winter forest soil respiration controlled by climate and microbial community composition. *Nature* **2006**, *439*, 711–714. [CrossRef] [PubMed]
34. Sorensen, P.O.; Templer, P.H.; Christenson, L.; Duran, J.; Fahey, T.; Fisk, M.C.; Groffman, P.M.; Morse, J.L.; Finzi, A.C. Reduced snow cover alters root-microbe interactions and decreases nitrification rates in an northern hardwood forest. *Ecology* **2016**, *97*, 3359–3368. [CrossRef] [PubMed]
35. Sorensen, P.O.; Templer, P.H.; Finzi, A.C. Contrasting effects of winter snowpack and soil frost on growing season microbial biomass and enzyme activity in two mixed-hardwood forests. *Biogeochemistry* **2016**, *128*, 141–154. [CrossRef]
36. Kaiser, K.; Kalbitz, K. Cycling downwards—dissolved organic matter in soils. *Soil Biol. Biochem.* **2012**, *52*, 29–32. [CrossRef]
37. Ni, X.; Liao, S.; Tan, S.; Wang, D.; Peng, Y.; Yue, K.; Wu, F.; Yang, Y. A quantitative assessment of amino sugars in soil profiles. *Soil Biol. Biochem.* **2020**, *143*, 107762. [CrossRef]

Article

Tree Fresh Leaf- and Twig-Leached Dissolved Organic Matter Quantity and Biodegradability in Subtropical Plantations in China

Jia-Wen Xu [1], Jing-Hao Ji [1], Dong-Nan Hu [2,*], Zhi Zheng [1] and Rong Mao [1,3,*]

[1] Key Laboratory of National Forestry and Grassland Administration on Forest Ecosystem Protection and Restoration of Poyang Lake Watershed, Nanchang 330045, China; xujiawen0823@163.com (J.-W.X.); jinuo1109057446@163.com (J.-H.J.); zhiz010211@gmail.com (Z.Z.)

[2] College of Forestry, Jiangxi Agricultural University, Nanchang 330045, China

[3] Wuyishanxipo Observation and Research Station of Forest Ecosystem, Fuzhou 335300, China

[*] Correspondence: dnhu98@163.com (D.-N.H.); maorong@jxau.edu.cn (R.M.)

Abstract: Extreme weather events often cause the input of fresh plant tissues into soils in forests. However, the interspecific patterns of tree fresh plant tissue-leached dissolved organic matter (DOM) characteristics are poorly understood. In this study, we collected fresh leaves and twigs of two broadleaf trees (*Liquidambar formosana* and *Schima superba*) and two coniferous trees (*Pinus massoniana* and *Pinus elliottii*) in subtropical plantations in China, and measured tree fresh tissue-leached DOM quantity and biodegradability. The interspecific patterns of fresh plant tissue-leached DOM production varied with organ types. Broadleaf tree leaves leached greater amounts of dissolved organic carbon (DOC), dissolved total nitrogen (DTN), and dissolved total phosphorus (DTP) than coniferous tree leaves, but an opposite pattern of DOC and DTN productions was observed between broadleaf and coniferous tree twigs. Regardless of tree species, leaves often leached greater quantities of DOC, DTN, and DTP than twigs. For both leaves and twigs, broadleaf tree tissue-leached DOM had greater aromaticity and lower biodegradability than coniferous tree tissue-leached DOM. Moreover, leaf-leached DOM had greater aromaticity and lower biodegradability than twig-leached DOM. In addition, DOM biodegradability negatively correlated with the initial aromaticity and DOC:DTN ratio, despite no relationship between DOM biodegradability and DOC:DTP ratio. These findings highlight the pivotal roles of leaf habit and organ type in regulating fresh tree tissue-leached DOM production and biodegradability and reveal that the substantial variations of fresh tissue-leached DOM biodegradability are co-driven by DOM aromaticity and N availability in subtropical plantations in China.

Keywords: decomposition; extreme weather event; leaching; plant growth form; stoichiometry

Citation: Xu, J.-W.; Ji, J.-H.; Hu, D.-N.; Zheng, Z.; Mao, R. Tree Fresh Leaf- and Twig-Leached Dissolved Organic Matter Quantity and Biodegradability in Subtropical Plantations in China. *Forests* **2022**, *13*, 833. https://doi.org/10.3390/f13060833

Academic Editors: Fuzhong Wu, Zhenfeng Xu and Wanqin Yang

Received: 29 April 2022
Accepted: 23 May 2022
Published: 27 May 2022

Publisher's Note: MDPI stays neutral with regard to jurisdictional claims in published maps and institutional affiliations.

1. Introduction

Dissolved organic matter (DOM), a small but labile fraction of soil organic matter, plays a crucial role in regulating key ecological processes in forests [1,2]. For example, DOM provides the main sources of energy and nutrients for heterotrophic microbes [3–6], and thus drives greenhouse gas emissions and soil organic matter formation in forests [7–9]. Moreover, DOM is an important vector controlling the transport of carbon (C) and other elements from terrestrial to aquatic systems [1,10,11]. Therefore, knowledge about DOM characteristics is necessary to fully understand forest ecosystem structure and function.

In forests, the leaching of soluble organic matter from plant litter, especially in the early stage of decomposition, is considered the primary source of DOM in soils [12,13]. It is well acknowledged that litter-leached DOM amounts are controlled by a variety of physical and chemical traits such as water holding capacity [14], soluble carbohydrates [15], and lignin concentration [16]. Likewise, litter-leached DOM biodegradability is co-regulated by C quality [3,17] and stoichiometric ratios between C and nutrients [18–21]. However, these previous studies have concentrated on senesced litter [4,16,22], and little is known about the production and biodegradability of DOM leached from fallen fresh plant tissues. In forests, extreme weather events such as hurricanes and storms often destroy the trees, increasing the input of fresh plant tissues into soils [23–25]. For example, typhoon-induced fine litterfall ranges from 2.21 to 5.12 Mg ha^{-1} year^{-1} in subtropical pine plantations [24]. Due to the substantial differences in physical and chemical properties between fresh and senesced plant tissues, the controls on amounts and biodegradability of fresh organ-leached DOM are unclear in these ecosystems. These knowledge gaps will limit our understanding of plant-mediated C and nutrient cycles in forests.

Here, we sampled the fresh leaves and twigs of four common afforestation species (broadleaf trees: *Liquidambar formosana* and *Schima superba*; coniferous trees: *Pinus massoniana* and *Pinus elliottii*) from subtropical plantations in southern China, and then conducted a 48 h leaching experiment to investigate the amounts of fresh tissue-leached dissolved organic carbon (DOC), dissolved total nitrogen (DTN), and dissolved total phosphorus (DTP). In this study, we used the amount of DOC in the leachates to indicate DOM production and assessed DOM aromaticity by measuring the specific ultraviolet absorbances at 254 nm (SUVA$_{254}$) and 350 nm (SUVA$_{350}$) [26,27]. Subsequently, we measured DOM biodegradability using a 28-day standard incubation experiment [17]. The main objectives of this study were to: (1) determine the effects of tree species and organ type on fresh tissue-leached DOM production and biodegradability, and (2) uncover the factors determining the variations of fresh tissue-leached DOM quantity and biodegradability in subtropical plantations.

2. Materials and Methods

2.1. Study Site

Fresh tree tissues were collected from plantations in the long-term Forest Restoration Experimental Station of Jiangxi Agricultural University (26°55′16″ N, 114°48′14″ E) located in the Luoxi Town, Taihe County, Jiangxi Province, southern China. The study site has a humid subtropical monsoon climate. Mean annual temperature is 18.6 °C, and mean annual precipitation is 1726 mm. These plantations have been established in 1991 and the main afforestation species are *L. formosana*, *S. superba*, *P. massoniana*, and *P. elliottii*. These selected four tree species have been widely planted for restoring degraded lands and maintaining ecosystem services in subtropical regions of China. For example, the area of *P. massoniana* plantations has reached 1.00×10^7 hm^2 according to the eighth National Forest Resource Survey in China. Mean tree height in the *L. formosana*, *S. superba*, *P. massoniana*, and *P. elliottii* plantations was 10.7 m, 9.8 m, 8.7 m, and 11.6 m, respectively, and mean diameter at breath height (DBH) in the *L. formosana*, *S. superba*, *P. massoniana*, and *P. elliottii* plantations was 16.2 cm, 20.3 cm, 17.9 cm, and 26.4 cm, respectively. The detailed description of the study site is shown in Xu et al. [28].

2.2. Plant Sampling and Measurement

In the study site, we randomly established six plots (10 m × 10 m) in the *L. formosana*, *S. superba*, *P. massoniana*, and *P. elliottii* plantations in June 2019. In each plot, we selected three individuals with similar tree height and DBH and then harvested thirty fully expanded and healthy leaves and ten twigs from the canopy of each targeted individual. For each organ type, plant materials collected from the same plot were mixed and divided into two subsamples. The first subsample was oven-dried at 65 °C to determine the initial moisture content and then ground to pass through 0.15 mm sieves for chemical analyses, and the second subsample was used for the leaching experiment. Plant organic C and total N concentrations were measured on a FlashSmart CHNS/O Elemental Analyzer (Thermo Fisher Scientific, Bremen, Germany), total P concentration was colorimetrically determined on an autoanalyzer (AA3, Seal Analytical, Germany) after acid digestion, total polyphenols were measured with the Folin–Ciocalteu method [29], soluble sugars and starch were measured by the anthrone colorimetry method [30], and tissue density was assessed by the procedures of Pérez-Harguindeguy et al. [31]. The initial chemical properties of tree tissues are shown in Table 1.

Tree tissue-leached DOM was extracted with a short-term leaching experiment [32]. For each tree organ, 3 g of fresh plant material was placed in 500 mL glass jars and soaked in 200 mL of deionized water in the dark at room temperature (about 20 °C) for 48 h. Subsequently, the leachates were filtered through 0.7 μm Whatman ™ GF/F glass microfiber filters (Little Chalfont, Buckinghamshire, UK) and used to measure DOM parameters. In the leachates, DOC and DTN concentrations were measured on a TOC analyzer (multi N/C 2100S, Analytik Jena, Jena, Germany), DTP concentration was colorimetrically measured on an autoanalyzer after peroxodisulfate oxidation [33], and the ultraviolet absorbances of DOM at 254 nm and 350 nm was measured with an ultraviolet-visible spectrophotometer (UV600SC, Jinghua Instruments, Shanghai, China). Plant-leached DOC, DTN, and DTP productions were obtained from the total amounts of DOC, DTN, and DTP in the leachates and the dry mass of fresh tree tissues. The specific ultraviolet absorbance at 254 nm ($SUVA_{254}$) and 350 nm ($SUVA_{350}$) were calculated by dividing the ultraviolet absorbance at 254 nm and 350 nm by the DOC concentration, respectively [26,27].

Plant-leached DOM biodegradability was assessed with a 28-day standard laboratory incubation experiment [17]. Prior to incubation, DOC concentration in the leachates was diluted to about 20 mg L^{-1} to avoid excessive microbial growth. Meanwhile, microbial inoculums were obtained by placing 5 g of fresh forest soils in 1000 mL of deionized water. Subsequently, 100 mL of diluted leachates per treatment was placed in a 500 mL glass jar, and 5 mL of microbial inoculums was added to the jar. In addition, six glass jars with 100 mL of deionized water and 5 mL of microbial inoculums was established as blanks. All glass jars were sealed and aerobically incubated in the dark at 20 °C for 28 days. By the end of incubation, DOC concentration in the leachates was determined on a TOC analyzer. Plant-leached DOM biodegradability was calculated from the difference between the initial and final DOC amounts and was expressed as the proportion of the initial DOC amount (%).

2.3. Statistical Analyses

All statistical analyses were performed with R version 4.1.1 [34], and the accepted significant level was set at $\alpha = 0.05$. Linear mixed models were used to assess the effects of tree species, organ type, and their interaction on DOM parameters with the 'nlme' package. For each tree species, paired *t*-test was used to examine the differences in DOM parameters between leaves and twigs. For each plant organ, one-way ANOVA with Tukey's HSD comparison was applied to determine the differences in DOM parameters among the tree species with the 'agricolae' package. Spearman's correlation analysis was used to examine the relationships between DOM parameters and initial properties of plant materials with the 'survey' package. A linear regression analysis was used to determine the relationships between DOM biodegradability and initial DOM properties with the 'abline' function.

Table 1. Initial properties of tree leaf and twig in subtropical plantations in southern China.

Species	Organ Type	Organic C	Total N	Total P	C:N Ratio	C:P Ratio	Polyphenols	Soluble Sugars	Starch	Tissue Density
		$mg\,g^{-1}$	$mg\,g^{-1}$	$\mu g\,g^{-1}$			$mg\,g^{-1}$	$mg\,g^{-1}$	$mg\,g^{-1}$	$g\,cm^{-3}$
L. formosana	Leaf	465 (3) B	15.7 (0.29) A	1338 (28) A	29.7 (0.4) C	348 (7) C	128.2 (1.9) A	66.5 (1.1) A	29.9 (1.1) A	0.782 (0.01) A
	Twig	551 (4) a	5.84 (0.07) a	704 (11) b	94.4 (1.6) b	784 (11) b	40.5 (0.7) b	37.5 (1.1) a	22.0 (0.9) a	0.567 (0.01) a
	p-value	<0.001	<0.001	<0.001	<0.001	<0.001	<0.001	<0.001	0.020	<0.001
S. superba	Leaf	482 (4) A	14.2 (0.17) B	1142 (25) B	34.1 (0.6) B	424 (12) B	120.1 (1.6) B	60.1 (1.6) B	26.7 (0.6) A	0.468 (0.01) C
	Twig	460 (4) d	4.12 (0.06) c	676 (7) b	111.8(2.0)a	681 (6) c	76.8 (0.7) a	34.9 (0.7) a	19.8 (1.1) ab	0.605 (0.02) a
	p-value	0.01	<0.001	<0.001	<0.001	<0.001	<0.001	<0.001	<0.001	0.001
P. massoniana	Leaf	475 (3) AB	13.5 (0.17) B	1377 (18) A	35.4 (0.6) B	345 (6) C	61.6 (1.0) D	52.1 (1.3) C	21.0 (0.7) B	0.693 (0.002) B
	Twig	516 (4) b	5.77 (0.06) a	583 (8) c	89.4 (0.9) b	885 (11) a	23.2 (0.5) c	29.1 (1.1) b	18.1 (0.8) b	0.456 (0.001) b
	p-value	0.01	<0.001	<0.001	<0.001	<0.001	<0.001	<0.001	0.016	<0.001
P. elliottii	Leaf	484 (3) A	8.8 (0.15) C	918 (22) C	55.3 (1.2) A	528 (13) A	84.7 (0.9) C	49.4 (1.4) C	20.0 (0.8) B	0.676 (0.002) B
	Twig	498 (5) c	5.27 (0.07) b	745 (9) a	94.7 (1.4) b	669 (10) c	21.3 (0.7) c	29.5 (1.1) b	16.4 (0.8) b	0.471 (0.001) b
	p-value	0.05	<0.001	<0.001	<0.001	<0.001	<0.001	<0.001	0.042	<0.001

The data in the parentheses are the standard errors of the means ($n = 6$). In the same column, the different uppercase letters indicate the significant differences among the tree leaf treatments ($p < 0.05$), and the lowercase letters indicate the significant differences among the tree twig treatments ($p < 0.05$). For the same tree species, *p*-value < 0.05 indicate the significant differences between leaf and twig.

3. Results

Tree species, organ type, and their interaction significantly affected DOC production, DTN production, DTP production, DOC:DTN ratio, and DOC:DTP ratio (Table 2). For each tree species, leaves leached greater amounts of DOC, DTN, and DTP than twigs (Table 3). Broadleaf tree (*L. formosana* and *S. superba*) leaves had greater DOC, DTN, and DTP productions than coniferous tree (*P. massoniana* and *P. elliottii*) leaves, whereas broadleaf tree twigs had lower DOC production than coniferous tree twigs (Table 3). Plant-leached DOC, DTN, and DTP productions showed a negative relationship with the initial tissue C:N and C:P ratios, but exhibited a positive relationship with polyphenols, soluble sugars, and starch (Table A1). In the leachates, tree leaves often had higher DOC:DTN and DOC:DTP ratios than tree twigs (Table 3). In addition, broadleaf tree leaves had greater DOC:DTN ratio in the leachates than coniferous tree leaves, although there was no significant difference between broadleaf and coniferous tree twigs (Table 3).

Table 2. Results (F-values) of linear mixed models on the effects of tree species (S), organ type (O), and their interaction on tree tissue-leached DOM properties and biodegradability in subtropical plantations in southern China.

Sources of Variation	DOC Production	DTN Production	DTP Production	DOC:DTN Ratio	DOC:DTP Ratio	SUVA$_{254}$	SUVA$_{350}$	DOM Biodegradability
S	527 ***	158 ***	426 ***	85 ***	120 ***	53 ***	87 ***	41 ***
O	2697 ***	1450 ***	1783 ***	481 ***	89 ***	60 ***	119 ***	27 ***
S × O	707 ***	417 ***	329 ***	67 ***	85 ***	0.35	10 ***	0.31

DOC, dissolved organic carbon; DTN, dissolved total nitrogen; DTP, dissolved total phosphorus; DOM, dissolved organic matter. ***, $p < 0.001$.

Table 3. Dissolved organic C (DOC), dissolved total N (DTN), and dissolved total P (DTP) productions, and stoichiometric ratios of tree tissue leachates in subtropical plantations in southern China.

Species	Organ Type	DOC Production	DTN Production	DTP Production	DOC:DTN Ratio	DOC:DTP Ratio
		mg g^{-1}	µg g^{-1}	µg g^{-1}		
L. formosana	Leaf	60.9 (1.4) A	323 (4) A	61.2 (1.5) A	188 (5) A	997 (29) B
	Twig	7.7 (0.4) b	80 (4) c	17.1 (0.6) a	97 (5) a	449 (7) c
	p-value	<0.001	<0.001	<0.001	<0.001	<0.001
S. superba	Leaf	35.7 (0.3) B	212 (4) B	46.4 (0.8) B	169 (3) B	770 (12) C
	Twig	8.3 (0.4) b	107 (4) b	11.5 (0.7) c	77 (3) b	726 (15) b
	p-value	<0.001	0.001	<0.001	<0.001	0.055
P. massoniana	Leaf	19.2 (0.3) C	161 (4) C	17.3 (0.6) C	120 (2) C	1114 (34) A
	Twig	14.0 (0.6) a	138 (6) a	14.0 (0.5) b	102 (3) a	998 (25) a
	p-value	0.001	<0.001	0.005	0.006	0.003
P. elliottii	Leaf	13.1 (0.3) D	134 (5) D	21.1 (0.8) C	98 (4) D	626 (22) D
	Twig	8.8 (0.6) b	114 (4) b	11.8 (0.6) c	77 (3) b	749 (18) b
	p-value	0.002	0.003	0.001	0.014	0.010

DOC, dissolved organic carbon; DTN, dissolved total nitrogen; DTP, dissolved total phosphorus. The data in the parentheses are the standard errors of the means ($n = 6$). In the same column, the different uppercase letters indicate the significant differences among the tree leaf treatments ($p < 0.05$), and the lowercase letters indicate the significant differences among the tree twig treatments ($p < 0.05$).

Tree species, organ type, and their interaction produced significant effects on SUVA$_{254}$ and SUVA$_{350}$ values (Table 2). Irrespective of tree species, leaves had higher SUVA$_{254}$ and SUVA$_{350}$ values in the leachates than twigs (Figure 1). For both leaves and twigs, broadleaf trees had greater SUVA$_{254}$ and SUVA$_{350}$ values in the leachates than coniferous trees (Figure 1).

Tree species and organ type independently produced a significant effect on DOM biodegradability (Table 2). Regardless of tree species, twig-leached DOM had greater biodegradability than leaf-derived DOM (Figure 1). For both leaves and twigs, broadleaf trees had lower DOM biodegradability than coniferous trees (Figure 1). In addition, tissue-leached DOM biodegradability was negatively related to the initial SUVA$_{254}$ value, SUVA$_{350}$ value, and DOC:DTN ratio, but exhibited no significant relationship with DOC:DTP ratio (Figure 2).

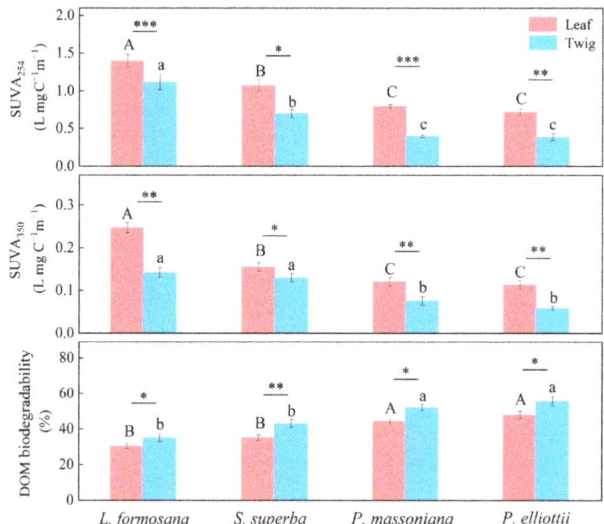

Figure 1. Optical properties and biodegradability of tree litter-derived dissolved organic matter (DOM) in subtropical plantations in southern China. The error bars are the standard error of the means ($n = 6$). Different uppercase letters indicate the significant differences among the tree leaf treatments ($p < 0.05$), and different lowercase letters indicate the significant differences among the tree twig treatments ($p < 0.05$). For each tree species, *, **, and *** indicate the significant differences between leaf and twig at the levels of $p < 0.05$, $p < 0.01$, and $p < 0.001$, respectively.

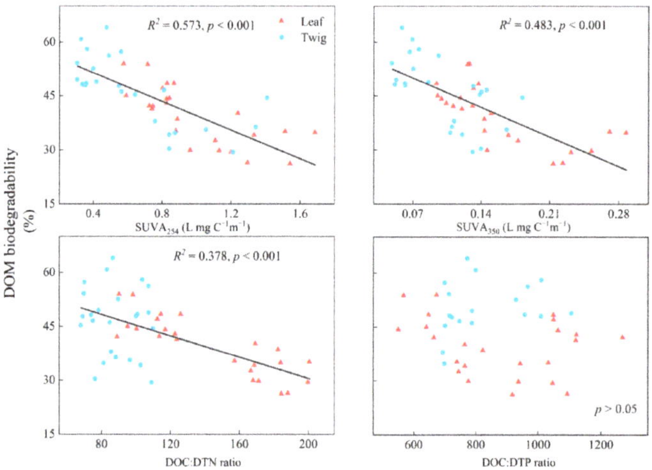

Figure 2. Relationships between plant-leached dissolved organic matter (DOM) biodegradability and the initial properties in subtropical plantations in southern China.

4. Discussion

The interspecific patterns of leaf- and twig-leached DOM production differed between broadleaf and coniferous trees. Specifically, broadleaf trees produced greater amounts of DOC and soluble nutrients in the leaf leachates but had lower DOC quantity in the twig leachates than coniferous trees (Table 3). Previous studies also observed greater amounts of DOM for broadleaf tree leaf litter than for coniferous tree leaf litter in temperate and subtropical forests [16,19,35]. In general, plant-leached DOM production is controlled by litter physical and chemical traits such as non-structural carbohydrates and lignin concentration [15,16,28,36]. Thus, these inconsistent results would be explained by the differences in interspecific patterns of leaf and twig traits between broadleaf and coniferous trees. These findings suggest that the interspecific patterns of fresh plant tissue-leached DOM production between broadleaf and coniferous trees is largely dependent on organ type in subtropical plantations.

In this study, the greater DOM production of broadleaf trees relative to coniferous trees would be explained by the following mechanisms. First, broadleaf trees have higher soluble sugars, starch, and polyphenols concentrations in fresh leaves than coniferous trees (Table 1). Second, compared with coniferous trees, broadleaf trees had greater specific leaf area and water holding capacity [37], which would enhance leaching of soluble compounds from fresh leaves. However, coniferous tree twigs had lower tissue density than broadleaf tree twigs (Table 1), which would favor the entry of waters into plant tissues [14]. Accordingly, DOC production was greater for coniferous tree twigs than for broadleaf tree twigs.

During leaching, broadleaf tree leaves had greater DTN and DTP productions than coniferous tree leaves, despite no significant difference between broadleaf and coniferous tree twigs (Table 3). These inconsistent interspecific patterns would be explained by the differences in nutrient concentrations in fresh tissues between broadleaf and coniferous trees. In forest soils, the soluble nutrient leached from plant litter is the primary nutrient source for microbial growth and activity [3,4,38]. Accordingly, the interspecific difference of fresh tissue-leached DTN and DTP productions would partly help explain the spatial variations of soil microbial-mediated ecological processes between broadleaf and coniferous tree plantations in subtropical regions.

Tree fresh tissue-leached DOC, DTN, and DTP productions negatively correlated with both C:N ratio and C:P ratio (Table A1). This result was contrary to the previous studies which did not observe any significant relationship between litter-leached DOM production and C:N:P stoichiometry [35,39]. These inconsistent findings indicate the contrasting roles of C:N:P stoichiometric ratios in regulating DOM production between fresh and senesced tree tissues. In plant tissues, the higher C:N and C:P ratios represent the lower metabolic activities and associated smaller amounts of non-structural compounds and nutrients [40,41]. In this study, plant tissues with higher C:N and C:P ratios were observed to have lower soluble sugars and starch (Table 1). Thus, fresh tree tissue-leached DOM production increased with decreasing C:N and C:P ratios.

For both leaves and twigs, broadleaf trees had greater $SUVA_{254}$ and $SUVA_{350}$ values in the leachates than coniferous trees (Figure 1), indicating that broadleaf tree tissue-leached DOM had greater amounts of aromatic compounds than coniferous tree tissue-leached DOM [26,27]. Moreover, both $SUVA_{254}$ and $SUVA_{350}$ values correlated negatively with C:N and C:P ratios, but correlated positively with soluble sugars, starch, and polyphenols (Table A1). Hagedorn and Machwitz [15] also found that the molar UV absorptivity at 285 nm of litter-derived DOM exhibited negative relationship with C:N and lignin:N ratios. In this study, broadleaf trees had higher soluble organic compounds in the leaves and twigs than coniferous trees (Table 1), and thus produced DOM with greater aromaticity in subtropical plantations.

Notably, broadleaf trees had lower fresh leaf- and twig-leached DOM biodegradability than coniferous trees during 28-day incubation (Figure 1). This interspecific pattern of fresh tissue-leached DOM biodegradability was contrary to the previous studies reporting greater senesced leaf-leached DOM biodegradability for broadleaf trees than for coniferous trees in forests [18,19,42]. These inconsistent results indicate that the interspecific variations of plant-leached DOM biodegradability differ between fresh and senesced tissues in forests. In general, DOM biodegradability is influenced by DOM aromaticity and C:N:P stoichiometry because microbial activity is limited by the availability of energy and nutrients [17,19,28,43]. The initial $SUVA_{254}$ value, $SUVA_{350}$ value, and DOC:DTN ratio were also observed to correlate negatively with DOM biodegradability in this study (Figure 2). In these forests, coniferous tree tissues often had lower aromatic C content and DOC:DTN ratio in the leachates than broadleaf tree tissues (Figure 1). Therefore, coniferous trees had higher fresh tissue-leached DOM biodegradability than broadleaf trees. These results imply that the interspecific variation of fresh tissue-leached DOM biodegradability is co-regulated by C quality and N availability in subtropical plantations.

Tree fresh leaves had greater DOC, DTN, and DTP productions and DOM aromaticity, but had lower DOM biodegradability than tree fresh twigs (Table 3 and Figure 1). Kammer and Hagedorn [44] also found that senescent leaves produced greater quantity of DOC with higher refractory components than senesced twigs in a temperate forest. In these forests, tree fresh leaves had higher soluble sugars and starch contents and lower tissue density than tree fresh twigs (Table 1). In contrast, tree leaves could allocate a greater amount of defensive compounds (i.e., condensed tannins and polyphenols) to resist herbivores than tree twigs due to the relatively higher nutrient concentrations [45]. In this study, higher polyphenol concentration was also observed for tree leaves than tree twigs (Table 1). Therefore, tree fresh leaves produced higher amounts of DOM containing greater aromatic compounds than tree fresh twigs in subtropical plantations. Due to the lower DOM aromaticity and DOC:DTN ratio, fresh tissue-leached DOM biodegradability was higher for tree twigs than for tree leaves in these plantations.

In forests, litter-leached DOM strongly influences soil organic matter formation and stabilization [1,9]. In these plantations, tree fresh leaves produced much greater amounts of DOM with higher biodegradability than tree senesced leaves [19]. These observations imply that tree fresh leaf-leached DOM would promote the accumulation of mineral-associated C fractions via the microbial pathway [9,38]. Considering that the intensity and frequency of extreme weather events exhibit an increasing trend as a result of global climate change [46], our findings highlight the crucial role of plant fresh tissue in regulating C budget in subtropical forest ecosystems.

5. Conclusions

In summary, tree fresh tissue-leached DOM production and biodegradability varied with leaf habit and organ type. During leaching, broadleaf tree fresh leaves produced greater amounts of DOM with higher aromaticity than coniferous tree fresh leaves, whereas broadleaf tree fresh twigs produced lower amounts of DOM with greater aromaticity than coniferous tree fresh twigs. Irrespective of organ type, broadleaf trees had higher tissue-leached DOM biodegradability than coniferous trees due to the lower DOC:DTN ratio and aromatic C content. Given the elevating incidences of extreme weather events such as hurricanes and storms and subsequent increased input of fresh plant tissues in the context of global climate change [46], these findings will help understand and predict the biogeochemical cycles in subtropical plantations in southern China.

Author Contributions: Investigation, supervision, writing—review and editing, J.-W.X.; investigation, methodology, data curation, J.-H.J.; methodology, supervision, writing—review and editing, D.-N.H.; investigation, methodology, data curation, Z.Z.; investigation, supervision, funding acquisition, writing—review and editing, R.M. All authors have read and agreed to the published version of the manuscript.

Funding: This study was supported by the Double Thousand Plan of Jiangxi Province (jxsq2018106044) to R.M.

Institutional Review Board Statement: Not applicable.

Informed Consent Statement: Not applicable.

Data Availability Statement: The data reported in this study are available on request from the corresponding authors. The data are not publicly available yet due to the authors are writing some other papers by mining these data.

Conflicts of Interest: The authors declare no conflict of interest.

Appendix A

Table A1. Spearman's correlation coefficients between tree tissue-leached DOM parameters and the initial chemical traits in subtropical plantations in southern China.

	C:N Ratio	C:P Ratio	Polyphenols	Soluble Sugars	Starch
DOC production	−0.891 ***	−0.701 ***	0.641 ***	0.746 ***	0.538 ***
DTN production	−0.870 ***	−0.716 ***	0.626 ***	0.709 ***	0.529 ***
DTP production	−0.845 ***	−0.627 ***	0.779 ***	0.870 ***	0.699 ***
SUVA$_{254}$ value	−0.543 ***	−0.475 **	0.705 ***	0.751 ***	0.758 ***
SUVA$_{350}$ value	−0.481 **	−0.483 **	0.781 ***	0.775 ***	0.849 ***

DOC, dissolved organic carbon; DTN, dissolved total nitrogen; DTP, dissolved total phosphorus; SUVA$_{254}$, the specific UV absorbance at 254 nm; SUVA$_{350}$, the specific UV absorbance at 350 nm. **, $p < 0.01$; ***, $p < 0.001$.

References

1. Kalbitz, K.; Sollinger, S.; Park, J.H.; Michalzik, B.; Matzner, E. Controls on the dynamics of dissolved organic matter in soils: A review. *Soil Sci.* **2000**, *165*, 277–304. [CrossRef]
2. Kalbitz, K.; Kaiser, K. Contribution of dissolved organic matter to carbon storage in forest mineral soils. *J. Plant Nutr. Soil Sci.* **2008**, *171*, 52–60. [CrossRef]
3. Marschner, B.; Kalbitz, K. Controls of bioavailability and biodegradability of dissolved organic matter in soils. *Geoderma* **2003**, *113*, 211–235. [CrossRef]
4. Joly, F.X.; Fromin, N.; Kiikkilä, O.; Hättenschwiler, S. Diversity of leaf litter leachates from temperate forest trees and its consequences for soil microbial activity. *Biogeochemistry* **2016**, *129*, 373–388. [CrossRef]
5. Jílková, V.; Jandová, K.; Cajthaml, T.; Kukla, J.; Jansa, J. Differences in the flow of spruce-derived needle leachates and root exudates through a temperate coniferous forest mineral topsoil. *Geoderma* **2022**, *405*, 115442. [CrossRef]
6. De Marco, A.; Spaccini, R.; Virzo De Santo, A. Differences in nutrients, organic components and decomposition pattern of *Phillyrea angustifolia* leaf litter across a low maquis. *Plant Soil* **2021**, *464*, 559–578. [CrossRef]
7. Schwesig, D.; Kalbitz, K.; Matzner, E. Mineralization of dissolved organic carbon in mineral soil solution of two forest soils. *J. Plant Nutr. Soil Sci.* **2003**, *166*, 585–593. [CrossRef]
8. Sarker, T.C.; Maisto, G.; De Marco, A.; Esposito, F.; Panico, S.C.; Alam, M.F.; Mazzoleni, S.; Bonanomi, G. Explaining trajectories of chemical changes during decomposition of tropical litter by ^{13}C-CPMAS NMR, proximate and nutrients analysis. *Plant Soil* **2018**, *436*, 13–28. [CrossRef]
9. Cotrufo, M.F.; Soong, J.L.; Horton, A.J.; Campbell, E.E.; Haddix, M.L.; Wall, D.H.; Parton, W.J. Formation of soil organic matter via biochemical and physical pathways of litter mass loss. *Nat. Geosci.* **2015**, *8*, 776–779. [CrossRef]
10. Neff, J.C.; Asner, G.P. Dissolved organic carbon in terrestrial ecosystems: Synthesis and a model. *Ecosystems* **2001**, *4*, 29–48. [CrossRef]
11. Jansen, B.; Kalbitz, K.; McDowell, W.H. Dissolved organic matter: Linking soils and aquatic systems. *Vadose Zone J.* **2014**, *13*, vzj2014.05.0051. [CrossRef]
12. Cleveland, C.C.; Neff, J.C.; Townsend, A.R.; Hood, E. Composition, dynamics, and fate of leached dissolved organic matter in terrestrial ecosystems: Results from a decomposition experiment. *Ecosystems* **2004**, *7*, 175–285. [CrossRef]
13. Soong, J.L.; Parton, W.J.; Calderon, F.; Campbell, E.E.; Cotrufo, M.F. A new conceptual model on the fate and controls of fresh and pyrolized plant litter decomposition. *Biogeochemistry* **2015**, *124*, 27–44. [CrossRef]
14. Xu, J.W.; Yang, N.; Shi, F.X.; Zhang, Y.; Wan, S.Z.; Mao, R. Bark controls tree branch-leached dissolved organic matter production and bioavailability in a subtropical forest. *Biogeochemistry* **2022**, *158*, 345–355. [CrossRef]
15. Hagedorn, F.; Machwitz, M. Controls on dissolved organic matter leaching from forest litter grown under elevated atmospheric CO$_2$. *Soil Biol. Biochem.* **2007**, *39*, 1759–1769. [CrossRef]
16. Don, A.; Kalbitz, K. Amounts and degradability of dissolved organic carbon from foliar litter at different decomposition stages. *Soil Biol. Biochem.* **2005**, *37*, 2171–2179. [CrossRef]
17. Pinsonneault, A.J.; Moore, T.R.; Roulet, N.T.; Lapierre, J.F. Biodegradability of vegetation-derived dissolved organic carbon in a cool temperate ombrotrophic bog. *Ecosystems* **2016**, *19*, 1023–1036. [CrossRef]
18. Ding, Y.D.; Xie, X.Y.; Ji, J.H.; Li, Q.Q.; Xu, J.W.; Mao, R. Tree mycorrhizal effect on litter-leached DOC amounts and biodegradation is highly dependent on leaf habits in subtropical forests of southern China. *J. Soils Sediment* **2021**, *21*, 3572–3579. [CrossRef]

19. Wu, P.P.; Ding, Y.D.; Li, S.L.; Sun, X.X.; Zhang, Y.; Mao, R. Carbon, nitrogen and phosphorus stoichiometry controls interspecific patterns of leaf litter-derived dissolved organic matter biodegradation in subtropical plantations of China. *iForest* **2021**, *14*, 80–85. [CrossRef]

20. Memoli, V.; De Marco, A.; Esposito, F.; Panico, S.C.; Barile, R.; Maisto, G. Seasonality, altitude and human activities control soil quality in a national park surrounded by an urban area. *Geoderma* **2019**, *337*, 1–10. [CrossRef]

21. De Marco, A.; Fioretto, A.; Giordano, M.; Innangi, M.; Menta, C.; Papa, S.; Virzo De Santo, A. C stocks in forest floor and mineral soil of two Mediterranean beech forests. *Forests* **2016**, *7*, 181. [CrossRef]

22. Chomel, M.; Baldy, V.; Guittonny, M.; Greff, S.; DesRochers, A. Litter leachates have stronger impact than leaf litter on Folsomia candida fitness. *Soil Biol. Biochem.* **2020**, *147*, 107850. [CrossRef]

23. Dale, V.H.; Joyce, L.A.; McNulty, S.; Neilson, R.P.; Ayres, M.P.; Flannigan, M.D.; Hanson, P.J.; Irland, L.C.; Lugo, A.E.; Peterson, C.J.; et al. Climate change and forest disturbances: Climate change can affect forests by altering the frequency, intensity, duration, and timing of fire, drought, introduced species, insect and pathogen outbreaks, hurricanes, windstorms, ice storms, or landslides. *BioScience* **2001**, *51*, 723–734. [CrossRef]

24. Xu, X. Nutrient dynamics in decomposing needles of *Pinus luchuensis* after typhoon disturbance in a subtropical environment. *Ann. For. Sci.* **2006**, *63*, 707–713. [CrossRef]

25. Yang, Q.P.; Xu, M.; Chi, Y.G.; Zheng, Y.P.; Shen, R.C.; Wang, S.L. Effects of freeze damage on litter production, quality and decomposition in a loblolly pine forest in central China. *Plant Soil* **2014**, *374*, 449–458. [CrossRef]

26. Weishaar, J.L.; Aiken, G.R.; Bergamaschi, B.A.; Fram, M.S.; Fujii, R.; Mopper, K. Evaluation of specific ultraviolet absorbance as an indicator of the chemical composition and reactivity of dissolved organic carbon. *Environ. Sci. Technol.* **2003**, *37*, 4702–4708. [CrossRef]

27. Hansen, A.M.; Kraus, T.E.C.; Pellerin, B.A.; Fleck, J.A.; Downing, B.D.; Bergamaschi, B.A. Optical properties of dissolved organic matter (DOM): Effects of biological and photolytic degradation. *Limnol. Oceanogr.* **2016**, *71*, 1015–1032. [CrossRef]

28. Xu, J.W.; Ding, Y.D.; Li, S.L.; Mao, R. Amount and biodegradation of dissolved organic matter leached from tree branches and roots in subtropical plantations of China. *For. Ecol. Manag.* **2021**, *484*, 118944. [CrossRef]

29. Yu, Z.; Dahlgren, R.A. Evaluation of methods for measuring polyphenols in conifer foliage. *J. Chem. Ecol.* **2000**, *26*, 2119–2140. [CrossRef]

30. Wang, F.C.; Chen, F.S.; Wang, G.G.; Mao, R.; Fang, X.M.; Wang, H.M.; Bu, W.S. Effects of experimental nitrogen addition on nutrients and nonstructural carbohydrates of dominant understory plants in a Chinese fir plantation. *Forests* **2019**, *10*, 155. [CrossRef]

31. Pérez-Harguindeguy, N.; Díaz, S.; Garnier, E.; Lavorel, S.; Poorter, H.; Jaureguiberry, P.; Bret-Harte, M.S.; Cornwell, W.K.; Craine, J.M.; Gurvich, D.E.; et al. New handbook for standardised measurement of plant functional traits worldwide. *Aust. J. Bot.* **2013**, *61*, 167–234. [CrossRef]

32. Schreeg, L.A.; Mack, M.C.; Turner, B.L. Nutrient-specific solubility patterns of leaf litter across 41 lowland tropical woody species. *Ecology* **2013**, *94*, 94–105. [CrossRef] [PubMed]

33. Ebina, J.; Tsutsui, T.; Shirai, T. Simultaneous determination of total nitrogen and total phosphorus in water using peroxodisulfate oxidation. *Water Res.* **1983**, *17*, 1721–1726. [CrossRef]

34. R Development Core Team. *R: A Language and Environment for Statistical Computing*; R Foundation for Statistical Computing: Vienna, Austria, 2021.

35. Hensgens, G.; Lechtenfeld, O.J.; Guillemette, F.; Laudon, H.; Berggren, M. Impacts of litter decay on organic leachate composition and reactivity. *Biogeochemistry* **2020**, *154*, 99–117. [CrossRef]

36. Chen, H.; Liu, X.J.; Blosser, G.D.; Rücker, A.M.; Conner, W.H.; Chow, A.T. Molecular dynamics of foliar litter and dissolved organic matter during the decomposition process. *Biogeochemistry* **2020**, *150*, 17–30. [CrossRef]

37. Li, S.L.; Xu, J.W.; Ding, Y.D.; Mao, R. Litter water-holding and water-loss characteristics of trees and ferns in the water conservation forests at the middle reaches of the Gan River. *J. Soil Water Conserv.* **2021**, *35*, 170–176.

38. Liang, C. Soil microbial carbon pump: Mechanism and appraisal. *Soil Ecol. Lett.* **2020**, *2*, 241–254. [CrossRef]

39. Bantle, A.; Borken, W.; Ellerbrock, R.H.; Schulze, E.D.; Weisser, W.W.; Matzner, E. Quantity and quality of dissolved organic carbon released from coarse woody debris of different tree species in the early phase of decomposition. *For. Ecol. Manag.* **2014**, *329*, 287–294. [CrossRef]

40. Tang, Z.Y.; Xu, W.T.; Zhou, G.Y.; Bai, Y.F.; Li, J.X.; Tang, X.L.; Chen, D.M.; Liu, Q.; Ma, W.D.; Xiong, G.M.; et al. Patterns of plant carbon, nitrogen, and phosphorus concentration in relation to productivity in China's terrestrial ecosystems. *Proc. Nat. Acad. Sci. USA* **2018**, *115*, 4033–4038. [CrossRef]

41. Sun, W.Z.; Shi, F.X.; Chen, H.M.; Zhang, Y.; Guo, Y.D.; Mao, R. Relationship between relative growth rate and C:N:P stoichiometry for the marsh herbaceous plants under water-level stress conditions. *Glob. Ecol. Conserv.* **2021**, *25*, e01416. [CrossRef]

42. Kiikkilä, O.; Kitunen, V.; Spetz, P.; Smolander, A. Characterization of dissolved organic matter in decomposing Norway spruce and silver birch litter. *Eur. J. Soil Sci.* **2012**, *63*, 476–486. [CrossRef]

43. Mao, R.; Li, S. Temporal controls on dissolved organic carbon biodegradation in subtropical rivers: Initial chemical composition versus stoichiometry. *Sci. Total Environ.* **2019**, *651*, 3064–3069. [CrossRef] [PubMed]

44. Kammer, A.; Hagedorn, F. Mineralisation, leaching and stabilisation of ^{13}C-labelled leaf and twig litter in a beech forest soil. *Biogeosciences* **2011**, *8*, 2195–2208. [CrossRef]

45. Gatehouse, J.A. Plant resistance towards insect herbivores: A dynamic interaction. *New Phytol.* **2002**, *156*, 145–169. [CrossRef] [PubMed]
46. Wang, F.M.; Liu, J.; Zou, B.; Neher, D.A.; Zhu, W.; Li, Z.A. Species-dependent responses of soil microbial properties to fresh leaf inputs in a subtropical forest soil in South China. *J. Plant Ecol.* **2014**, *7*, 86–96. [CrossRef]

Article

Seven-Year Changes in Bulk Density Following Forest Harvesting and Machine Trafficking in Alberta, Canada

David H. McNabb [1,*] and Andrei Startsev [2,†]

[1]   Forest Soil Science Ltd., 8775 Strathearn Crescent NW, Edmonton, AB T6C 4C5, Canada
[2]   Alberta Environmental Centre, Alberta Research Council, Vegreville, AB T9C 1T4, Canada; andrei.startsev@gmail.com
*   Correspondence: dhmcnabb@telus.net
†   Retired.

Abstract: Processes responsible for natural recovery of compacted forest soils are poorly understood, making estimating their recovery problematic. Bulk density was measured over 7 years at nine boreal forest sites in Alberta, Canada, where harvest-only and three skidding treatments were installed (~10,000 samples). Air and soil temperatures, soil moisture and redox potential, and snow depth were also measured on the harvest-only and adjacent seven-cycle skid trail. Significant increases in bulk density occurred when the soil water potential was wetter than −25 kPa. After 1 year, an additional significant increase in bulk density of 0.03 Mg m^{-3} was measured across all treatments, soil depths, and sites. The increase is attributed to the soil mechanics process of rebound and disruption of soil biological processes. By year 7, the secondary increase in bulk density had recovered in trafficked soil, but not on the harvest-only area. Some soil freezing had no effect on bulk density, which was moderated by the depth of the snowpack. The array of soil physical processes, soil texture, water supply, mechanics of water freezing in soil, and weather required to make soil freezing an effective decompacting agent did not occur. The shrink–swell process was not relevant because the soils remained wet. As a result, the bulk density of the trafficked soil failed to recover after 7 years to a depth of 20 cm. The freeze–thaw process as a decompaction agent is far more complex than commonly assumed, and its effectiveness cannot be assumed because soil temperatures below 0 °C are measured.

Keywords: soil compaction; natural soil recovery; decompaction processes; freeze–thaw process; boreal forests; sampling bulk density

Citation: McNabb, D.H.; Startsev, A. Seven-Year Changes in Bulk Density Following Forest Harvesting and Machine Trafficking in Alberta, Canada. *Forests* **2022**, *13*, 553. https://doi.org/10.3390/f13040553

Academic Editors: Fuzhong Wu, Zhenfeng Xu and Wanqin Yang

Received: 1 March 2022
Accepted: 27 March 2022
Published: 31 March 2022

Publisher's Note: MDPI stays neutral with regard to jurisdictional claims in published maps and institutional affiliations.

1. Introduction

The potential to reduce soil productivity and adversely affect other ecological functions of soil by trafficking soil with ground-based equipment has long been a concern in agriculture and forestry [1]. More than 150 papers published between 1970 and 1977 reported that soil compaction affected the growth of agricultural crops and forest trees [2]. Compaction of forest soils was the focus of 26 papers with only 2 papers failing to report compaction-reduced tree growth. These papers drew attention to the most obvious examples of agroforestry machines impacting soils and crop productivity. At the time, indiscriminate areal trafficking of soil and a lack of awareness of the problem were important issues. Concentrating the trafficking to a few skid trails reduced the impact [3], and other options to manage soil compaction were promoted [4]. New literature reviews continue to summarize the expanding range of impacts soil compaction causes and options for their management [5–7], and evaluate a wider array of machines and operational practices to reduce soil impacts [8,9], but the potential for long-term term natural recovery of anthropogenically compacted soils remains speculative [10].

The ecological consequences of compaction on forest soils are complex. Soil compaction causes an increase in bulk density and a related decrease in air-filled porosity as a function of an increasing number of machine passes [11]. These changes in soil affect a wide range of other soil properties important for the growth of roots [12], including soil resistance to root penetration [13], soil aeration [14], soil water availability, and movement of water into and through soil [15,16]. The potential for soil compaction to affect multiple, interrelated soil properties has complicated our ability to relate changes in tree growth on compacted soil to a single-soil property. This complexity also makes it difficult to define specific values of bulk density that limit root and tree growth [4]. Values of natural and compacted bulk densities are also significantly correlated [17]. Particle size distribution and organic matter content are important soil properties responsible for these differences [18]. A newer alternative is to use relative density instead of bulk density as a measure of the severity of soil compaction. Relative density is the ratio of measured bulk density divided by the bulk density obtained using a standardized compaction or consolidation test [19,20]. Relative density has been related to differences that soil compaction has on tree growth for different sites and soils in British Columbia, Canada [21].

The effectiveness of natural decompaction processes to restore compacted soil to its original condition is uncertain because of the range of factors and interactions involved [22]. As a result, efforts to estimate soil recovery are based on studying a chronosequence of similarly treated sites. Unfortunately, the history of older sites may not be well documented, and the ability to visually identify compacted and adjacent control soils on older sites becomes less dependable [4]. Finally, the estimated time for compacted soil to recover in chronosequence studies is based on regression analyses of when the difference between the bulk density of samples selected to represent compacted and control soils becomes zero. Estimates of forest soil recovery based on this methodology have ranged from about two decades in northern climates [23] to 50 to 70 years in France [24] and more than 32 to 70 years in the western US [25,26]. Surface soils consistently recover faster than subsoil [25,26]. In a recent review of the natural recovery of skid trails, DeArmond et al. [10] identified 18 of 64 studies where the physical properties of the surface soil to a depth of 5 to 15 cm had recovered, but recovery was primarily limited to the less traveled skid trails.

More reliable measures of soil recovery should be obtained if bulk density is repeatedly measured on the same site over time, but short-term reports of soil recovery have been inconsistent. For example, Labelle and Jaeger [27] reported two loams in New Brunswick, Canada, which indicated no natural recovery after 5 years. On the other hand, Page-Dumroese et al. [28] reported a possible minor recovery of compacted, coarse-textured soil in a 0–10 cm layer after 5 years when a significance level of $\alpha = 0.10$ was used. They did note that the number of samples required to estimate a 15% difference in bulk density on most of the sites was generally greater than the number of samples collected. For repeated measurement of bulk density over time to be dependable, many high-quality bulk density samples must be collected at each location and time. The standard agroforestry method of measuring bulk density by the core method recommends using large-volume rings with a diameter of at least 75 mm [29]. Validation of statistically significant differences in bulk density as a measure of soil recovery will be more dependable if the monitoring of the soil and climatic environment helps confirm that specific natural recovery processes were active [30].

Our objective was to quantify changes in bulk density over a 7-year period following forest harvesting and controlled skidding on boreal forest soils in west-central Alberta. We assumed that cold winters would make the freeze–thaw process an important factor contributing to the natural decompaction of these soils. Weather stations to measure air and soil temperatures and snow depth were installed at each installation to confirm when such events occurred.

2. Materials and Methods

2.1. Sites

We are reporting changes in bulk density over 7 years for 9 sites that were operationally trafficked in western–central Alberta [6]. The primary objective of the study was to assess the effects that differences in soil wetness and trafficking intensity had on soil bulk density, air-filled porosity, and other changes in soil physical properties. Changes in bulk density, air-filled porosity, infiltration rate, and pore-size distribution immediately after trafficking have been reported for the original 14 sites [11,15,16]. Changes in soil aeration and soil morphology up to 4 years after harvesting have also been reported [14].

Although 14 sites were originally established, funding was only available to measure 9 of the first 10 sites over 7 years (Figure 1). One large cutblock contained two research installations, but only one was instrumented with a weather station for this phase of the project. The locations of the study sites are shown in Figure 1.

Figure 1. West-central Alberta location of nine sample sites where soil bulk density was measured six times over 7 years. The distance between sites 7 and 8 is about 330 km.

Prior to skidding, four replicated plots, 40 m on a side, were established about 50 m from the temporary harvest road (Figure 2). This location allowed controlled skidding of each replication and space for the skidder to realign the trees for decking with butts at roadside. Skidding cycles of 3, 7, and 12 empty and loaded passes were assigned to each plot. The same skid trail layout was used for all replications and sites. Control areas were skidded from the sides of the plot or down the center of the 12-cycle plot.

Except for site 2, the sites were harvested with a tracked feller-buncher operating parallel to a temporary harvest road. The feller-buncher made one pass across the site at a spacing of about 12 m. Bunches of whole trees were skidded to the roadside and decked with the butts of the trees at the edge of the road. Skidders were two-axle grapple skidders that weighed 16–17 mg, the width of tires ranged were between 0.8 and 1.2 man, and wider tires were on the rear of skidders (Table 1). Trees at site 2 were felled and processed at the stump with a harvester and log-transported with a small three-axle forwarder (Valmet 540). The compacted slash mat was thin, and the resulting soil compaction was not different from skidders [11].

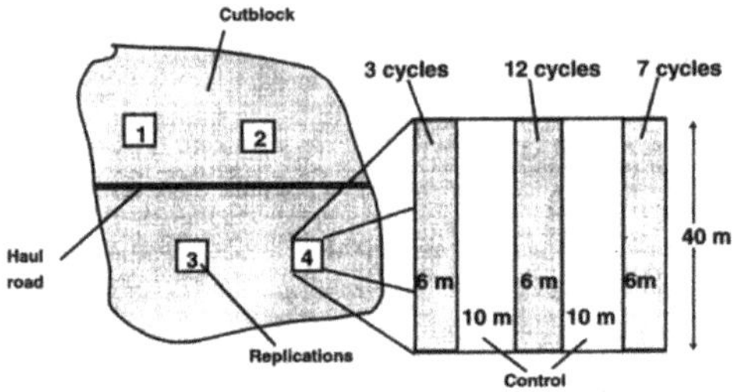

Figure 2. Illustration of location and location of each replication at each site.

Table 1. Soil organic matter and particle size distribution (n = 4), soil classification, and soil water content and tensiometer water potential (n = 64) at the time of skidding [11].

Site	Depth	Organic Matter	Particle Size Distribution			Texture	Classification	Soil Water When Skidded		TireTrack Width
			Clay	Silt	Sand					
	cm		kg kg^{-1}					kg kg^{-1}	kPa	m
1	5	0.023	0.16	0.5	0.34	Silt loam/loam	Orthic/Gleyed Gray Luvisol	0.264	−13	0.8/1.1
	10	0.036	0.13	0.52	0.35	Silt loam/loam		0.218	−15	
	20	0.006	0.16	0.43	0.41	Loam		0.181	−15	
2	5	0.025	0.17	0.64	0.19	Silt loam	Gleyed Gray Luvisol	0.251	−15	0.7
	10	0.028	0.19	0.59	0.22	Silt loam		0.234	−7	
	20	0.015	0.23	0.59	0.18	Silt loam		0.259	−8	
3	5	0.031	0.12	0.53	0.35	Silt loam	Orthic Gray Luvisol	0.251	−6	0.8/1.1
	10	0.026	0.12	0.51	0.37	Silt loam		0.234	−7	
	20	0.014	0.15	0.48	0.37	Loam		0.259	−7	
4	5	0.024	0.21	0.43	0.36	Loam	Orthic Gray Luvisol	0.337	−9	1.1
	10	0.031	0.2	0.4	0.4	Loam		0.295	−6	
	20	0.01	0.23	0.38	0.39	Loam		0.271	−7	
6	5	0.027	0.15	0.44	0.41	Loam	Brunisolic Gray Luvisol	0.293	−23	0.8
	12	0.026	0.13	0.42	0.45	Loam		0.234	−30	
	20	0.008	0.17	0.39	0.44	Loam		0.231	−38	
7	5	0.033	0.22	0.56	0.22	Silt loam	Eluviated Dystric Brunisol	0.26	−7	1.1
	10	0.027	0.21	0.55	0.24	Silt loam		0.23	−13	
	20	0.012	0.28	0.52	0.2	Silty clay loam/clay loam		0.189	−10	
8	5	0.025	0.29	0.47	0.27	Clay loam	Orthic/Gleyed Gray Luvisol	0.448	−40	1.1
	10	0.022	0.31	0.46	0.32	Clay loam		0.306	−37	
	20	0.016	0.24	0.44	0.27	Silt loam		0.232	−26	
9	5	0.058	0.18	0.55	0.28	Silt loam	Orthic Gray Luvisol	0.229	−13	0.8/1.1
	10	0.026	0.15	0.54	0.27	Silt loam		0.184	−16	
	20	0.016	0.18	0.55	0.24	Silt loam		0.184	−15	
10	5	0.028	0.25	0.5	0.25	Silt loam/loam	Orthic/Gleyed Gray Luvisol	0.278	−7	0.8
	10	0.015		0.47	0.27	Loam		0.222	−11	
	20	0.01	0.33	0.43	0.24	Clay loam		0.219	−9	

2.2. Bulk Density

We are reporting changes in soil bulk density through the first 7 years (1994–2001) of the study. Bulk density was measured using stainless steel rings (7.6 cm outside diameter and 3 cm height); the wall thickness of the rings was 0.16 cm. Rings were pushed into the soil with a T-bar handle and a driving head [11]. Soil cores were collected from the midpoint of the 5, 10, and 20 cm depths and at random preselected locations in each skidding corridor and control. For each remeasurement period, the sample point was shifted 0.5 m along the corridor in the direction of machine travel from the previous measurement point and always in the same direction; the relative position across the corridor remained the same. Bulk density was measured within 2–3 days after each site was felled and skidded (year 0), and 1, 2, 3, 4 and 7 years later in late summer and early fall. A maximum of 192 samples were collected from each site and time (3 soil depths per point, 4 points per treatment, 4 treatments per replication, and 4 replications per site). Only a few samples were either lost, labels unidentifiable, and bags broken or removed from the database because a value was an outlier (4 + standard errors). The database contains approximately 10,000 values of bulk density.

Site 10 was not measured in year 4, and the last sampling date was year 6 because of a lack of funds in other years. When the data for Site 10 was excluded, nearly 80% of the resultant mean bulk densities changed by less than 0.01 Mg m^{-3}. Thus, the effect that these two missing sample periods had on our analyses is assumed to be minor. We use values of bulk density collected immediately after skidding as the reference bulk density to assess whether there were significant changes in bulk density from 0, 3, 7, and 12 cycles of soil trafficking in subsequent years.

2.3. Soil Climate

The most representative replicate at each site, based on soil and terrain, was instrumented with a multiplexed CR10 data logger (Campbell Scientific Inc., Logan, UT, USA) to measure air and soil temperature in the control soil and adjacent 7-cycle skid trail. The distance between soil measurement points in these two treatments was between 6 and 8 m. Thermocouples were inserted horizontally at the interface between the LFH layer and mineral soil (0 cm depth) and at 5, 10, 20, and 50 cm depths. Soil and air temperatures were recorded every second hour. Snow depth was measured over the harvest-only treatment with a sonic range sensor (Campbell Scientific Inc.).

2.4. Statistical Analyses

The analysis of bulk density data was performed using SAS 6.11 [31], a doubly repeated mixed model with time and depth as repeated factors [32]. This step was followed by analyses of effects sliced by the interacting factor and differences among least square means for significant effects. Analysis of effects was conducted starting from higher-order interactions according to Milliken and Johnson [33].

3. Results

3.1. Soils and Weather and Soil Climate

The soils were all medium textured, ranging from sandy loam to clay loam (Table 1). The similarity in textures, organic matter content, and classifications reflect their common glacial till origins. Soil organic matter was low, averaging 0.036 kg kg^{-1} at the 5 cm depth and a third of that amount at 20 cm (Table 1). The sites are in the Upper and Lower Cordilleran Ecoregions of western Alberta [34]. The values of soil water content were obtained from each bulk density sample, and the soil water potential was measured with a handheld tensiometer next to each bulk density sample [11].

Differences in the depth of winter snowpack dominated the weather data over the 7 years that data were collected. The depth of snow cover was deeper the first three winters (1994–1997) than during the last four winters (1997–2001). Data to illustrate the differences between a heavy and light snowpack winter are shown in Figure 3. Although the sites were separated by up to 330 km, the confidence limits for snow depth within a year showed that the differences were small within the Upper and Lower Cordilleran Ecoregions of western Alberta [34].

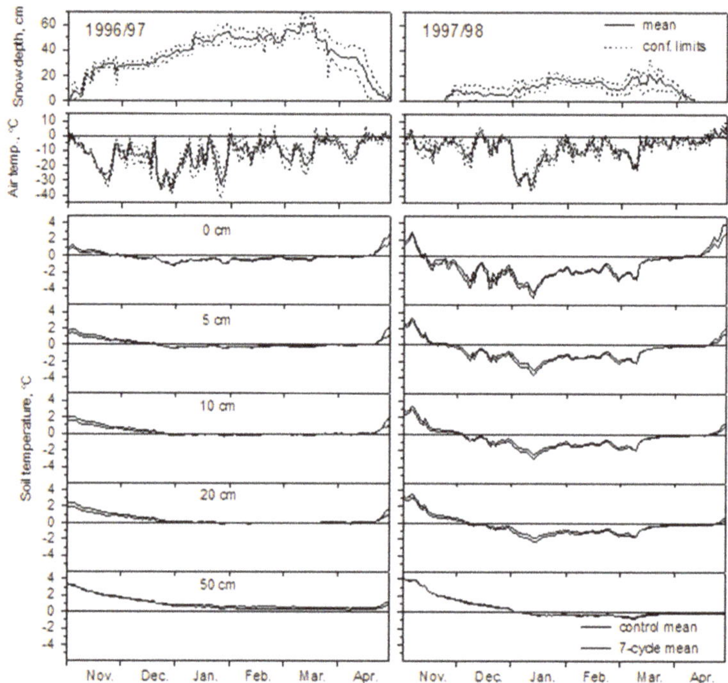

Figure 3. Snow depth and daily minimum air temperatures for a winter with normal snowfall (1996–1997) and below normal snowfall (1997–1998). Values are the mean (solid lines) and the confidence limits (dashed lines) for the 9 sites each winter.

Periodic cycles of colder air temperatures had a more pronounced effect on soil temperatures when snow depth was low (Figure 3). A snow depth of about 10 cm was insufficient to prevent soil temperatures from dropping to near −4 °C at 5 cm and −2 °C at the 20 cm depth when daily minimum air temperatures were between −25 and −33 °C (3–14 January 1998). A snow cover of at least 40 cm prevented soils from dropping below −1 °C at the surface despite daily minimum air temperatures of −30 °C and colder, which occurred four times during the winter of 1996–1997. Whether the soil water froze either time will be an important part of our discussion.

Snow depth was measured over the undisturbed litter layer of the harvest-only treatment. Most of the forest floor was compacted or displaced on the adjacent seven-cycle treatment. Nevertheless, the differences in soil temperature between the control and trafficked soil were less than 1 °C (Figure 3). Thorud and Duncan [35] previously reported that the removal of the forest floor had less effect on soil freezing than did the depth of snow cover.

3.2. Skidding Traffic Increases Bulk Density

The mean values of bulk density and standard errors for each treatment, depth, site, and year are shown in Table 2. The only obvious trend in standard errors was values decreasing with depth—0.046, 0.039, and 0.033 Mg m^{-3} for depths of 5, 10, 20 cm, respectively.

Table 2. Soil bulk density data for 9 sites, 4 treatments, 3 depths, and 6 sampling periods over 7 years. Soil depth is the average depth of a soil core with a height of 3 cm. The sample size was 16 and the standard errors are in parentheses.

Site/Year	Soil Depth, 5 cm				Soil Depth, 10 cm				Soil Depth, 20 cm			
	Control	3 Cycle	7 Cycle	12 Cycle	Control	3 Cycle	7 Cycle	12 Cycle	Control	3 Cycle	7 Cycle	12 Cycle
								$Mg\ m^{-3}$				
Site 1												
0	1.13 (0.04)	1.29 (0.04)	1.33 (0.04)	1.29 (0.04)	1.34 (0.04)	1.46 (0.04)	1.44 (0.04)	1.44 (0.03)	1.46 (0.02)	1.46 (0.03)	1.41 (0.04)	1.47 (0.03)
1	1.20 (0.04)	1.24 (0.04)	1.35 (0.04)	1.36 (0.03)	1.37 (0.06)	1.52 (0.04)	1.51 (0.03)	1.49 (0.03)	1.48 (0.04)	1.61 (0.02)	1.58 (0.02)	1.58 (0.02)
2	1.12 (0.04)	1.07 (0.05)	1.14 (0.04)	1.19 (0.04)	1.28 (0.03)	1.30 (0.05)	1.36 (0.05)	1.34 (0.04)	1.39 (0.03)	1.39 (0.03)	1.38 (0.03)	1.36 (0.03)
3	1.09 (0.04)	1.12 (0.07)	1.27 (0.05)	1.26 (0.04)	1.29 (0.03)	1.40 (0.05)	1.46 (0.04)	1.44 (0.04)	1.46 (0.03)	1.46 (0.04)	1.50 (0.04)	1.45 (0.02)
4	1.20 (0.05)	1.1 (0.05)	1.30 (0.04)	1.28 (0.05)	1.48 (0.03)	1.45 (0.05)	1.42 (0.05)	1.43 (0.06)	1.58 (0.02)	1.58 (0.04)	1.54 (0.03)	1.51 (0.04)
7	1.14 (0.03)	1.09 (0.04)	1.14 (0.04)	1.15 (0.05)	1.22 (0.03)	1.19 (0.04)	1.28 (0.03)	1.27 (0.04)	1.27 (0.03)	1.24 (0.05)	1.29 (0.03)	1.15 (0.06)
Site 2												
0	1.29 (0.05)	1.34 (0.05)	1.40 (0.04)	1.46 (0.04)	1.37 (0.04)	1.38 (0.04)	1.41 (0.03)	1.42 (0.04)	1.29 (0.05)	1.40 (0.04)	1.38 (0.03)	1.41 (0.04)
1	1.20 (0.04)	1.39 (0.04)	1.34 (0.05)	1.42 (0.03)	1.23 (0.04)	1.41 (0.04)	1.43 (0.03)	1.39 (0.04)	1.30 (0.04)	1.35 (0.05)	1.43 (0.03)	1.42 (0.04)
2	1.18 (0.04)	1.26 (0.06)	1.33 (0.03)	1.35 (0.05)	1.30 (0.05)	1.38 (0.04)	1.45 (0.03)	1.40 (0.04)	1.33 (0.04)	1.40 (0.05)	1.42 (0.03)	1.44 (0.04)
3	1.31 (0.05)	1.36 (0.05)	1.31 (0.03)	1.43 (0.03)	1.29 (0.04)	1.38 (0.04)	1.40 (0.03)	1.44 (0.05)	1.29 (0.03)	1.38 (0.05)	1.45 (0.03)	1.42 (0.04)
4	1.24 (0.05)	1.32 (0.05)	1.37 (0.03)	1.38 (0.03)	1.31 (0.05)	1.40 (0.04)	1.43 (0.03)	1.43 (0.04)	1.22 (0.04)	1.33 (0.04)	1.44 (0.04)	1.42 (0.05)
7	1.03 (0.06)	1.16 (0.05)	1.27 (0.03)	1.25 (0.03)	1.13 (0.04)	1.31 (0.04)	1.35 (0.03)	1.30 (0.03)	1.16 (0.05)	1.27 (0.05)	1.31 (0.04)	1.33 (0.03)
Site 3												
0	1.19 (0.06)	1.27 (0.04)	1.39 (0.04)	1.37 (0.03)	1.31 (0.03)	1.42 (0.03)	1.47 (0.04)	1.48 (0.03)	1.35 (0.03)	1.47 (0.03)	1.50 (0.04)	1.51 (0.03)
1	1.28 (0.04)	1.27 (0.05)	1.27 (0.06)	1.35 (0.04)	1.33 (0.03)	1.37 (0.03)	1.39 (0.05)	1.45 (0.03)	1.40 (0.04)	1.44 (0.03)	1.45 (0.04)	1.54 (0.02)
2	1.22 (0.04)	1.35 (0.05)	1.33 (0.06)	1.43 (0.04)	1.36 (0.04)	1.45 (0.03)	1.45 (0.02)	1.52 (0.03)	1.47 (0.03)	1.49 (0.03)	1.53 (0.03)	1.59 (0.03)
3	1.28 (0.03)	1.24 (0.07)	1.35 (0.06)	1.44 (0.03)	1.33 (0.03)	1.48 (0.05)	1.44 (0.03)	1.50 (0.03)	1.40 (0.03)	1.54 (0.05)	1.50 (0.04)	1.60 (0.03)
4	1.17 (0.05)	1.27 (0.03)	1.35 (0.06)	1.41 (0.03)	1.31 (0.04)	1.37 (0.04)	1.43 (0.03)	1.45 (0.04)	1.37 (0.03)	1.44 (0.04)	1.51 (0.03)	1.58 (0.02)
7	1.32 (0.05)	1.48 (0.06)	1.42 (0.06)	1.46 (0.05)	1.49 (0.03)	1.50 (0.04)	1.56 (0.05)	1.53 (0.04)	1.50 (0.02)	1.52 (0.03)	1.59 (0.04)	1.61 (0.03)
Site 4												
0	1.21 (0.04)	1.38 (0.05)	1.39 (0.03)	1.42 (0.04)	1.39 (0.02)	1.39 (0.04)	1.46 (0.03)	1.47 (0.03)	1.38 (0.04)	1.43 (0.04)	1.46 (0.02)	1.50 (0.02)
1	1.27 (0.03)	1.33 (0.06)	1.46 (0.03)	1.41 (0.04)	1.39 (0.03)	1.49 (0.02)	1.49 (0.03)	1.57 (0.04)	1.39 (0.03)	1.46 (0.03)	1.51 (0.03)	1.57 (0.03)
2	1.30 (0.03)	1.30 (0.03)	1.36 (0.05)	1.27 (0.06)	1.35 (0.03)	1.40 (0.03)	1.43 (0.04)	1.54 (0.02)	1.41 (0.03)	1.42 (0.03)	1.54 (0.04)	1.53 (0.03)
3	1.27 (0.05)	1.33 (0.04)	1.38 (0.05)	1.39 (0.04)	1.35 (0.04)	1.46 (0.03)	1.52 (0.02)	1.52 (0.03)	1.38 (0.04)	1.47 (0.03)	1.49 (0.02)	1.53 (0.03)
4	1.27 (0.04)	1.28 (0.03)	1.38 (0.04)	1.36 (0.05)	1.34 (0.03)	1.43 (0.04)	1.50 (0.02)	1.46 (0.09)	1.39 (0.02)	1.49 (0.03)	1.50 (0.03)	1.57 (0.02)
7	1.24 (0.04)	1.28 (0.05)	1.33 (0.04)	1.32 (0.05)	1.27 (0.03)	1.33 (0.04)	1.42 (0.03)	1.45 (0.04)	1.29 (0.04)	1.32 (0.03)	1.40 (0.03)	1.43 (0.03)
Site 6												
0	1.08 (0.06)	1.16 (0.06)	1.31 (0.05)	1.14 (0.05)	1.22 (0.05)	1.34 (0.05)	1.30 (0.07)	1.32 (0.04)	1.31 (0.03)	1.43 (0.04)	1.45 (0.04)	1.41 (0.06)
1	1.05 (0.04)	1.20 (0.06)	1.13 (0.05)	1.17 (0.07)	1.31 (0.04)	1.35 (0.04)	1.33 (0.05)	1.36 (0.04)	1.34 (0.05)	1.42 (0.02)	1.47 (0.03)	1.45 (0.03)
2	1.22 (0.04)	1.18 (0.05)	1.30 (0.05)	1.23 (0.05)	1.35 (0.03)	1.29 (0.05)	1.42 (0.04)	1.30 (0.03)	1.37 (0.02)	1.37 (0.03)	1.40 (0.04)	1.31 (0.03)
3	1.09 (0.05)	1.14 (0.05)	1.11 (0.05)	1.11 (0.05)	1.25 (0.04)	1.27 (0.04)	1.27 (0.06)	1.30 (0.05)	1.39 (0.03)	1.41 (0.02)	1.41 (0.03)	1.49 (0.03)
4	1.08 (0.04)	1.08 (0.04)	1.12 (0.06)	1.19 (0.06)	1.26 (0.04)	1.22 (0.05)	1.32 (0.04)	1.28 (0.04)	1.41 (0.02)	1.47 (0.03)	1.50 (0.02)	1.42 (0.05)
7	1.24 (0.04)	1.24 (0.04)	1.22 (0.05)	1.22 (0.04)	1.35 (0.02)	1.37 (0.02)	1.38 (0.03)	1.34 (0.04)	1.36 (0.03)	1.37 (0.02)	1.43 (0.04)	1.36 (0.06)

Table 2. *Cont.*

Site/Year	Soil Depth, 5 cm				Soil Depth, 10 cm				Soil Depth, 20 cm			
	Control	3 Cycle	7 Cycle	12 Cycle	Control	3 Cycle	7 Cycle	12 Cycle	Control	3 Cycle	7 Cycle	12 Cycle
						$Mg\ m^{-3}$						
Site 7												
0	1.00 (0.05)	1.12 (0.05)	1.18 (0.04)	1.22 (0.06)	1.25 (0.06)	1.23 (0.03)	1.28 (0.05)	1.36 (0.05)	1.43 (0.03)	1.45 (0.05)	1.44 (0.05)	1.47 (0.05)
1	1.05 (0.04)	1.20 (0.06)	1.21 (0.06)	1.30 (0.07)	1.30 (0.05)	1.35 (0.05)	1.42 (0.06)	1.43 (0.04)	1.43 (0.05)	1.49 (0.02)	1.56 (0.03)	1.50 (0.04)
2	1.14 (0.05)	1.27 (0.04)	1.23 (0.07)	1.33 (0.08)	1.33 (0.05)	1.39 (0.05)	1.39 (0.06)	1.46 (0.04)	1.48 (0.04)	1.53 (0.03)	1.51 (0.04)	1.55 (0.03)
3	1.03 (0.04)	1.20 (0.06)	1.14 (0.06)	1.27 (0.05)	1.32 (0.05)	1.45 (0.04)	1.38 (0.06)	1.44 (0.07)	1.51 (0.02)	1.50 (0.04)	1.49 (0.04)	1.53 (0.04)
4	0.99 0.04)	1.21 (0.06)	1.19 (0.06)	1.35 (0.06)	1.27 (0.05)	1.44 (0.04)	1.48 (0.05)	1.47 (0.04)	1.44 (0.04)	1.47 (0.04)	1.56 (0.02)	1.51 (0.03)
7	1.27 (0.05)	1.27 (0.06)	1.29 (0.06)	1.31 (0.05)	1.35 (0.03)	1.37 (0.05)	1.47 (0.05)	1.49 (0.05)	1.40 (0.05)	1.39 (0.04)	1.48 (0.03)	1.48 (0.04)
Site 8												
0	1.15 (0.06)	1.15 (0.07)	1.31 (0.06)	1.28 (0.05)	1.33 (0.05)	1.36 (0.03)	1.45 (0.05)	1.38 (0.02)	1.39 (0.04)	1.43 (0.03)	1.45 (0.02)	1.37 (0.03)
1	1.41 (0.04)	1.37 (0.03)	1.49 (0.04)	1.39 (0.04)	1.55 (0.03)	1.51 (0.04)	1.55 (0.02)	1.51 (0.03)	1.53 (0.03)	1.49 (0.03)	1.49 (0.02)	1.47 (0.02)
2	1.38 (0.05)	1.37 (0.06)	1.50 (0.03)	1.40 (0.04)	1.52 (0.03)	1.55 (0.04)	1.59 (0.02)	1.54 (0.02)	1.58 (0.03)	1.56 (0.03)	1.59 (0.03)	1.54 (0.02)
3	1.15 (0.07)	1.33 (0.06)	1.36 (0.04)	1.36 (0.06)	1.44 (0.05)	1.54 (0.03)	1.50 (0.04)	1.50 (0.04)	1.54 (0.03)	1.47 (0.04)	1.52 (0.03)	1.49 (0.03)
4	1.48 (0.03)	1.44 (0.04)	1.38 (0.05)	1.39 (0.04)	1.56 (0.03)	1.56 (0.03)	1.46 (0.05)	1.59 (0.02)	1.54 (0.03)	1.50 (0.04)	1.48 (0.02)	1.51 (0.02)
7	1.49 (0.05)	1.50 (0.07)	1.43 (0.04)	1.49 (0.04)	1.52 (0.03)	1.61 (0.04)	1.51 (0.03)	1.56 (0.02)	1.61 (0.03)	1.55 (0.03)	1.50 (0.02)	1.55 (0.02)
Site 9												
0	1.24 (0.05)	1.27 (0.05)	1.29 (0.06)	1.32 (0.07)	1.30 (0.04)	1.29 (0.03)	1.38 (0.04)	1.36 (0.03)	1.35 (0.04)	1.34 (0.03)	1.42 (0.03)	1.44 (0.04)
1	1.24 (0.04)	1.29 (0.03)	1.32 (0.04)	1.29 (0.08)	1.29 (0.05)	1.33 (0.04)	1.44 (0.03)	1.34 (0.07)	1.32 (0.03)	1.39 (0.02)	1.47 (0.04)	1.45 (0.03)
2	1.28 (0.05)	1.38 (0.02)	1.36 (0.05)	1.41 (0.03)	1.29 (0.04)	1.35 (0.04)	1.41 (0.04)	1.40 (0.05)	1.31 (0.03)	1.40 (0.04)	1.46 (0.04)	1.48 (0.03)
3	1.22 (0.08)	1.06 (0.14)	1.35 (0.04)	1.08 (0.15)	1.28 (0.09)	1.17 (0.12)	1.38 (0.02)	1.24 (0.14)	1.37 (0.08)	1.34 (0.08)	1.49 (0.03)	1.29 (0.13)
4	1.30 (0.05)	1.30 (0.04)	1.39 (0.04)	1.43 (0.03)	1.27 (0.03)	1.33 (0.03)	1.40 (0.04)	1.49 (0.03)	1.30 (0.03)	1.33 (0.04)	1.44 (0.03)	1.49 (0.03)
7	1.34 (0.04)	1.41 (0.03)	1.45 (0.04)	1.31 (0.03)	1.37 (0.04)	1.39 (0.02)	1.50 (0.04)	1.44 (0.05)	1.43 (0.03)	1.43 (0.04)	1.46 (0.06)	1.57 (0.05)
Site 10												
0	1.09 (0.05)	1.30 (0.04)	1.30 (0.03)	1.31 (0.03)	1.18 (0.04)	1.35 (0.03)	1.41 (0.02)	1.42 (0.03)	1.27 (0.03)	1.42 (0.03)	1.44 (0.02)	1.39 (0.03)
1	1.08 (0.03)	1.28 (0.03)	1.36 (0.03)	1.35 (0.04)	1.12 (0.03)	1.37 (0.03)	1.39 (0.02)	1.46 (0.03)	1.28 (0.03)	1.43 (0.03)	1.50 (0.02)	1.42 (0.03)
2	1.10 (0.04)	1.32 (0.03)	1.33 (0.04)	1.34 (0.03)	1.19 (0.03)	1.36 (0.03)	1.40 (0.03)	1.46 (0.03)	1.32 (0.02)	1.45 (0.02)	1.49 (0.02)	1.44 (0.03)
3	1.12 (0.04)	1.32 (0.03)	1.39 (0.03)	1.32 (0.04)	1.19 (0.05)	1.43 (0.03)	1.43 (0.02)	1.45 (0.02)	1.29 (0.02)	1.44 (0.03)	1.50 (0.02)	1.44 (0.03)
4	Site was not sampled											
7	1.04 (0.03)	1.16 (0.03)	1.22 (0.06)	1.05 (0.06)	1.13 (0.03)	1.38 (0.03)	1.42 (0.02)	1.41 (0.02)	1.19 (0.03)	1.37 (0.03)	1.40 (0.03)	1.38 (0.02)

The main effects model of the ANOVA show trafficking, depth, site, and year are all highly significant ($p < 0.0000$, Table 3). Soil depth is the dominant variable because bulk density increases rapidly with increasing depth (Table 4). The relationship is consistent for the entire period (depth and year interaction was not significant). These upland forest soils typically have a thin mineral topsoil with the B horizon beginning within 6 to 16 cm of the surface [36].

Table 3. Analyses of Variance of soil bulk density collected over a sampling period of 7 years (mixed ANOVA with depth and year as repeated factors).

Source of Variance	Degrees of Freedom	Type III F	Pr > F	Significance Level
Traffic Cycles	3	32.71	0.0000	****
Depth	2	801.82	0.0000	****
Site	8	13.91	0.0000	****
Year	5	11.26	0.0000	****
Traffic Cycles * Depth	6	2.36	0.0300	*
Traffic Cycles * Site	24	1.32	0.1720	
Traffic Cycles * Year	6	1.49	0.1976	
Depth * Site	16	20.42	0.0000	****
Depth * Year	10	2.71	0.0027	**
Site * Year	39	14.91	0.0000	****
Traffic Cycles * Depth * Site	48	1.43	0.0442	*
Traffic Cycles * Depth * Year	30	0.78	0.7903	
Traffic Cycles * Site * Year	117	1.21	0.0889	
Depth * Site * Year	78	1.98	0.0000	****

* Significance level, Pr < 0.05; ** Significance level, Pr < 0.01; **** Significance level, Pr < 0.0001.

Table 4. Summary of descriptive statistics for standard erros at the 5, 10, and 20 cm d.

Statistic	5 cm Depth	10 cm Depth	20 cm Depth	Combined Depths
Mean	0.0463	0.0391	0.0333	0.0395
Median	0.0441	0.0363	0.0310	0.0370
Standard Error	0.0010	0.0010	0.0008	0.0006
Standard Deviation	0.0151	0.0147	0.0117	0.149
Kurtosis	14.9592	15.7848	20.2396	13.4195
Skewness	2.71672	2.9093	3.1446	2.5526
Range	0.1288	0.1270	0.1106	0.1328
Minimum	0.0205	0.0170	0.0165	0.0165
Maximum	0.1493	0.1440	0.1272	0.1493
Count	210	210	210	630

Only the three cycles of skidding treatment caused a significant increase in bulk density ($p < 0.0000$, Table 3, Figure 4). An increase in bulk density between 3 and 7 cycles occurred but was not significant. A factor contributing to the sustained increase between 3 and 7 cycles on these wet soils was that samples for bulk density were collected from predetermined points within the 6 m wide skidding corridor [11]. Therefore, some of the increase in bulk density between 3 and 7 cycles is attributed to increased areal coverage of the skidding corridor. By seven cycles, the traffic pattern had stabilized into a well-defined skid trail along the 40 m transect. Bulk density increases faster when compaction is only measured in wheel tracks [27].

Site was a significant factor ($p < 0.0000$, Table 4) because increases in bulk density from three cycles of skidding were only significant at six of the nine sites. After 12 cycles of skidding, the average increase in bulk density when trafficking at the 5, 10, and 20 cm depths were 0.161, 0.106, and 0.084 Mg m^{-3}, respectively. The percentage increases in bulk density with a soil depth were 14.2%, 8.2%, and 6.2%, respectively. The low increase in bulk density is partly attributed to the use of wider tires on the pressure harvesting machines (Table 1). Compaction of the wettest of these soils reduced their low air-filled porosity until the soils were essentially saturated at the higher skidding cycles [11]. Essentially saturated soils contained trapped air at a high degree of saturation. Trapped air prevents a further increase in bulk density because water is incompressible and trafficking does not allow time for the water in the soil to drain [11,37]. Sites 6 and 8 were not significantly compacted and had soil water potentials lower than -25 kPa (Table 1). The increase in bulk density for these soils was less than 6% despite the soils having an average air-filled porosity of $0.32 \text{ m}^3 \text{ m}^{-3}$ after three cycles [11].

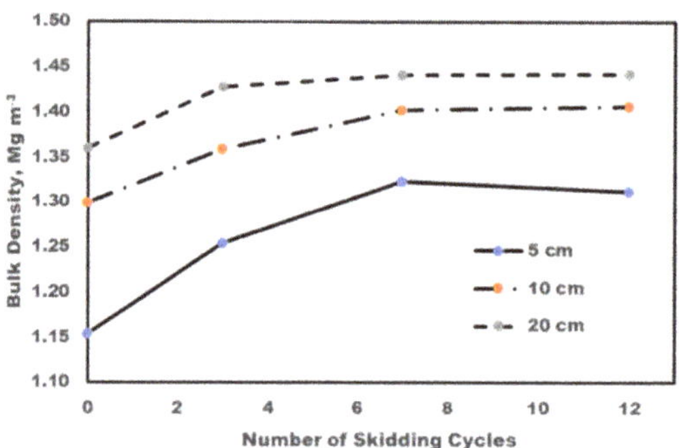

Figure 4. Bulk density measured immediately after harvesting and skidding (year 0) as a function of the number of skidding cycles.

3.3. Postharvest Changes in Bulk Density

Changes in bulk density occurred in the 7 years postharvest (Figure 5). The first change was a significant increase in bulk density after 1 year. Year of sampling was a significant source of variation ($p < 0.0000$, Table 3) and was consistent across all levels of trafficking, including the harvest-only control. The interactions of traffic cycles and site, and trafficking cycles and year were not significant.

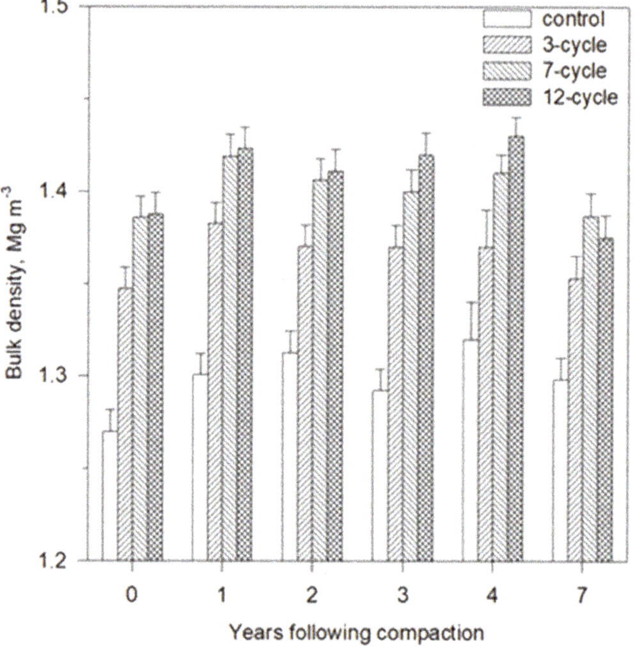

Figure 5. Mean bulk density of harvest-only (control) and 3, 7, and 12 skid cycles of soil trafficking over a period of 7 years. Each value of bulk density is the mean for the 5, 10, and 20 cm depths measured at nine sites ($n = 144$; Table 2).

The significant increase in bulk density after 1 year was approximately 0.03 Mg m^{-3} (Figure 5). This is a 2.4% average increase in bulk density, which is about 25% of the original increase as a percentage. The postharvest increase in bulk density of trafficked soil persisted for three winters when the snowpacks were deep (Figure 3). The bulk density samples for year 4 were collected in the fall of 1997 prior to the first of four winters when the snowpack was thin. By year 7, the bulk density of the three trafficked treatments had decreased to values similar to those measured immediately following trafficking. During the intervening time period, these sites had potentially undergone a minor amount of freezing (Figure 3). However, the bulk density of the harvest-only soil had not returned to the values measured immediately after harvesting.

The site and year interaction ($p < 0.0000$) and the depth and year interaction ($p < 0.0027$) were both significant (Table 3). Sites 1 and 8 had the greatest year-to-year variation (Table 2). The year-to-year variation in soil wetness was greater on these sites than the other sites. The sustained wetness on some sites also caused changes in soil morphology and decreased soil drainage class within the first 3 to 4 years [14], the time period when two-thirds of the data were collected. Wet or dry soils can sometimes make it more difficult to consistently collect high-quality soil cores with a standardized protocol [28,30]. Field staff also reported that wet soil made the collection of core samples on some sites more difficult.

The interactions that are not statistically significant are also relevant to understanding the dynamics of natural recovery processes. Traffic cycles and year ($p < 0.1976$), and traffic cycles, depth, and year interactions ($p < 0.7903$) are most notable (Table 4). Their lack of significance indicates that soil to a sampling depth of 22 cm behaved as a homogeneous unit over the entire 7-year period. This is a further indication that the soil freezing, if it occurred, was ineffective, and postharvest soil biological processes were not sufficiently effective to loosen the soil compacted by skidding at any depth.

3.4. Measuring Bulk Density

The mean of the standard errors for bulk density for the three depths in Table 2 is 0.039 Mg m^{-3} with a standard deviation of 0.0149 Mg m^{-3} (Table 3). The median value is 0.037 Mg m^{-3}. The only obvious tread in standard deviation was values decreasing with depth (0.046, 0.039, and 0.033 Mg m^{-3} for depths 5, 10, 20 cm, respectively). The inclusive value for kurtosis is 13.39, and the data are positively skewed to the right (2.54) for all depths. Hence, the variability of the values of bulk density is not described by a bell-shaped distribution curve, but by the widening of the lower portion of the curve [38] (Figure 6). Standard errors > 0.065 Mg m^{-3} represent 3.6% of the samples. Half of these values, six of the seven values > 0.095 Mg m^{-3}, and 75% of all values occurred on Site 9 in year 3 (Table 2). Site 9 was the only site where a small amount of gravel was encountered [11]. Nevertheless, the mean standard error at Site 9 for year 0 was 0.042 Mg m^{-3}, including a maximum value of 0.069 Mg m^{-3}. At that time, a small amount of gravel was not regarded as a problem or likely to have a measurable effect on the value of bulk density. The sampling protocol required that sampling rings be pressed into the soil by body weight. Roots were normally the dominant factor causing a high discard rate of potential bulk density sampling points. At Site 9, small gravel caused more sample locations to be abandoned or cores discarded during trimming. Nevertheless, a temporary relaxation of the commitment to the sampling protocol on a difficult-to-sample site is most likely responsible for this one-time deviation. This deviation reinforces the need for a strict adherence to a sampling protocol.

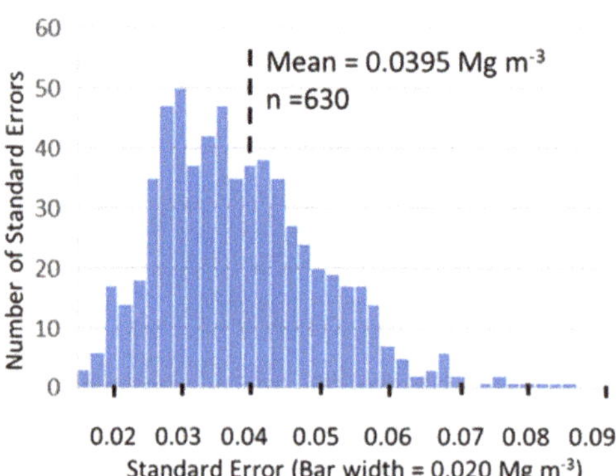

Figure 6. Histogram of all the standard errors reported in Table 2. Seven values between 0.095 and 0.150 g m^{-3} are not shown. Each value of standard error is based on 16 individual soil cores (Table 2).

4. Discussion

4.1. Measured Changes in Bulk Density

Ampoorter et al. [39], using a meta-analysis of data from several soil compaction studies including this one, found the number of skidding cycles was the only factor responsible for an increase in bulk density, but the relationship was weak. Soil water content was the only common measure of soil wetness considered and was not significant. However, other research has confirmed wet soils are more susceptible to compaction, and 60% to 90% of the increase in bulk density occurs in the first skidding cycle [27,40]. Soil water content alone is not a reliable variable to evaluate compactability of soil because the relationship between soil water content and maximum bulk density primarily depends on soil texture [17,41]. Higher compacted bulk density can only be achieved at lower soil water contents in course-textured soils [42]. Field measurements of soil water potential is a more reliable predictor of when a soil is most susceptible to compaction [11]. Soil resistance to compression increases as soil water potential becomes more negative [43].

Labelle and Jaeger [27] reported a slight increase in bulk density of soil in wheel tracks of forwarders lasting 12 to 24 months at two sites in New Brunswick, Canada. However, a statistically significant increase in the bulk densities of harvested and/or skidded soils after trafficking is unprecedented. Nevertheless, the statistical power of our data set and the consistency of the increase across all treatments and soil depths for at least 4 years make it difficult to assume that the increase is an artifact of sampling, natural variability, or an anomaly [44,45]. We believe that a combination of factors is contributing to the initial postharvest significant increase in bulk density, as well as a similar decrease in bulk density in the skidding treatments by year 7. In this situation, natural processes of decompaction [22] were ineffective. Instead, we considered postharvest changes in soil ecology [46], and the soil mechanics of deformation [41] as plausible factors responsible for an increase in bulk density after year 1 (Figure 5).

We hypothesize that the dynamics and intensity of the below-ground biological processes in conjunction with the protection of the soil provided by a mature forest canopy and intact forest floor allow mature forests to develop a slightly less dense soil. First, the upper 20 cm of forest soils commonly contain more than 80% of the roots in forested ecosystems [47]. To maintain this high root density, complex and dynamic processes of root mortality and regeneration are necessary, which are site and soil environment specific [48]. The root system also requires an equally active root rhizosphere biota to maintain its efficiency. As a result, a substantial proportion of annual gross primary production of forests is allocated to below ground processes [49]. These processes had not been disturbed on our sites for at least a century.

Clearcut forest harvesting removed the predominantly coniferous tree canopy, causing massive root mortality, stopping new root regeneration, and exposing the forest floor to more cyclic climatic variations. At the same time, the soil environment was also conducive to accelerated decomposition of organic matter [50]. As a result, a combination of root mortality, minimal root regeneration and rhizosphere maintenance, and accelerated rate of decomposition is most likely responsible for the natural consolidation of our soils. This consolidation resulted in the small significant increase in bulk density measured after year 1 (Figure 5). These combinations of processes would have occurred across all treatments but are solely responsible for the increase in bulk density in the harvest-only soil.

The increase in bulk density of trafficked soil after 1 year is attributed to a soil mechanics process. In soil mechanics, a small decrease in bulk density occurs whenever a consolidating force applied to a soil is removed [41]. The decrease in bulk density is referred to as rebound. Rebound is an instantaneous process that occurs once a load is removed from a soil and can only be measured in laboratory consolidation tests. Rebound is attributed to the elastic properties of soil [51]. The amount and kind of clay minerals in a soil are the primary determinants of the amount of rebound [52]. Consolidation tests of sieved agricultural soils have found that rebound results in a 2% to 5% decrease in bulk density after a consolidating load is removed. Rebound with similar decreases in bulk density has been reported for undisturbed forest soils [53]. Root systems of trees also increase soil strength [54] and would increase rebound because the root network would provide additional elastic resistance after a wheel passes.

For bulk density to increase in all trafficked treatments and depths at year 1, the soil apparently returned to its most dense condition, which occurred when the soil was under the wheel/track. The massive disruption of soil ecological processes [50] would have adversely impacted the biological elasticity of roots and other biota. It is this disruption that allowed the trafficked soil to 'reconsolidate' after 1 year. The reconsolidation is assumed to occur from the soil collapsing back into the same structural arrangement of soil particles and aggregates when under the wheel/track. The envisioned process is likened to expansion and contraction of a surface soil from the formation of ice lenses during a freeze–thaw cycle [55]. The increase was consistent across the three skidding treatments and three soil depths, which persisted for at least 4 years in the skidding treatments (Figure 5).

The values of bulk density at year 7 in all skidding treatments decreased to approximately the values measured immediately after trafficking (Figure 5). The most plausible explanation for the decrease in the values of bulk density for only the trafficked soil is that vegetation regenerated enough root growth and soil biota recovery to exploit the pore spaces created by rebound. A survey of vegetation cover of these sites at year 4 found the 12-cycle treatment had a vegetation cover of 80%, decreasing with less trafficking to the harvest-only area with a cover of 57% (Startsev, unpublished). Unpublished data on water infiltration and pore size distribution measured at year 5 at a few of these sites also inferred a slight improvement in macroporosity may have started. These data are supportive of the small decrease in bulk density measured by year 7 (Figure 5).

Soil on harvest-only areas failed to recover the porosity lost during the first year (Figure 5). More time is needed for successional development of a more mature forest canopy and forest floor. This development will support and protect the associated higher level of gross primary productivity found in soils under mature forests [48]. The process is generally slower in conifer dominated boreal forest ecosystems than elsewhere [49]. Other than the assumed role in reversing the rebound effect of soil compaction, biological processes had no other obvious role in restoring these compacted soils after 7 years.

4.2. Freeze–Thaw (F–T) as a Decompaction Process

Soil temperature and snow depth were measured on these sites because deep snow insulates the soil, which could limit the effectiveness of the F–T process [35]. The thermal conductivities of air, water, and ice are 0.025, 0.555, and 2.24 W m^{-1} K^{-1}, respectively [56]. The thermal conductivities of snow range between 0.10 and 0.5 W m^{-1} K^{-1}, which is primarily a function of density. In seasonal snowpacks, thermal conductivity also increases with the weathering of the snow grains [57]. The thermal conductivity of snow is 5 to 20 lower than that of mineral soils [56]. Hence, snow is a high-quality insulation that slows the loss of heat from the soil, which slows or prevents soil from freezing.

Our data confirmed the insulating potential of snow during the first 3 years of monitoring (Figure 3), and that a snowpack of at least 40 cm kept mineral soil temperatures close to zero despite cycles of air temperatures dropping to -30 to $-20\,^\circ$C [56]. Thinner snowpacks were measured in 1997/1998, and later years only resulted in soil temperatures decreasing to -4 to $-2\,^\circ$C to a depth of 20 cm. The low soil temperatures were not sustained when air temperatures returned to above $-20\,^\circ$C. The warming of the soil is due to the heat released from melting ice in the soil [56].

When subzero soil temperatures are sufficient to contribute to soil decompaction is a complex question. Soil water must first undergo the process of nucleation, which is a structural transformation of water, before ice crystals can begin to form and grow [58]. The process only occurs in soil when water is supercooled. The nucleation temperature can be a few degrees below $0\,^\circ$C in a pure clay–water mixture [59]. In natural fine-textured soils, nucleation generally occurs at soil temperatures $<-1\,^\circ$C, and initially occurs on the chemically inert surfaces of grains of sand [60] and high silt soils [61]. The growth of ice crystals next to sand grains pushes the finer soil particles away as the ice crystals grow, which can leave clean grains of sand visible on the surfaces of soil peds. Nucleation generally occurs at the surface of the same sand grains and locations when soil undergoes repeated cycles of subzero temperatures [55].

Without an external supply of water, the water/ice ratio decreases as ice crystals grow. For a high silt soil with 75% silt and 25% clay, Azmatch et al. [61] found about half the soil water had changed to ice at a temperature of $-0.65\,^\circ$C. At a temperature of $-9.0\,^\circ$C, approximately 25% of the soil water remained unfrozen. Surface chemistry of clay minerals slows the decrease in the water/ice ratio as the clay content and its chemical activity increase [58]. Soil must be nearly saturated if only in situ soil water is available for growing ice crystals that will be large enough to decrease bulk density [62]. In this situation, the maximum volume expansion of a saturated soil remains less than about 3% at a soil temperature of $-6.7\,^\circ$C. Such conditions are not common unless the soils are poorly drained. The partial freezing of soil water, including those high in silt, can increase the tensile strength of soil by 100-fold at a temperature of only $-0.65\,^\circ$C [61]. Therefore, the measurement of subzero soil temperatures between -2 and $0\,^\circ$C or encountering frozen surface soils in the field cannot automatically be interpreted as an indicator that the F–T process may be decompacting the soil (Figure 3).

The environmental conditions for the F–T process to effectively decompact soil requires a substantial amount of soil water moving upward from deeper in the soil [58]. When the freezing front is stationary, water and vapor moving upward from lower soil layers causes ice crystals to grow into ice lenses and heave the soil upward [60,63,64]. The freezing front will advance deeper into the soil when the heat loss to the atmosphere exceeds the supply of water and heat to the freezing front from deeper in the soil. This advance can be triggered by colder air temperatures, decreasing unsaturated hydraulic conductivity of the nonfrozen soil because the deeper soil no longer contains as much water, or a combination of the two. Freezing of drier soil deeper in the profile will only produce small ice crystals within soil pores that cannot effectively decompact the soil [62]. This is a critical factor limiting the effectiveness of the F–T to decompact subsoils.

A sustained cycle of cold temperatures is more effective at cooling the soil profile. Abrupt decreases in soil temperature do not penetrate soil as deeply as slowly changing soil temperatures of the same amplitude occurring over a long period of time [65]. Snow cover, even a thin one, also moderates the temperature fluctuations (Figure 3) [66]. A sustained balancing of the heat loss to the atmosphere and input of heat from deeper in the soil is called the zero-curtain effect [67]. The depth of the zero curtain in our soils was about 50 cm. The most notable feature of the zero-curtain effect was a synchronized decrease in soil temperature at all depths during the cold cycle starting on 3 January and persisting to 14 January 1998. The effect caused a slight change in the amplitude of soil temperature with increasing depth, and the absence of the normal time lag associated with the peak decreased in soil temperature with increasing depth [65]. This response is indicative of soil cooling as a result of the simultaneous growth of in situ ice crystals at all depths. Otherwise, water flowing upward in the soil during the cold cycle would have increased the volume of ice forming near the surface, and soil temperatures deeper in the soil would not have decreased as much or as quickly with increasing soil depth. As air temperatures increased at the end of the cold cycle, the recovery of soil temperatures was also a synchronized response. This response occurred because the source of most of the heat warming the soil at these depths originated from the latent heat of fusion released from the in situ melting of ice crystals. Our interpretation is also supported by the low saturated hydraulic conductivity of less than 2×10^{-6} m sec^{-1} in these fine-textured soils starting at a depth of 10 cm [14]. The actual rate would have been 100 or more times slower than this value because of the lower viscosity of subzero water and the formation of ice crystals blocking the larger pore spaces. The zero-curtain effect persisted until early March. The occurrence of the zero-curtain effect under seasonal snowpacks may be a limited phenomenon in northern temperate forests. The effect would not occur in the absence of a temperature moderating snowpack or a sustained cycle of cold air temperatures occurring before the snowpack was established [56]. It would not have been nearly as obvious without a temporary large decrease in air temperatures previously discussed (Figure 3).

The F–T process is potentially most effective in surface soils where the cycles of freezing soil temperatures are greater and occur more frequently [55,60]. The potential for ice formation to decompact soil is reduced because ice normally forms horizontal lenses in the soil that collapse when the ice melts [55]. In forestry, this phenomenon is responsible for frost heaving small germinates and new planted seedlings out of the ground [60]. In soil engineering, most of the adverse impacts of F–T on roads occur from one to three F–T cycles; additional cycles cause minimal additional adverse impacts [68,69]. The F–T process in subsoils is seldom effective because cycles of F–T are fewer, unsaturated permeability much lower, and external water supply insufficient. While the surface soils of severely compacted temporary forest roads on high clay content soil have been partly loosed to depths of 10–15 cm, the subsoil remains compact and nonforested after 28 years [70]. The subsoil in two early wagon roads crossing the prairies in southeastern Alberta have also remained compacted after 80 and 100 years [71]; these soils are assumed to have frozen every year.

4.3. Shrink–Swell (S–S) as a Decompaction Process

As a decompaction process, S–S primarily depends on the presence of plastic clay minerals [22]. However, soil must undergo numerous cycles of wetting and drying for evapotranspiration and precipitation cycles to be effective. The clearcut harvesting and mostly intact forest floor on these sites had severely limited evapotranspiration. Five of the sites were imperfectly or poorly drained, and the other four had an argillic subsoil horizon prior to harvesting [14]. As a result, the excess soil moisture reduced soil aeration in the compacted and harvest-only soils on four of these sites, and all sites were moist to wet when sampled in late fall. The future effectiveness of the S–S process in these soils will be poor because other forest soils in the center of the study area have a low plasticity index [70]. A low plasticity index is indicative of soils with lesser amounts of expandable clay minerals [41,59].

4.4. Management Implications

Trafficking and rebound collapse caused a 16.6% increase in bulk density at 5 cm, decreasing to 8.6% at 20 cm for the significantly compacted soil (Table 2). The increase in bulk density only decreased the macropore space [15], but these are the pore sizes primarily responsible for gas diffusion and air permeability [72] and water permeability on these sites [15]. These boreal forest soils commonly have poor internal drainage because they are finer-textured soils and the sites have low slope classes [73]. The undisturbed soil at six of these nine sites had mottles within the top 20 cm [1]. Harvesting and trafficking caused four of these sites to exhibit morphological changes, and two of these sites changed from imperfectly drained to poorly drained soil after 3–4 years because of prolonged anaerobic soil conditions.

Mechanical site preparation is commonly used to ensure the prompt reforestation of conifers in harvested areas across the boreal forests of Alberta. Disc trenchers and small mounders mounted on wheeled skidders are most common; on wetter soils, larger mounds are built with tracked excavators using mounding buckets. Hence, the sustained poor drainage and loss of air-filled porosity [11,14,15] will likely require a switch to the more aggressive and expensive mechanical site preparation practices on moderately well-drained soils and other more poorly drained areas.

The skidders trafficking soils in this study were in the 16–17 Mg weight class; new skidders are in the 22 + Mg class. New machines have the ability to compact soil to a higher bulk density across a wider range of soil wetness. These changes will drive the need to use aggressive forms of site preparation on more areas and more frequently.

The biological consequences of a postharvest increase in bulk density are unlikely to be a long-term ecological concern on most boreal forest sites in the region. The 25% increase in postharvest bulk density was important on these sites because of the dynamic changes occurring in air-filled porosity and redox potential [11,14]. The postharvest increase in bulk density is also a short-term issue (Figure 5) but has the potential to adversely affect the prompt reforestation of some sites at a critical time. Whether the current change in drainage class is a long-term issue is probably unlikely. The data show that changes in redox potential is dynamic and sensitive to the establishment of a new vegetation cover [14]. Therefore, a long-term change in drainage class in these soils from the machines used is probably unlikely. However, changes in machine systems and climate could result in a different outcome because natural drainage class is the dominant factor affecting site productivity of *Picea glauca* (Moench) Voss [74].

4.5. Measuring Bulk Density

Bulk density is an increasingly important soil property for measuring the ecological consequences of soil compaction, assessing soil health [75], and quantifying soil carbon storage. Hence, the collection of bulk density samples using the core (cylinder) method requires a higher level of precision and quality control. Page-Dumroese et al. [28] also reported these types of issues affect the monitoring of field trials and operational practices in forestry.

The standard errors decreased with depth, which is attributed to more natural soil perturbations closer to the surface and differences in morphological soil development (Tables 1 and 4). Nevertheless, kurtosis and skewness were consistent for the three depths. The variance in standard errors widened at the base and was shewed to the right. The range of values may have only started to decrease at 20 cm. The consistency of these results suggests that this distribution should be regarded as a normal soil behavior (Figure 6) [59]. The largest standard errors mostly occurred at Site 9 at year 3 and were attributed to site conditions and a probable lapse in diligence in discarding some soil cores (Figure 6). These extreme values were obvious, but other sources of errors may not be during the collection of soil cores.

The widening of the base of the standard error histogram suggests two sources of variation are likely (Figure 6). First is the natural variability of soil bulk density. These data suggest that the minimum is about 0.02 Mg m^{-3}. McNabb and Boersma [53] reported standard errors between 0.023 and 0.059 Mg m^{-3} (n = 19–23) for four forest soils collected at a higher-quality control standard (7–12 cm depth) because the cores were used in soil mechanics tests. The lowest values were for three Andisols that were quite uniform, and the highest value was for an older Xeric Haplohumult with more spatial and profile variability.

Forest soils likely have higher natural variability than soils in other ecosystems because of large root decay, windthrow, and other types of perturbations that occur close to the soil surface.

The second source of variation in standard error is the quality control used during the collection of soil cores. These two errors are not necessarily additive but are responsible for widening of the base of the distribution of the standard errors (Figure 6). Grossman and Reinsch [29] concluded without evidence that mass had less effect on bulk density than sample volume. D2937-17 [76] recommends a core volume of approximately 940 cm^3, which is the size of the mold used in the standard compaction test, but 7.5 cm diameters have worked well in most agroforestry tests of cylinder sizes [77,78]. A slide-hammer assembly with a removable sleeve with a diameter of about 7.5 cm has long been regarded as a superior design for agroforestry applications [29,79,80]. However, Grossman and Reinsch [29] provided an equation to calculate whether a hammer-driven core sampler is a good design. The specification is the ratio of the cross-sectional area of steel/material in the sampler (i.e., inside wall thickness of the cylinder) relative to its outside diameter. For double-walled samplers, these measurements include the sleeve and driving cylinder. The ratio of the area in material(s) in the sample cylinder to the outside area of the cylinder should not be greater than 10% to 15%. Hvorslev [81] established these values, and they have been part of the ASTM standard for decades. For our stainless steel rings with a wall thickness of 0.16 cm, the ratio is 8.3%. For a double-walled sampler with a diameter of 7.5 cm and a total wall thickness of 0.6 cm, the area ratio is 27%, and for a smaller sampler, 5 cm double-walled sampler with a wall thickness estimated at 0.5 cm, the ratio is about 32%.

The higher ratio requires more of the energy applied to a driven core sampler to displace soil so that the cutter edge can advance deeper. As a result, thick-walled samplers are far more likely to fracture and loosen dry or compacted soils, compact wetter cohesive soils, or limit the entry of soil into the ring. Woody roots in forest soils compound this problem when a driven core sampler temporarily bounces on an uncut root. In cases where the ring does not fill with soil, Gross and Reinsch [29] and Hao et al. [82] propose measuring the height of the unfilled ring with a ruler or filling the void with beads to correct for soil volume. This is an unacceptable practice and becomes a major source of unknown error in the measurement of bulk density. Soil engineers have long measured the height of soil within a sampling cylinder and compared it with the depth that the cylinder penetrates the soil as a measure of how severely the soil core had been disturbed [81]. As a result, the North American project to evaluate soil health lists Blake and Hartge [80] as their preferred methods for measuring bulk density [75].

Blake [79,80] considered the volume changes in soil from shattering or compaction to be a genuine problem. D2937-17 [76] provides a design specification for a driven sampler with a thin-wall sampler to reduce these errors. The design can be scaled to a 7.5 cm diameter ring. The design includes a space in the driving head for the expansion of a sample if it shatters. If the driving head is removed prior to extracting the cylinder, the elevation of the soil inside the ring can be compared with the surrounding soil. This is a critical practice to ensure that only high-quality soil cores are extracted from the soil profile. A tolerance for error should be established, and samples failing to meet this standard should be abandoned while still in the ground. D1937-17 [76] also provides helpful advice on the soil and conditions when the drive-cylinder method should not be used as well as several recommendations when it may not be applicable.

5. Conclusions

This paper reiterates the point that significant compaction measured as an increase in bulk density is related to soil water potential, which can be measured in the field with a handheld penetrometer. The value applies to a specific weight class of forestry machines. Heavier machines will compact soil to a higher bulk density and do so across a wider range of soil wetness. Hence, the applicable value of soil water potential will be lower because heavier machines can significantly compact soil at a lower soil water content.

Harvesting and soil trafficking resulted in a slight but statistically significant increase in bulk density after 1 year on these sites. Older mature forests are viewed as providing a deeper forest floor and mature canopy that protects natural processes that have tended to loosen soil over an extended period of time, which cannot be maintained after harvesting. The bulk density of all trafficked soil increased by the same small amount, which is attributed to soil rebound. These are natural soil mechanical and biophysical responses to a soil perturbation; they have always happened. Only the size and quality of our database allowed us to document them.

This project confirms that small, significant increases in bulk density can have site- and soil-specific ecological consequences. The ecological consequence on these sites was that compaction prolonged anaerobic soil conditions in soils with partially impaired drainage. The dynamics of the process indicate that the shift will be temporary. Reforestation will shift the water balance of these sites back to a less anaerobic environment, but in the interim, more aggressive forms of site preparation will be necessary to promptly reforest the sites.

Monitoring of air and soil temperatures found that subzero soil temperature is not indicative of a freeze–thaw process that could decompact soil. As a decompaction process, the freeze–thaw process is complex with exacting requirements of soil texture and mineralogy, soil permeability, temperature gradients in the profile, and external supply of water. These requirements were not met, and the bulk density of trafficked soil remained unchanged. A snow depth of 40 cm in these ecotypes effectively moderated soil temperatures with and without an intact forest floor.

Recent changes simplifying the measurement of bulk density using the core method and videos showing the ease with which sample rings can be hammered into the ground are misleading. The collection of high-quality core samples for bulk density and other physical tests has never been easy. When bulk density is used as an indicator of soil health, monitoring the sustained impact of soil compaction on ecosystem services, and soil carbon storage, the precision of the values of bulk density obtained by the core method becomes paramount. The protocol for the collection of bulk density using the core method must be updated with an emphasis on the use of better practices and specifications for the core method and a rigorous commitment to a quality control and assurance program in the field.

Author Contributions: Conceptualization, D.H.M.; methodology, D.H.M.; formal analysis, A.S.; investigation, all; resources, D.H.M.; data curation, A.S.; writing, D.H.M.; visualization, D.H.M.; supervision, all; project administration, all; funding acquisition, all. All authors have read and agreed to the published version of the manuscript.

Funding: The data collection and original publications were performed while both authors were employed by the Alberta Environmental Centre, Alberta Environment at Vegreville, Alberta, which became part of the Alberta Research Council toward the end of the data collection period. Research sites and forest harvesting equipment for the field trials were provided by Canadian Forest Products Ltd., Weldwood of Canada Ltd. (Hinton Division), Weyerhaeuser Canada Ltd. (Grande Prairie), Sundance Forest Products Ltd., Sunpine Forest Products Ltd., Millar Western Industries Ltd., and Alberta Newsprint Company. These companies also provided direct funding for the project via the Alberta Forest Development Trust, Government of Alberta. Funding for the first 3 years was also provided by Foothills Model Forest (now fRI Research), Hinton, Alberta. The analyses and preparation of a draft report of these data were initially made possible by a grant from the Forest Resources Improvement Association of Alberta (FRIAA), Open Funds Initiative, Government of Alberta, to ForestSoil Science Ltd. Final paper and interpretations are personal contributions.

Institutional Review Board Statement: Not applicable.

Data Availability Statement: Data may be available from the senior author.

Acknowledgments: The authors thank Susan Paquin and Rod Kusiek for their technical and laboratory support and as crew leaders for the collection of soil samples, and the field crew; Michelle Hiltz for advice on statistical analyses; and reviewers.

Conflicts of Interest: The funders had no role in the design of the study; in the collection, analyses, and interpretation of data; in the writing of the manuscript; or in the decision to publish the results. DM has also done research and development on soil restoration implements and practices, which has resulted in Canada Patent No 2586933, and U.S. Patent No 8176993 B2, a 'Ripper Plough for Soil Tillage'.

References

1. Rosenberg, N.R. Response of plants to the physical effects of soil compaction. *Adv. Agron.* **1964**, *16*, 181–196. [CrossRef]
2. Greacen, E.L.; Sands, R. Compaction of forest soils: A review. *Aus. J. Soil Res.* **1980**, *18*, 163–189. [CrossRef]
3. Froehlich, H.A.; Aulerich, D.D.; Curtis, R. *Designing Skidtrails Systems to Reduce Soil Impacts from Tractive Logging Machines*; Research Paper 44; Oregon State University: Corvallis, OR, USA, 1981; p. 15.
4. Froehlich, H.A.; McNabb, D.H. Minimizing soil compaction in Pacific Northwest forests. In *Forest Soils and Treatment Impacts. Proceedings of the 6th North American Forest Soils Conference Knoxville, TN, USA, 19–23 June 1983*; Stone, E.L., Ed.; The University of Tennessee: Knoxville, TN, USA, 1984; pp. 159–192.
5. Hamza, M.A.; Anderson, M.A. Soil compaction in cropping systems: A review of the nature, causes, and possible solutions. *Soil Till. Res.* **2005**, *82*, 121–145. [CrossRef]
6. Horn, J.; Vossbrink, J.; Peth, S.; Becker, S. Impact of modern forest vehicles on soil physical properties. *For. Ecol. Manag.* **2007**, *248*, 56–63. [CrossRef]
7. Cambi, M.; Certini, G.; Neri, F.; Marchi, E. The impact of heavy traffic on forest soils: A review. *For. Ecol. Manag.* **2015**, *338*, 124–138. [CrossRef]
8. Alakukku, L.; Weisskofp, P.; Chamen, W.C.T.; Tijink, F.G.J.; van der Linde, J.P.; Pires, S.; Sommer, C.; Spoor, G. Prevention strategies for field traffic-induced subsoil compaction: A review: Part 1. Machine/soil interactions. *Soil Till. Res.* **2003**, *73*, 145–160. [CrossRef]
9. Chamen, W.C.T.; Alakukku, L.; Pires, S.; Sommer, C.; Spoor, G.; Tijink, F.; Weisskofp, P. Prevention strategies for field traffic-induced subsoil compaction: A review: Part 2. Equipment and field practices. *Soil Till. Res.* **2003**, *73*, 161–174. [CrossRef]
10. DeArmond, D.; Ferraz, J.B.S.; Higuchi, N. Natural recovery of skid trails: A review. *Can. J. For. Res.* **2021**, *51*, 948–961. [CrossRef]
11. McNabb, D.H.; Startsev, A.D.; Nguyen, H. Soil wetness and traffic level effects on bulk density and air-filled porosity of compacted boreal forest soils. *Soil Sci. Soc. Am. J.* **2001**, *65*, 1238–1247. [CrossRef]
12. da Silva, A.P.; Kay, B.C. Estimating the least limiting water range of soils from properties and management. *Soil Sci. Soc. Am. J.* **1997**, *61*, 877–883. [CrossRef]
13. Zou, C.; Penfold, C.; Sands, R.; Misra, R.K.; Hudson, I. Effects of air-filled porosity, soil matric potential and soil strength on primary root growth of radiata pine seedlings. *Plant Soil* **2001**, *236*, 105–115. [CrossRef]
14. Startsev, A.D.; McNabb, D.H. Effects of compaction on aeration and morphology of boreal forest soils in Alberta, Canada. *Can. J. Soil Sci.* **2009**, *89*, 45–56. [CrossRef]
15. Startsev, A.D.; McNabb, D.H. Skidder traffic effects on water retention, pore size distribution, and van Genuchten parameters of boreal forest soils. *Soil Sci. Soc. Am. J.* **2001**, *65*, 224–231. [CrossRef]
16. Startsev, A.D.; McNabb, D.H. Effects of skidding on soil infiltration in west central Alberta. *Can. J. Soil Sci.* **2000**, *80*, 617–624. [CrossRef]
17. Howard, R.F.; Singer, M.J.; Frantz, G.A. Effects of soil properties, water content, and compactive effort on the compaction of selected California forest and range soils. *Soil Soc. Soc. Am. J.* **1981**, *45*, 231–236. [CrossRef]
18. Zhao, Y.; Krzic, M.; Bulmer, C.E.; Schmidt, M.G. Bulk Density of British Columbia Forest Soils from the Proctor Test: Relationships with Selected Physical and Chemical Properties. *Soil Sci. Soc. Am. J.* **2008**, *72*, 442–452. [CrossRef]
19. Carter, M.R. Relative measures of soil bulk density to characterize compaction in tillage studies on fine sandy loams. *Can. J. Soil Sci.* **1990**, *70*, 425–433. [CrossRef]
20. Håkansson, I.; Lipiec, J. A review of the usefulness of relative bulk density values in studies of soil structure and compaction. *Soil Till. Res.* **2000**, *53*, 71–85. [CrossRef]
21. Zhao, Y.; Krzic, M.; Bulmer, C.E.; Schmidt, M.G.; Simard, S.W. Relative bulk density as a measure of compaction and its influence on tree height. *Can. J. For. Res.* **2010**, *40*, 1724–1734. [CrossRef]
22. Dexter, A.R. Amelioration of soil by natural processes. *Soil Till. Res.* **1991**, *20*, 87–100. [CrossRef]
23. Corns, I.G.W. Compaction by forestry equipment and effects on coniferous seedling growth on four soils in the Alberta foothills. *Can. J. For. Res.* **1988**, *18*, 75–84. [CrossRef]
24. Mohieddinne, H.; Brasseur, B.; Spicher, F.; Gallet-Moron, E.; Buridant, J.; Kobaissi, A.; Horen, H. Physical recovery of forest soil after compaction by heavy machines, revealed by penetration resistance over multiple decades. *For. Ecol. Manag.* **2019**, *449*, 117472. [CrossRef]
25. Wert, S.; Thomas, B.R. Effects of skid roads on diameter, height, and volume growth in Douglas-fir. *Soil Sci. Soc. Am. J.* **1981**, *45*, 629–632. [CrossRef]

26. Froehlich, H.A.; Miles, D.W.R.; Robbins, R.W. Soil bulk density recovery on compacted skid trails in Central Idaho. *Soil Sci. Soc. Am. J.* **1985**, *49*, 1015–1019. [CrossRef]

27. Labelle, E.R.; Jaeger, D. Soil compaction caused by cut-to-length forest operations and possible short-term natural rehabilitation of soil density. *Soil Sci. Soc. Am. J.* **2011**, *75*, 2314–2329. [CrossRef]

28. Page-Dumroese, D.S.; Jurgensen, M.F.; Tiarks, A.E.; Ponder, F., Jr.; Sanchez, F.G.; Fleming, R.L.; Kranabetter, J.M.; Powers, R.F.; Stone, D.M.; Elioff, J.D.; et al. Soil physical property changes at the North American Long-Term Soil Productivity study sites: 1 and 5 years after compaction. *Can. J. For. Res.* **2006**, *36*, 551–564. [CrossRef]

29. Grossman, R.B.; Reinsch, T.G. 2.1 Bulk density and linear extensibility. In *Methods of Soil Analysis, Part 4: Physical Methods, 5.4*; Dane, J.H., Topp, G.C., Eds.; Soil Science Society of America Inc.: Madison, WI, USA, 2002; pp. 201–228. [CrossRef]

30. Goutal, N.; Soivin, P.; Ranger, J. Assessment of the natural recovery rate of soil specific volume following forest soil compaction. *Soil Sci. Soc. Am. J.* **2012**, *76*, 1426–1435. [CrossRef]

31. SAS Institute. *SAS System for Linear Models*, 3rd ed.; SAS Institute Inc.: Cary, NC, USA, 1991.

32. Littell, R.C.; Milliken, G.A.; Stroup, W.W.; Wolfinger, R.D. *SAS System for Mixed Models*; SAS Institute Inc.: Cary, NC, USA, 1996.

33. Milliken, G.A.; Johnson, D.E. *Analysis of Messy Data: Designed Experiments*; Chapman Hall: London, UK, 1994; Volume 1.

34. Strong, W.L. *Ecoregions and Ecodistricts of Alberta*; Alberta Forestry, Lands, and Wildlife: Edmonton, AB, Canada, 1992; Volume 1.

35. Thorud, D.B.; Duncan, D.P. Effects of snow removal, litter removal, and soil compaction on soil freezing and thawing in a Minnesota oak stand. *Soil Sci. Soc. Am. Proc.* **1972**, *36*, 153–157. [CrossRef]

36. Siltanen, R.M.; Apps, M.J.; Zoltai, S.C.; Mair, R.M.; Strong, W.L. *A Soil Profile and Organic Carbon Data Base for Canadian Forest and Tundra Mineral Soils*; Natural Resources Canada: Edmonton, AB, Canada, 1997.

37. McNabb, D.H. Soil failures under rigid-tracked forestry machines as a function of slope and soil wetness. In *Exceeding the Vision: Forest Mechanization in the Future, Proceedings of the 52nd International Symposium on Forestry Mechanization, Sopron, Hungary/Forchtenstein, Austria, 6–9 October 2019*; Czupy, I., Ed.; University of Sopron Press: Sopron, Hungary, 2019; pp. 340–349, ISBN 978-963-334-343-2.

38. Wikipedia.org. Kurtosis, Updated 6 February 2022. Available online: https://en.wikipedia.org/w/index/php?title=Kurtosis&oldid=1070274736 (accessed on 12 February 2022).

39. Ampoorter, E.; de Schrijver, A.; van Nevel, L.; Hermy, M.; Verheyen, K. Impact of mechanized harvesting on compaction of sandy and clayey forest soils: Results of a meta-analysis. *Ann. For. Sci.* **2012**, *69*, 533–542. [CrossRef]

40. Williamson, J.R.; Neilson, W.A. The influence of forest site on rate and extent of soil compaction and profile disturbance of skid trails during ground-based harvesting. *Can. J. For. Res.* **2000**, *30*, 1196–1205. [CrossRef]

41. Das, B.M. *Fundamentals of Geotechnical Engineering*, 4th ed.; Cengage Learning: Stamford, CT, USA, 2013.

42. Krzic, M.; Bulmer, C.E.; Teste, F.; Dampier, L.; Rahman, S. Soil properties influencing compactability of forest soils in British Columbia. *Can. J. Soil Sci.* **2004**, *84*, 219–226. [CrossRef]

43. McNabb, D.H.; Boersma, L. Nonlinear model for compressibility of partly saturated soils. *Soil Sci. Soc. Am. J.* **1996**, *60*, 333–341. [CrossRef]

44. Casler, M.D. Fundamentals of experimental design: Guidelines for designing successful experiments. *Agron. J.* **2015**, *107*, 692–705. [CrossRef]

45. Campbell, K.G.; Thompson, Y.M.; Guy, S.O.; McIntosh, M.; Glaz, B. Is, or is not, the two great ends of Fate: Errors in agronomic research. *Agron. J.* **2015**, *107*, 718–729. [CrossRef]

46. Marshall, V.G. Impacts of forest harvesting on biological processes in northern forest soils. *For. Ecol. Manag.* **2000**, *133*, 43–60. [CrossRef]

47. Jackson, R.B.; Canadell, J.; Ehleringer, J.R.; Mooney, H.A.; Sala, O.E.; Schultze, E.D. A global analysis of root distributions for terrestrial biomes. *Oecologia* **1996**, *108*, 389–411. [CrossRef]

48. Grier, C.C.; Logan, R.S. Old-growth Pseudotsuga menziesii communities of a western Oregon watershed: Biomass distribution and production budgets. *Ecol. Monogr.* **1979**, *47*, 373–400. [CrossRef]

49. Waring, R.H.; Schlesinger, W.H. *Forest Ecosystems Concepts and Management*; Academic Press, Inc.: New York, NY, USA, 1985.

50. Startsev, N.A.; McNabb, D.H.; Startsev, A.D. Soil biological activity in recent clearcuts in Alberta. *Can. J. Soil. Sci.* **1998**, *78*, 69–78. [CrossRef]

51. Kirchhof, G. Plastic Properties. In *Encyclopedia of Soil Science*, 2nd ed.; Lai, R., Ed.; Taylor and Francis: New York, NY, USA, 2006; pp. 1311–1313.

52. Stone, J.A.; Larson, W.E. Rebound of five one-dimensional compressed granular soils. *Soil Sci. Soc. Am. J.* **1980**, *44*, 819–822. [CrossRef]

53. McNabb, D.H.; Boersma, L. Evaluation of the relationship between compressibility and shear strength of Andisols. *Soil Sci. Soc. Am. J.* **1993**, *57*, 923–929. [CrossRef]

54. Mickovski, S.B.; Hallett, P.D.; Bransby, M.F.; Davies, M.C.R.; Sonnenberg, R.; Bengough, A.G. Mechanical reinforcement of soil by willow roots: Impacts of root properties and root failure system. *Soil Sci. Soc. Am. J.* **2009**, *73*, 1276–1285. [CrossRef]

55. Kay, B.C.; Grant, C.D.; Groenevelt, P.H. Significance of ground freezing on soil bulk density under zero tillage. *Soil Sci. Soc. Am. J.* **1985**, *49*, 973–978. [CrossRef]

56. Zhang, T. Influence of the seasonal snow cover on the ground thermal regime: An overview. *Am. Geophys. Union* **2005**, *43*, RG4002. [CrossRef]

57. Smith, M.; Jamieson, B. A new set of thermal conductivity measurements. In Proceedings of the International Snow Science Workshop 2014, Banff, AB, Canada, 29 September–3 October 2014; pp. 507–510. [CrossRef]
58. Chalmers, B.; Jackson, K.A. *Experimental and Theoretical Studies of Mechanism of Frost Heaving*; Research Paper 199; Cold Regions Research and Engineering Laboratory: Hanover, NH, USA, 1970; p. 23.
59. Mitchell, J.K.; Soga, K. *Fundamentals of Soil Behavior*, 3rd ed.; John Wiley & Sons, Inc.: Hoboken, NJ, USA, 2005.
60. Heidmann, L.J. *Frost Heaving of Tree Seedlings: A Literature Review of Causes and Possible Control*; USDA Forest Serv. Gen. Tec. Rep. RM-21; Rocky Mountain Forest and Range Experiment Station: Fort Collins, CO, USA, 1976; p. 10.
61. Azmatch, T.C.; Sego, D.C.; Arenson, L.U.; Biggar, K.W. Tensile strength of frozen soils using for-point bending test. In Proceedings of the 63th Canadian Geotechnical Conference, Calgary, AB, Canada, 12–16 September 2010; pp. 436–442.
62. Dagesse, D.F. Freezing-induced bulk soil changes. *Can. J. Soil Sci.* **2010**, *90*, 389–401. [CrossRef]
63. Groenevelt, P.H.; Kay, B.D. On the Interaction of Water and Heat Transport in Frozen and Unfrozen Soils: II. The liquid phase. *Soil Sci. Soc. Am. J.* **1994**, *38*, 400–404. [CrossRef]
64. Sheng, D.; Axelsson, K.; Knutsson, S. Frost heave due to ice lens formation in freezing soils 1. Theory and verification. *Nord. Hydrol.* **1995**, *26*, 125–146. [CrossRef]
65. Jury, W.A.; Gardner, W.R.; Gardner, W.H. *Soil Physics*, 5th ed.; John Wiley & Sons, Inc.: New York, NY, USA, 1991.
66. Hinkel, K.M.; Outcalt, S.I. Identification of Heat-Transfer Processes during Soil Cooling, Freezing and Thaw in Central Alaska. *Permafr. Periglac. Processes* **1994**, *5*, 217–235. [CrossRef]
67. Oulcalt, S.I.; Nelson, F.E.; Hinkel, K.M. The Zero-Curtain Effect: Heat and mass transfer across an isothermal region in freezing soil. *Water Resour. Res.* **1990**, *26*, 1509–1516.
68. Eigenbroad, B.D. Effects of cyclic freezing and thawing on volume changes and permeabilities of soft fine-grained soils. *Can. Geotech. J.* **1996**, *33*, 529–537. [CrossRef]
69. Viklander, P. Permeability and volume changes in till due to cyclic freeze/thaw. *Can. Geotech. J.* **1988**, *35*, 471–477. [CrossRef]
70. McNabb, D.H. Tillage of compacted haul roads and landings in the boreal forests of Alberta, Canada. *For. Ecol. Manag.* **1994**, *66*, 179–194. [CrossRef]
71. Gillund, G.; Kennedy, B. Historical wagon trails: An evaluation of the durability of subsoil compaction in Alberta. In Proceedings of the 32nd Annual Alberta Soil Science Workshop, Grande Prairie, AB, Canada, 13–15 March 1995; pp. 201–206.
72. Schjønning, P.; Lamandé, M.; Berisso, F.E.; Simojoki, A.; Alakukku, L.; Andreasen, R.R. Gas diffusion, non-Darcy air permeability, and computed tomography images of a clay subsoil affected by compaction. *Soil Sci. Soc. Am. J.* **2013**, *77*, 1977–1990. [CrossRef]
73. Bechingham, J.D.; Corns, I.G.W.; Archibald, J.H. *Field Guide to Ecosites of West-Central Alberta*; Special Report—Field Guide (NoFC—Edmonton); Northern Forest Centre: Edmonton, AB, USA, 1996.
74. Yang, G.G.; Klinka, K. Use of synoptic variables in predicting white spruce site index. *For. Ecol. Manag.* **1996**, *80*, 95–105. [CrossRef]
75. Norris, C.E.; Bean, G.M.; Cappellazzi, S.B.; Cope, M.; Greub, K.L.H.; Liptzin, D.; Rieke, E.L.; Tracy, P.W.; Morgan, C.L.S.; Honeycutt, C.W. Introducing the North American project to evaluate soil health measurements. *Agron. J.* **2020**, *112*, 3195–3215. [CrossRef]
76. *ASTM D2937-17*; Standard Test Method for Density of soil in place by the drive-cylinder method. ASTM International: West Conshohocken, PA, USA, 2017; p. 8.
77. Terry, T.A.; Cassel, D.K.; Wollum, A.G., II. Effects of soil sample size and included root and wood on bulk density in forested soils. *Soil Sci. Soc. Am. J.* **1981**, *45*, 135–138. [CrossRef]
78. Page-Dumroese, D.S.; Jurgensen, M.F.; Brown, R.E.; Mroz, G.D. Comparison of Methods for determining bulk densities of rocky forest soils. *Soil Sci. Soc. Am. J.* **1999**, *63*, 379–383. [CrossRef]
79. Blake, G.R. Bulk density. In *Methods of Soil Analysis Part 1 Physical and Mineralogical Properties, Including Statistics of Measurement and Sampling*; Black, D.A., Ed.; Soil Science Society of America Inc.: Madison, WI, USA, 1965; pp. 374–390. [CrossRef]
80. Blake, G.R.; Hartge, K.H. Bulk density. In *Physical and Mineralogical Methods*, *5.1*, 2nd ed.; Klute, A., Ed.; Soil Science Society of America Inc.: Madison, WI, USA, 1986; pp. 363–375. [CrossRef]
81. Hvorslev, M.J. *Subsurface Exploration and Sampling of Soils for Civil Engineering Purposes: Report on a Research Project of the Committee on Sampling and Testing, Soil Mechanics and Foundations Division, American Society of Civil Engineers*; Waterways Experiment Station: Vicksburg, MS, USA, 1949.
82. Culley, J.L.B. Density and compressibility. In *Soil Sampling and Methods of Analysis*; Carter, M.R., Ed.; CRC Press: Boca Raton, FL, USA, 1993; pp. 529–539.

Article

The Changes in Soil Microbial Communities across a Subalpine Forest Successional Series

Zhihui Wang [1], Yi Bai [1], Jianfeng Hou [1], Fei Li [1], Xuqing Li [1], Rui Cao [1,2], Yuyue Deng [1,3], Huaibin Wang [1,4], Yurui Jiang [1,5] and Wanqin Yang [1,*]

[1] School of Life Science, Taizhou University, Taizhou 318000, China; 17816890031@163.com (Z.W.); baiyi@tzc.edu.cn (Y.B.); mehoujianfeng@163.com (J.H.); flissrs@yeah.net (F.L.); xuqing.li@outlook.com (X.L.); cr_leshan@163.com (R.C.); dengyuyue96@gmail.com (Y.D.); sichuanwhb@163.com (H.W.); jiangyrchina@163.com (Y.J.)

[2] Long-Term Research Station of Alpine Forest Ecosystems, Institute of Ecology & Forestry, Sichuan Agricultural University, Chengdu 611130, China

[3] School of Geography and Ecotourism, Southwest Forestry University, Kunming 650224, China

[4] School of Life Science, China West Normal University, Nanchong 650224, China

[5] College of Forestry, Northwest A&F University, Xianyang 712100, China

* Correspondence: scyangwq@163.com; Tel.: +86-576-8866-0339

Citation: Wang, Z.; Bai, Y.; Hou, J.; Li, F.; Li, X.; Cao, R.; Deng, Y.; Wang, H.; Jiang, Y.; Yang, W. The Changes in Soil Microbial Communities across a Subalpine Forest Successional Series. *Forests* **2022**, *13*, 289. https://doi.org/10.3390/f13020289

Academic Editor: Luís González

Received: 27 December 2021
Accepted: 10 February 2022
Published: 11 February 2022

Publisher's Note: MDPI stays neutral with regard to jurisdictional claims in published maps and institutional affiliations.

Abstract: Knowledge regarding changes in soil microbial communities with forest succession is vital to understand soil microbial community shifts under global change scenarios. The composition and diversity of soil microbial communities across a subalpine forest successional series were therefore investigated in the Wanglang National Nature Reserve on the eastern Qinghai-Tibet Plateau, China. The calculated diversity indices of soil bacteria (8.598 to 9.791 for Shannon-Wiener, 0.997 to 0.974 for Simpson, 4131 to 4974 for abundance-based coverage estimator (ACE) and 3007 to 3511 for Species richness indices), and ACE (1323 to 921) and Species richness (1251 to 879) indices of soil fungi decreased from initial to terminal succession stages, but Shannon-Wiener and Simpson of soil fungi indices varied slightly with forest succession. Meanwhile, the composition and structure of soil microbial communities varied markedly with forest succession. The relative abundance of the dominant bacterial phyla (*Acidobacteria*, *Firmicutes* and *Actinobacteria*) and fungal taxa (*Mortierellomycota*, *Rozellomycota* and unassigned phylum clade GS01) varied considerably with forest succession. However, regardless of successional stage, *Proteobacteria* and *Acidobacteria* dominated soil bacterial communities and *Ascomycota* and *Basidiomycota* dominated soil fungal communities. Moreover, the changes in soil microbial diversity with forest succession were significantly affected by soil pH, soil organic carbon, soil temperature, altitude, and non-woody debris stock. Importantly, soil pH was the dominant driver of soil microbial community shift with forest succession. In conclusion, the forests at different succession stages not only conserve same microbial populations, but also nurse unique microbial diversity across the forest succession series; and the biodiversity of soil bacterial and fungal communities has differential responses to forest succession.

Keywords: bacterial community; forest succession; forest variables; fungal community; non-woody debris; soil biodiversity

1. Introduction

Soil microbial communities play irreplaceable roles in driving the biogeochemical cycles of carbon and nutrients and maintaining site productivity and energy flow [1–3]. As a result, investigations on the changes in soil microbial community structure and diversity with environmental factors can provide key evidence for understanding the mechanisms driving bioelement cycles, managing forest ecosystems and conserving soil biodiversity. The present consensus is that the composition, structure and diversity of soil microbial communities can be driven by combinations of biotic and abiotic factors [4–6], making the responses of soil microbial communities to environmental changes relatively complex. In particular, the changes in forest community structure could simultaneously alter soil properties [7], microclimate [8], and the input and qualities of above- and below plant biomass [9,10]; these changes could then lead to soil microbial community shifts. For instance, Yan et al. [11] have concluded that more complex forest community structure can provide more diverse food sources for soil microbes and increase the heterogeneity of the soil environment, thereby increasing the biodiversity of the soil microbiota. *Proteobacteria*, *Actinobacteria* and *Acidobacteria* were important members of soil bacterial community, *Proteobacteria* represent the most metabolically diverse group of anoxygenic chlorophototrophs [12]. The vast majority of *Actinobacteria* are important saprophytes capable of breaking down a wide range of plant and animal debris in the process of decomposition [13]. *Acidobacteria* is generally acidophilic, oligotrophic and difficult to cultivate [14]. Forests with abundant plant species diversity have relatively higher amounts of root exudates and leaf litter, and alter soil properties such as nutrient regimes, temperature and moisture dynamics and pH, leading to an increase in the substrate quality and quantity for soil microbes and thus affecting the structure and diversity of soil microbial communities [15]. However, soil microbial community shifts across the forest successional series have not been fully investigated.

Forest succession is an ecological process in which one forest community evolves into another due to the forces of nature or humans [16–18]. Global change-driven forest succession [19,20] may modify key soil processes and, in particular, have important impacts on the structure and functioning of soil microbial communities [21]. As a result, full investigation on the changes in soil microbial communities with forest succession is vital to understand the shifts in soil microbial communities under global change scenarios. In theory, forest succession could influence the shifts in soil microbial communities in at least the following ways. First, the quality and quantity of woody and non-woody debris on the forest floor vary greatly with forest succession [22], and the related changes in substrate quality and quantity alter the composition and diversity of soil microbial communities [23]. Second, soil temperature and moisture can vary greatly with forest succession owing to the changes in canopy structure and understory cover [24,25]; both soil temperature and moisture are key factors in determining the composition and structure of soil microbial communities [15,26]. Third, the changes in soil microbial community structure can be driven by soil properties (e.g., pH, nutrient concentration and ecological stoichiometry) [5,27,28], which are sensitive to forest succession [29,30]. Although changes in soil microbial communities with forest types have been widely reported [31–33], the shifts in soil microbial communities in response to changes in soil properties, plant debris stock, and soil temperature and moisture across the forest successional series remain unknown.

The subalpine forests distributed on the eastern Qinghai-Tibet Plateau and in the upper reaches of Yangtze River play paramount roles in conserving water and soil, supporting biodiversity, responding to climate change, and participating in the global carbon cycle [34]. As affected by long-term natural disturbances such as earthquakes, debris flows and snowslides, and commercial logging of natural forests, subalpine forest communities of different successional stages have been widely observed in the subalpine forest region [34,35]. To date, changes in woody and non-woody debris stocks across a subalpine forest successional series [22] and changes in soil microbial diversity with gap size [36] and along environmental gradients [37] in the subalpine forest region have been documented. Nevertheless, little information is available on soil microbial community shifts across the subalpine forest successional series.

Based on the importance of forest succession on soil microbial community diversity and composition and little information on successional series in the subalpine forest region, we proposed the following research questions: (1) would soil microbial diversity change with forest succession? (2) would the composition of soil microbial community be richer with succession? (3) what role would forest and soil variables play in the effect of succession on microbial community change? To answer these questions, we investigated the shifts in soil microbial community composition and diversity across the subalpine forest successional series and the key drivers of those shifts. We hypothesized that (1) soil microbial diversity would increase from initial to terminal stages of forest succession; (2) the relative abundance of dominant taxa in soil bacterial and fungal communities would vary markedly among different succession stages, and soil bacterial and fungal communities would respond differentially to forest succession; and (3) non-woody debris stock would dominate the composition and diversity of soil microbial communities across the forest successional series. The objectives of this study were to elucidate the important function of forest succession on the diversity of soil microbial community and explore the key forest variables driver factors of those shifts. Those results help us better understand that maintaining different succession stages of forest communities through moderate disturbances is beneficial to the conservation of soil biodiversity at the forest region level.

2. Material and Methods

2.1. Field Description

The study region is located in Wanglang National Nature Reserve (32°49′–33°02′ N, 103°55′–104°10′ E; 2300–4983 m a.s.l.), Pingwu County, Sichuan, southwestern China, which is distributed on the eastern Qinghai-Tibet Plateau and in the upper reaches of Yangtze River. The area has a semihumid climate typical of the Danba-Songpan region. The annual precipitation is 859.9 mm, and the annual mean temperature is 2.9 °C, with maximum and minimum mean temperatures of 12.7 °C (in July) and −6.1 °C (in January), respectively [35]. The subalpine forest consists of abundant plant species, including mosses, herbs, shrubs, deciduous broadleaved trees and coniferous trees.

2.2. The Method of Selection of Study Plots and Their Characteristics

Wanglang subalpine coniferous region seriously damaged by long-term the commercial logging of natural forests below 2700 m since the 1950s and natural disturbance such as the 1976 Songpan-Pingwu earthquake [38,39]. Forest logging in the area ceased in 1962. Therefore, the Wanglang Nature Reserve has formed many secondary forests in different stages of succession and primary forests that have never been destroyed [35,38–40]. We considered their distribution based on forest disturbance history and natural recovery process and interviews with local supervisors, We defined the forests of S1 to S5 succession stages by referring to the succession division of Zhang et al. [35], and defined the primary coniferous forests which have never been destroyed to S6 succession stage. According to Zhang et al. [35], the S1 successional stage was dominated by shrubs (e.g., *Salix* spp., *Ulmus pumila* L. and *Hippophae rhamnoides* L.); the S2 successional stage was dominated by the deciduous broadleaved species *Betula* spp. and *Populus* spp.; the S3 successional stage was characterized by mixed forests dominated by deciduous broadleaved species (e.g., *Betula* spp.) and coniferous species (e.g., *Abies faxoniana* Rehd.); the S4 successional stage was dominated by middle-age coniferous species with a diameter at breast height (DBH) less than 20 cm (e.g., *Larix gmelinii* (Rupr.) Kuzen. and *Sabina saltuaria* Rehd.); the S5 successional stage was dominated by mature coniferous plants with a DBH less than 40 cm and more than 20 cm (e.g., *A. faxoniana* Rehd.); and the S6 succession stage was characterized as primary coniferous forest dominated by *Picea purpurea* Mast (Table S1). The traits of the six successional stages, including the coordinates, altitudes, directions, canopy covers, soil names and dominant species are shown in Table S1.

2.3. Soil Sampling Method

Soils across the subalpine forest successional series were sampled in August 2019. Replicate plots (10 m × 20 m) were established in each successional stage as follows: 12 plots in S1, 9 plots in S2, 9 plots in S3, 6 plots in S4, 6 plots in S5 and 6 plots in S6. After removing the litter layers, nine soil samples in each plot were randomly collected from the top 20 cm of the soil profiles by a stainless steel corer (5-cm diameter) and mixed. Moreover, the topsoil temperature (T) of each plot was measured by a portable soil thermometer (CEM,

DT-131, Shenzhen, China). The roots and stones were removed from the soil samples, which were then divided into two subsamples. One subsample was stored in a bag on ice, immediately transported to a laboratory, and stored at $-80\,^{\circ}\text{C}$ for high-throughput sequencing for microbial community diversity analysis [31]. The other subsample was transported to the laboratory and air-dried for soil property analysis.

2.4. Soil Agrochemical and Chemical Properties Analyses

After being passed through a 1-mm sieve, the air-dried soil samples were used for chemical analyses. Soil pH was measured at a soil: water ratio of 1:2.5 (*m:v*) with a digital pH meter (FE20K, Mettler-Toledo, Greifensee, Switzerland). The concentrations of total nitrogen (TN) in the soils were determined from a milled sample by combustion at $950\,^{\circ}\text{C}$ using an elemental analyzer (Vario EL III, Elementar, Langenselbold, Germany). Soil organic carbon (SOC) was measured by the potassium dichromate oxidation method [41]. The C:N ratio (C:N) was calculated as the mass ratio of SOC and TN [42].

2.5. Soil DNA Extraction, PCR Amplification and High-Throughput Sequencing

Total soil DNA was extracted from 0.5 g soil samples using the FastDNA Spin Kit for Soil (MP Biomedicals, Santa Ana, CA, USA) according to the instructions of the manufacturer. The V4 hypervariable region of the 16S rDNA [43] and the internal transcribed spacer (ITS1 and ITS2) region of rDNA [44] were selected as sequencing targets for bacteria and fungi, respectively. The library was built using the NEBNext® Ultra™ DNA Library Prep Kit for Illumina (NEB, Ipswich, MA, USA) at Novogene Company (Beijing, China). Sequencing was performed using a paired_end sequencing strategy with barcode connectors based on the Illumina MiSeq PE250 platform (Illumina Inc., San Diego, CA, USA). The barcode sequences were spliced with FLASH (v 1.2.7) software to obtain the raw tags [45]. The clean tags were obtained by quality filtering the raw tags using QIIME (v1.9.0) software [46,47]. Chimeric sequences were removed using the UCHIME algorithm [48]. Operational taxonomic unit (OTU) clustering for all the effective tags with 97% identity was performed by UPARSE (v 7.0.1001) software after discarding singletons [49]. The SSUrRNA database of SILVA132 [50] and Mothur software, and the UNIT (v 7.2) database [51] and QIIME (v 1.9.0) [52] software were selected as species annotation analysis for 16S rDNA and ITS, respectively.

2.6. Statistical Analyses

Alpha diversity indices of soil microbial communities (Shannon-Wiener, Simpson, abundance-based coverage estimator (ACE) and species richness) were calculated using QIIME (v 1.9.0) software [53]. To identify soil bacterial and fungal taxa differentially represented among the six forest successional stages, differentially abundant taxa were selected using LEfSe (v 1.0) software at a linear discriminant analysis (LDA) score of 4 [54]. Analysis of similarities (ANOSIM) was performed based on the Bray–Curtis distance algorithm [55] to identify the significance of the differences in the microbial communities among different succession stages. A redundancy analysis (RDA) was performed to screen the environmental factors that influence species differences at the phylum level among different stages of succession. Differential the abundance of soil microbial phyla composition, alpha diversity indices, and the concentrations of soil SOC, TN, C:N, temperature and pH, the stocks of woody debris (WD) and non-woody debris (NWD) among successional stages were tested by Kruskal-Wallis test and Wilcoxon rank-sum test. These tests were completed with IBM SPSS Statistics v. 20 (IBM Corporation, New York, NY, USA). Differences were considered significant at the 0.05 significance level. The metrics of the ANOSIM and RDA were determined using the vegan package [56] in R (v 2.15.3) software [57]. The R^2 and *p* value of the RDA, indicating the effects of each environmental factor on species distribution, were calculated using the Envfit function of the vegan package [56].

3. Results

3.1. Changes in Soil Properties and Stocks of Woody Debris and Non-Woody Debris with Succession

The stock of woody debris (WD), soil concentrations of SOC, TN and C:N showed a significant tendency of increase from the S1 to S6 succession stages (all $p < 0.05$). On the contrary, the stock of non-woody debris (NWD), soil temperature (T) and pH were observed significantly higher in initial than terminal stages of succession (all $p < 0.05$; Table 1). Especially, the S4 succession stage had the lowest soil concentrations of SOC and TN and the highest soil C:N and soil pH (Table 1).

Table 1. Soil physicochemical properties and the stocks of woody debris and non-woody debris among different succession stages.

Succession Stage	SOC (g kg^{-1})	TN (g kg^{-1})	C:N	T (°C)	pH	WD (Mg ha^{-1})	NWD (Mg ha^{-1})
S1	54.61 ± 37.09 b	5.70 ± 3.16 b	9.19 ± 1.81 c	12.96 ± 1.35 a	6.51 ± 0.41 a	31.73 ± 7.31 c	1.31 ± 0.30 a
S2	50.22 ± 25.20 b	5.16 ± 2.21 b	9.62 ± 1.76 bc	12.21 ± 2.04 ab	6.16 ± 0.62 ab	24.49 ± 10.34 c	0.70 ± 0.11 ab
S3	74.07 ± 42.92 ab	6.75 ± 0.37 b	10.93 ± 1.73 abc	10.54 ± 1.48 bc	6.54 ± 0.36 ab	51.49 ± 5.90 bc	0.82 ± 0.19 ab
S4	33.63 ± 9.45 b	2.34 ± 0.30 c	14.60 ± 4.62 ab	11.75 ± 1.31 ab	6.78 ± 0.28 a	36.07 ± 5.75 c	1.07 ± 0.12 a
S5	85.25 ± 54.12 ab	7.05 ± 4.08 bc	11.84 ± 0.84 ab	11.47 ± 0.33 ab	6.40 ± 0.32 ab	82.46 ± 20.38 a	0.43 ± 0.08 c
S6	136.93 ± 60.44 a	9.86 ± 3.40 a	13.42 ± 1.94 a	9.97 ± 0.99 c	5.90 ± 0.51 b	77.27 ± 21.97 a	0.80 ± 0.23 ab

SOC: the concentrations of soil organic carbon; TN: the concentrations of total nitrogen; C:N: the mass ratio of SOC and TN; T: soil temperature; pH: soil pH; WD: the stock of woody debris; NWD: the stock of non-woody debris; Values followed by different lowercase letters mean significant difference ($p < 0.05$) among six successional series based on Wilcoxon rank-sum test. S1 to S6 represent successional stages from 1 to 6, respectively.

3.2. Changes in the Sequence Data and Soil Microbial Alpha Diversity

A total of 3,004,539 and 2,845,500 high-quality effective sequences of bacteria and fungi were acquired across the subalpine forest successional series; the sequences were clustered with 97% identity into 13,684 and 12,233 OTUs, respectively. Among these OTUs, 3490 bacterial and 761 fungal OTUs were common in soil microbial communities at six successional stages (Figure 1). Soil bacterial and fungal communities at the initial succession stage (S1) had the most independent OTUs (1225 and 1572, respectively) (Figure 1). The alpha diversity of soil bacterial community at the initial successional stage (S1) was significantly higher than that at the terminal successional stages (S5 and S6) ($p < 0.05$; Figure 3.3), although there was not a remarkable difference in fungal alpha diversity among the six forest successional stages according to these two indices ($p > 0.05$; Figure 3). However, ACE and Species richness indices in fungal communities were significantly higher at the S1 succession stage than at the S6 succession stage ($p < 0.05$; Section 3.3 and Figure 3).

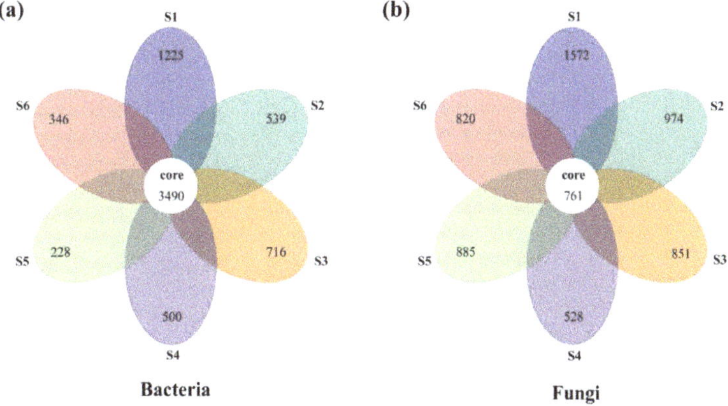

Figure 1. Flower diagram of the common and unique bacterial (**a**) and fungal (**b**) OTUs across a subalpine forest succession series. S1 to S6 represent successional stages from 1 to 6, respectively.

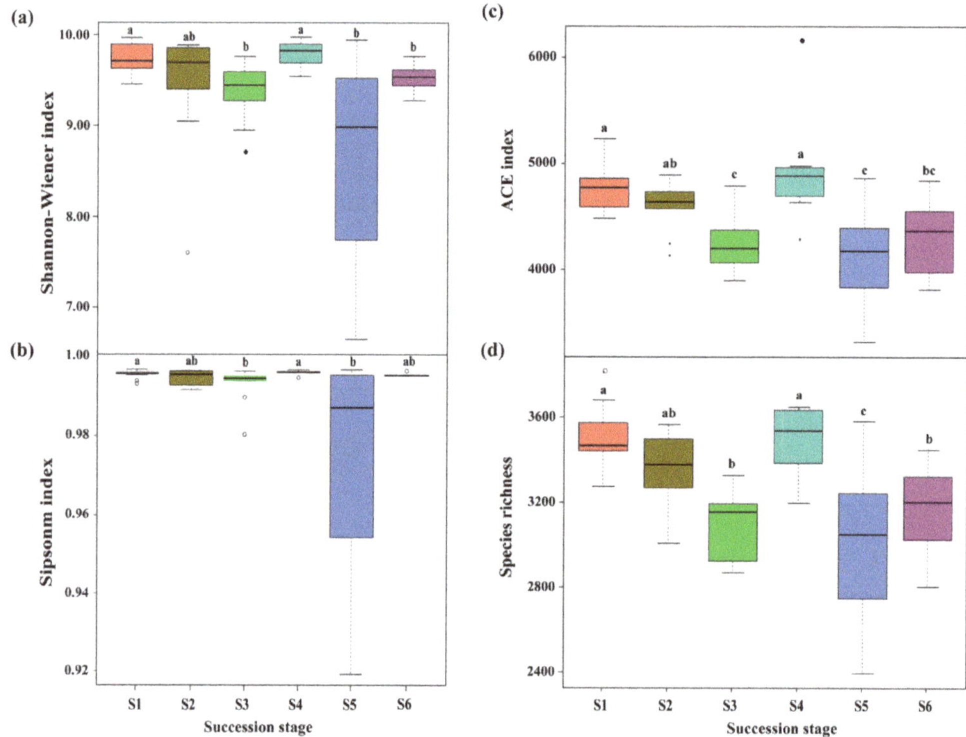

Figure 2. Differences in bacterial alpha diversity across a subalpine forest succession series. Calculations based on the OTUs at 97% sequence identity. Shannon-Wiener index (**a**), Simpson index (**b**), ACE index (**c**) and Species richness (**d**) represent the indices of alpha diversity. Values followed by different lowercase letters mean significant difference ($p < 0.05$) among six successional series based on Wilcoxon rank-sum test. S1 to S6 represent successional stages from 1 to 6, respectively.

3.3. Composition and Structure of Soil Microbial Community

At the phylum level, soil bacterial communities at the six successional stages were dominated by *Proteobacteria* (37.7–45.2%), *Acidobacteria* (15.3–27.5%) and *Actinobacteria* (5.0–10.4%) (Table 2). The relative abundance of eight of the 10 most abundant phyla varied markedly ($p < 0.05$) with forest succession, except for that of *Proteobacteria* and *Verrucomicrobia* (Figure 4a). In particular, higher relative abundance of *Firmicutes* ($p < 0.001$), Tenericutes ($p < 0.01$) and *Acidobacteria* ($p < 0.05$) were observed at terminal succession stages (S5 or S6) (Figure 4b; Table S2). In contrast, the relative abundance of *Chloroflexi* ($p < 0.01$), *Actinobacteria* ($p < 0.01$) and *Rokubacteria* ($p < 0.01$) were markedly higher at the initial successional stages (S1, S2 and S3) (Figure 4a; Table S2). Soil fungal communities across the subalpine forest successional series were dominated by *Ascomycota* (27.7–46.4%), *Basidiomycota* (13.5–31.8%) and *Mortierellomycota* (2.0–12.5%) (Table 3). The relative abundance of the other phyla totaled less than 2% (Table 3). Among the fungal phyla, clade GS01 was not enriched in the S4 and S6 successional stages, and *Blastocladiom* was not enriched in the S4 successional stage (Table 3). Furthermore, the significantly higher relative abundance of *Mortierellomycota* ($p < 0.05$), *Rozellomycota* ($p < 0.01$) and clade GS01 ($p < 0.05$) were observed at the terminal succession stage (S5 or S6) (Figure 4b; Table S2).

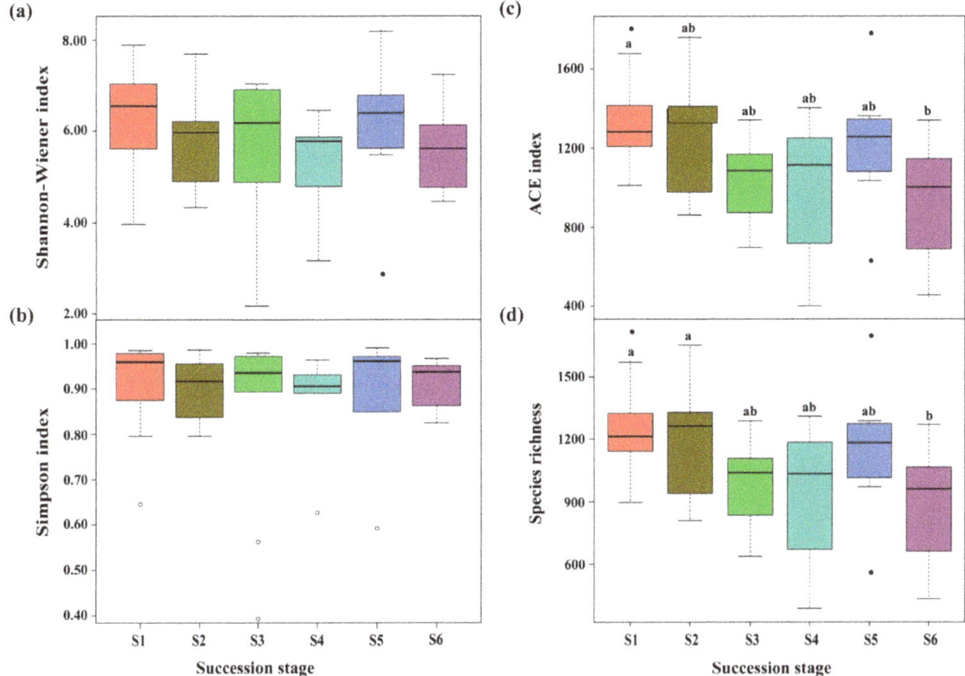

Figure 3. Differences in fungal alpha diversity across a subalpine forest succession series. Calculations based on the OTUs at 97% sequence identity. Shannon-Wiener index (**a**), Simpson index (**b**), ACE index (**c**) and Species richness (**d**) represent the indices of alpha diversity. Values followed by different lowercase letters mean significant difference ($p < 0.05$) among six successional series based on Wilcoxon rank-sum test. S1 to S6 represent successional stages from 1 to 6, respectively.

Table 2. Percentage composition of the dominant phyla of soil bacterial community among different succession stages.

Bacterial Phyla (%)	S1	S2	S3	S4	S5	S6
Proteobacteria	42.16	42.75	37.66	45.5	40.26	45.19
Acidobacteria	21.11	20.48	18.64	22.59	15.34	27.46
Actinobacteria	9.33	10.02	10.45	7.34	5.03	8.23
Firmicutes	3.65	3.58	8.84	2.85	12.33	0.95
Bacteroidetes	4.83	4.71	2.77	4.38	2.91	4.21
Gemmatimonadetes	2.39	2.65	4.88	4.88	2.34	2.70
Chloroflexi	3.69	3.46	3.99	3.14	1.98	2.91
Verrucomicrobia	3.64	4.55	1.75	1.68	1.73	3.44
Rokubacteria	2.79	3.58	3.72	2.57	1.44	1.57
Tenericutes	1.13	0.06	2.15	0.37	13.51	0.00
Others	5.27	4.16	5.16	4.71	3.14	3.34

The abundance of each taxon in relation to the abundance of all the taxa was calculated as the relative abundance (%) (based on the data of sequence analysis). S1 to S6 represent successional stages from 1 to 6, respectively.

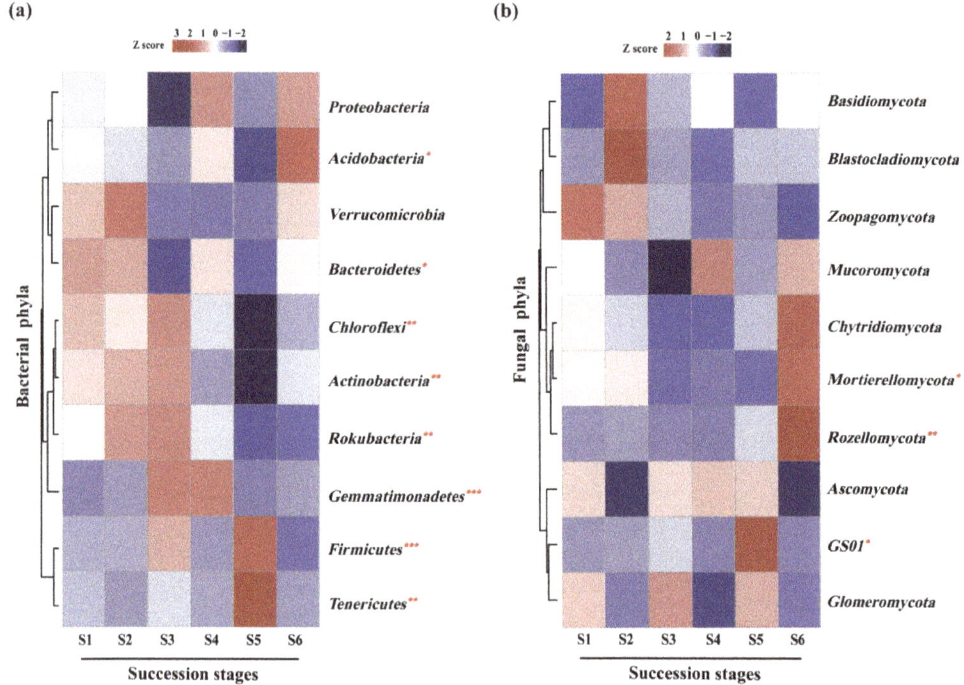

Figure 4. Relative abundance and significance analyses of the dominant soil bacterial (**a**) and fungal (**b**) communities at phylum level across a subalpine forest succession series. Significant effect: * $p < 0.05$, ** $p < 0.01$, *** $p < 0.001$. Kruskal-Wallis test. S1 to S6 represent successional stages from 1 to 6, respectively.

Table 3. Percentage composition of the dominant phyla of soil fungal community among different succession stages.

Fungal Phyla (%)	S1	S2	S3	S4	S5	S6
Ascomycota	45.29	27.93	45.07	46.40	45.78	27.74
Basidiomycota	13.53	31.80	18.75	21.95	13.77	22.28
Mortierellomycota	7.25	7.62	2.00	2.87	2.00	12.54
Rozellomycota	0.47	0.74	0.18	0.09	1.53	6.66
Chytridiomycota	0.17	0.13	0.04	0.04	0.12	0.30
Mucoromycota	0.12	0.09	0.04	0.17	0.09	0.15
Zoopagomycota	0.07	0.05	0.03	0.01	0.02	0.01
Glomeromycota	0.08	0.04	0.09	0.02	0.08	0.03
GS01	0.01	0.01	0.03	0.00	0.13	0.00
Blastocladiomycota	0.00	0.03	0.01	0.00	0.01	0.01
Others	33.01	31.57	33.77	28.44	36.46	30.27

The abundance of each taxon in relation to the abundance of all the taxa was calculated as the relative abundance (%) (based on the data of sequence analysis). S1 to S6 represent successional stages from 1 to 6, respectively.

Furthermore, based on an LDA score of 4, linear discriminant analysis effect size (LEfSe) indicated that there were a large number of significantly different taxa (biomarkers) among different succession stages (Figure 5). In total, 38 and 31 taxa (including the phylum, class, order, family, genus and species levels) were selected as biomarkers for soil bacterial and fungal communities, respectively, along the succession gradients (Figure 5).

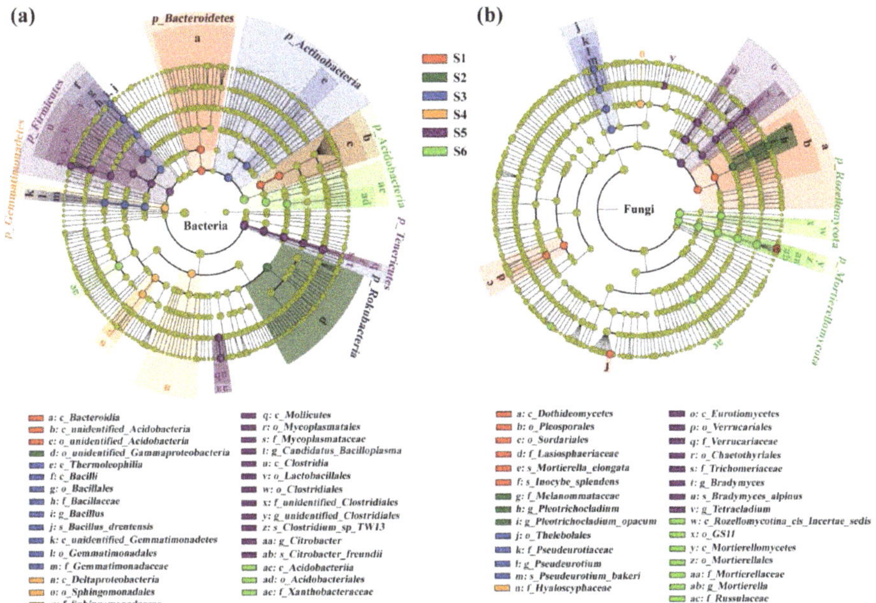

Figure 5. Cladograms generated by linear discriminant analysis effect size (LEfSe) indicating differences in the bacterial (**a**) and fungal (**b**) taxa across a subalpine forest succession series. Different colored bars indicate taxa were enrichment in the corresponding succession stages. S1 to S6 represent successional stages from 1 to 6, respectively.

ANOSIM test verified that the composition of soil bacterial and fungal communities was substantially distinct among the six successional stages (all $p = 0.001$; Figure 6). Specifically, ANOSIM showed that there were significant differences in soil bacterial community structure among different succession stages, except between S1 and S2, between S1 and S4, and between S3 and S4. Slight differences in soil fungal community composition were observed between S1 and S2, between S3 and S5 and between S4 and S5, but significant differences were found between the remaining successional stages (Table S3).

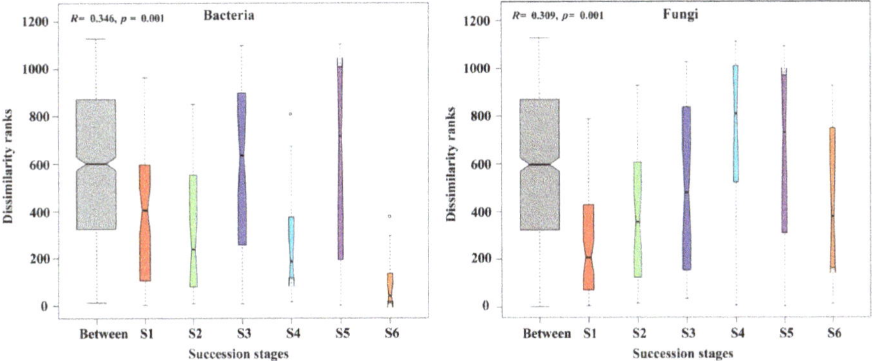

Figure 6. Boxplot of analysis of similarities (ANOSIM) based on the Bray–Curtis distances of samples for the dissimilarities of bacterial and fungal communities across a subalpine forest succession series. S1 to S6 represent successional stages from 1 to 6, respectively.

3.4. Relationships of Soil Microbial Diversity with Forest Variables

According to the analysis of detrended correspondence analysis (DCA) (Table S4), RDA was selected to analyze the relationships of soil microbial diversity with second forest variables based on the analysis of variance inflation factor (VIF) (Table S5). RDA confirmed that the variations in soil bacterial and fungal communities across the successional series were associated with forest variables (Figure 7). Among the forest variables, soil pH had an extremely significant effect on soil bacterial ($p < 0.01$) and fungal community composition ($p < 0.001$; Tables 4 and 5). Meanwhile, soil bacterial diversity was also significantly affected by SOC, soil T, the stock of non-woody debris (NWD) and altitude ($p < 0.05$; Table 4), while soil fungal community composition was also significantly influenced by soil temperature (T) ($p < 0.05$; Table 5). In particular, among dominant bacterial phyla, soil pH was significantly and positively correlated with the relative abundance of *Gemmatimonadetes* and *Firmicutes*, but markedly and negatively correlated with the abundance of *Proteobacteria*, *Bacteroidetes* and *Verrucomicrobia* (Figure 7a). The stock of NWD and soil temperature (T) was prominently and positively correlated with the abundance of most of the bacterial phyla, but observably and negatively correlated with the abundance of *Gemmatimonadetes*, *Firmicutes* and *Tenericutes*. On the contrary, the correlation of altitude and most of soil bacterial community composition was negative (Figure 7a). In addition, the relative abundance of *Proteobacteria* and *Bacteroidetes* were also significantly and positively correlated with the concentration of SOC (Figure 7a). Among dominant fungal phyla, soil pH was notably positively correlated with the abundance of *Rozellomycota* and *Ascomycota* not the other dominant phyla (Figure 7b). In contrast, soil temperature (T) was significantly and positively correlated with the abundance of the most abundant phyla, except for *Rozellomycota* and *Basidiomycota* (Figure 7b).

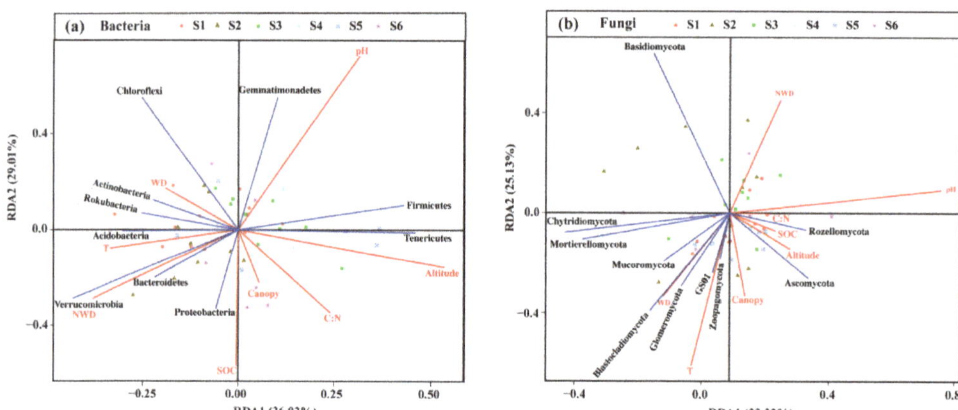

Figure 7. Redundancy analysis (RDA) about the effect of forest variables on the differences in the compositions of soil bacterial (**a**) and fungal (**b**) communities across a subalpine forest succession series. SOC: the concentrations of soil organic carbon; C:N: the mass ratio of SOC and soil total nitrogen; T: soil temperature; pH: soil pH; WD: the stock of woody debris; NWD: the stock of non-woody debris. S1 to S6 represent successional stages from 1 to 6, respectively.

Table 4. Significance of the effects of forest variables on bacterial community composition among different succession stages.

	Forest Variable							
	SOC	C:N	T	pH	WD	NWD	Altitude	Canopy
RDA1	−0.142	0.637	−0.963	0.543	−0.828	−0.841	0.991	0.187
RDA2	−0.990	−0.770	−0.268	0.840	0.561	−0.541	−0.128	−0.982
r^2	0.135	0.072	0.066	0.298	0.028	0.125	0.152	0.021
Pr(>r)	0.039 *	0.202	0.225	0.002 **	0.527	0.047 *	0.026 *	0.642

SOC: the concentrations of soil organic carbon; C:N: the mass ratio of SOC and soil total Nitrogen; T: soil temperature; pH: soil pH; WD: the stock of woody debris; NWD: the stock of non-woody debris; Significant effect: * $p < 0.05$, ** $p < 0.01$.

Table 5. Significance of the effects of forest variables on fungal community composition among different succession stages.

	Forest Variables							
	SOC	C:N	T	pH	WD	NWD	Altitude	Canopy
RDA1	0.973	0.999	−0.389	0.994	−0.763	0.570	0.934	0.205
RDA2	−0.232	0.023	−0.921	0.107	−0.647	0.821	−0.358	−0.979
r^2	0.016	0.012	0.145	0.332	0.074	0.094	0.032	0.040
Pr(>r)	0.683	0.749	0.026 *	<0.001 ***	0.153	0.105	0.495	0.422

SOC: the concentrations of soil organic carbon; C:N: the mass ratio of SOC and soil total nitrogen; T: soil temperature; pH: soil pH; WD: the stock of woody debris; NWD: the stock of non-woody debris; Significant effect: * $p < 0.05$, *** $p < 0.001$.

4. Discussion

The composition and diversity of soil bacterial and fungal communities varied significantly with forest succession. The alpha diversity of soil bacteria and fungi decreased from initial to terminal succession stages except for Shannon-Wiener and Simpson indices of soil fungal community, which was in contrast to our first hypothesis. *Proteobacteria* and *Acidobacteria* dominated soil bacterial communities, and *Ascomycota* and *Basidiomycota* dominated soil fungal communities in all six successional stages. However, a significant difference was observed in the relative abundance of dominant soil microbial phyla among the six successional stages. Meanwhile, the ANOSIM results demonstrated that soil microbial community structure was substantially different among the six successional stages. These results were consistent with our second hypothesis. RDA showed that soil pH had the greater effect on the differences in the composition of soil communities across the forest successional series, while non-woody debris stock was correlated with only the composition of soil bacterial community, partly supporting our third hypothesis. Generally, these results confirm that forests in different successional stages not only nursed the same dominant microbial communities but also support unique biomarkers and that the biodiversity in soil bacterial and fungal communities differed greatly in response to forest vegetation succession.

4.1. Changes in the Diversity of Soil Microbial Community with Forest Succession

Soil microbial diversity demonstrated a global tendency to decrease from initial to terminal succession stages; an exception to this finding was demonstrated by the Shannon-Wiener and Simpson indices for the fungal communities, which was supported by the findings of several studies [58,59]. However, the changes in soil bacterial and fungal alpha diversity varied greatly with forest succession from those reported previously (Table 6). A considerable increase in bacterial alpha diversity and an increase followed by a decrease in fungal diversity were found with forest succession in different regions [25,30,32]. Meanwhile, Zhang et al. [31] and Liu et al. [25] observed a trend of decreasing first and then increasing. In addition, the alpha diversity of soil bacterial and fungal communities did not show significant changes with succession [30,42,59]. As succession progressed, the aboveground plant community changed significantly. However, at middle and late successional stages, the changes in plant species and the intensification of competition led to a loss of soil nutrients and a decrease in the diversity of soil microbial communities [25]. Moreover, the decreased alpha diversity index may be due to the different growth strategies of microbes; in particular, at initial succession stages, the soil microbial population lives in harsh and unpredictable conditions and has a high reproductive rate but a low survival rate, resulting in high diversity but low abundance. At terminal succession stages, the microbial population lives in a favorable and predictable environment, with a low reproductive rate but a high survival rate, resulting in increased competitiveness and a stable population number [31]. In addition, the differences in soil bacterial and fungal alpha diversity may have occurred since soil microorganisms have differing nutrient preferences [60].

Table 6. Comparisons of forest soil microbial community alpha diversity index across forest succession series in different regions.

Alpha Diversity Index	Bacteria Tendency	Fungi Tendency	Reference
Shannon_Wiener, Simpson	decreased, then increased	ns	This study
ACE, Specie richness	decreased	decreased	
Shannon_Wiener, Chao1, Simpson	increased	increased, then decreased	[32]
ACE, Chao 1, Shannon	decreased, then increased	–	[31]
ACE, Chao 1	–	increased, then decreased	[42]
Shannon_Wiener	ns	–	
Chao 1, Shannon	increased	increased, then decreased	[25]
OTUs, Chao 1	increased	increased, then decreased	[30]
Shannon	ns	ns	
OTUs	decreased	ns	
Shannon	decreased	increased, then decreased	[59]
Chao 1	ns	ns	
Shannon	increased	–	[61]
ACE, Chao 1, Shannon, Sobs	decreased, then increased	decreased, then increased	[62]

ns: no significance; –: no data.

4.2. Changes in Soil Microbial Community Composition with Forest Succession

Proteobacteria, *Acidobacteria* and *Actinobacteria* were the dominant bacterial phyla in the six forest successional stages, but their relative abundance differed significantly, which was consistent with the findings of a growing number of studies in different ecosystems [62–64]. Fierer et al. [65] analyzed the extent to which these phyla were dominant in many soil bacterial communities based on phylogenetic analysis. The abundance of *Proteobacteria* increased with increasing vegetation successional stage, and this result was generally consistent with findings in New Zealand [66]. These findings suggested that *Proteobacteria* likely play a functional role in the restoration of soil [67]. Many soil *Proteobacteria* are copiotrophic and become abundant when labile substrates are available during secondary forest succession [65]. However, a higher abundance of *Actinobacteria* was observed at initial successional stages in our study since *Acidobacteria* are oligotrophic and prefer nutrient-poor environments. Notably, *Planctomycetes*, the dominant phylum in other forest succession ecosystems [31,68,69], was not dominant in these subalpine successional forests. This phenomenon suggested that serious water and soil erosion during subalpine forest succession might limit habitat availability for *Planctomycetes*, which is mainly present in oceans, marine sediments, freshwater lakes and wastewater [70].

Ascomycota and *Basidiomycota* were two dominant phyla in soil fungal communities among the six successional stages, but slight changes were observed in their abundance among the stages. This result was consistent with the findings of previous studies [25,71] but was contradictory to the observation in a *Pinus yunnanensis* forest [42]. Meanwhile, the abundance of *Mortierellomycota*, *Rozellomycota* and clade GS01 at terminal succession stages (S5 and S6) were higher than those in the other stages. Generally, the members of *Ascomycota* are common in extreme environments, while *Mortierellomycota*, *Rozellomycota* or clade GS01 members may have a strong preference for host plants in infertile soil, as these fungi exhibited environmental selection pressure and dispersal limitation with forest succession. Therefore, the abundance of these taxa in soil fungal communities increased with forest succession, indicating the increased accumulation of soil nutrients and maturation of the ecosystem.

The structure of soil bacterial and fungal communities differed substantially among the six successional stages based on ANOSIM analyses. This result was also observed in previous research [32]. However, other observations showed a lack of significant difference of community structure among successional stages [72]. These results suggested that the development pattern is consistent between plant and microbial communities and showed that above- and underground communities follow a synchronous succession process [73]. This further verified the importance of plant succession in microbial community structure.

4.3. The Roles of Forest Variables in Microbial Community Composition and Forest Succession

Our observations showed that the composition of soil microbial communities across the successional series was significantly affected by forest variables (e.g., soil pH, soil organic carbon, non-woody debris stock, altitude and soil temperature), which was consistent with the findings of previous studies [61,74,75]. Soil pH is a critical factor that affects the structure of soil bacterial and fungal communities [76–78]. Landesman et al. [27] have confirmed that the changes in tree species caused by secondary succession can lead to the acidification of forest soil, which was in line with our first result. Meanwhile, the members of *Acidobacteria*, *Mortierellomycota* and *Rozellomycota* [79,80], can physiologically tolerate acidic pH levels [81,82]. Therefore, the increase in the number of these microbial communities at terminal succession stages could be explained by the acidification of soil pH in this period. In addition to soil pH, soil temperature was also strongly correlated with the characteristics of soil microbial communities [83,84]. Soil temperature can not only directly accelerate microbial metabolic rates and biochemical processes [85–87] but also indirectly affect activity levels by changing the temperature dependency of the community [88]. In addition, plant successional processes, including nutrient and water uptake, as well as the initiation, branching, and orientation of root growth, also result in changes in soil temperature [89]. These factors could result in microbial communities performing variably over the succession period in response to soil temperatures.

Our results also found that the composition of soil bacterial communities, but not the composition of soil fungal communities, was significantly correlated with SOC and non-woody debris stock across the forest successional series, which was inconsistent with the findings of some previous studies [90,91]. Soil bacterial and fungal communities utilize soil organic carbon at different rates [92,93]. Variations in NWD among successional stage and vegetation type evoke different microbial responses to the quantitative and qualitative differences in plant compounds [94]. Meanwhile, our results showed that the soil concentration of SOC increased with plant community succession, which may be due to the decomposition of plant NWD. The concentration of soil organic matter strongly affects the structure and function of soil microbial communities [95,96]. Our results further demonstrated that distinct soil and vegetation properties could induce different soil microbial composition across forest succession series.

5. Conclusions

The diversity and composition of soil microbial communities varied greatly with forest succession. We summarized the following points:

(1)   The alpha diversity of soil bacteria, and ACE and Species richness indices of fungi significantly decreased with plant community succession, while there were slight changes in Shannon-Wiener and Simpson indices of soil fungi, indicating that soil bacterial and fungal communities had differential responses to forest succession.

(2)   The relative abundance of the most dominant phyla exhibited prominent differences among the six successional stages, although *Proteobacteria* and *Acidobacteria*, and *Ascomycota* and *Basidiomycota* were the dominant phyla of bacteria and fungi, respectively, regardless of succession level. The structure of soil microbial communities showed an obvious separation across the forest successional series.

(3)   Soil pH significantly affected the composition of soil microbial communities, while the NWD stock was correlated with only the change in soil bacterial community composition with forest succession, suggesting that the factors driving the changes in the structure and diversity of soil bacterial and fungal communities with forest succession differed greatly in the subalpine forest ecosystem.

To sum up, forest communities at different successional stages not only share dominant soil bacterial and fungal taxa but also support significantly different dominant groups of soil bacteria and fungi with community succession. The results imply that maintaining different succession stages of forest communities through moderate disturbances is beneficial to the conservation of soil biodiversity at the forest region level.

Supplementary Materials: The following supporting information can be downloaded at: https://www.mdpi.com/article/10.3390/f13020289/s1, Table S1: Site characteristics across a successional gradient in the subalpine forest on eastern Qinghai-Tibet Plateau; Table S2: Discrepancy of relative abundance of the dominant phyla of soil bacterial and fungal community composition among different forest succession stages; Table S3: The analysis of similarities (ANOSIM) among different forest succession stages based on the distance algorithm of Bray–Curtis; Table S4: The detrended correspondence analysis (DCA) based on based on the OTUs data at 97% sequence identity; Table S5: The detection of collinearity of forest variables by the analysis of Variance Inflation Factor (VIF).

Author Contributions: W.Y.: Conceptualization, funding acquisition, project administration, resources, supervision, validation and writing—review and editing. Z.W.: conceptualization, data curation, investigation, methodology and writing—original draft. F.L.: funding acquisition and investigation. Z.W., Y.B., J.H., F.L., X.L., Y.D. and Y.J.: investigation. Z.W., R.C. and H.W.: methodology. All authors have read and agreed to the published version of the manuscript.

Funding: This work was supported by the National Natural Science Foundation of China (grant numbers 32071554, 32001139).

Institutional Review Board Statement: Not applicable.

Informed Consent Statement: Not applicable.

Data Availability Statement: All data generated or analyzed during this study are included in this published article. Data in Table 6 is available through Zhang et al. 2016, Chai et al. 2019, Zhao et al. 2020, Li et al. 2020, Liu et al. 2020 (25), Liu et al. 2020 (62), Shang et al., 2021, Wang et al. 2021.

Acknowledgments: The authors of this study appreciate the field assistance of Wanglang National Nature Reserve involved in the initial sampling assignments.

Conflicts of Interest: The authors declare that they have no known competing financial interests or personal relationships that could have appeared to influence the work reported in this paper.

References

1.   Bodelier, P. Toward Understanding, Managing, and Protecting Microbial Ecosystems. *Front. Microbiol.* **2011**, *2*, 80. [CrossRef] [PubMed]

2.   Van Der Heijden, M.G.A.; Bardgett, R.D.; Van Straalen, N.M. The unseen majority: Soil microbes as drivers of plant diversity and productivity in terrestrial ecosystems. *Ecol. Lett.* **2008**, *11*, 296–310. [CrossRef] [PubMed]

3.   He, Z.; Yu, Z.; Huang, Z.; Davis, M.; Yang, Y. Litter decomposition, residue chemistry and microbial community structure under two subtropical forest plantations: A reciprocal litter transplant study. *Appl. Soil Ecol.* **2016**, *101*, 84–92. [CrossRef]

4. Lamarche, J.; Bradley, R.L.; Hooper, E.; Shipley, B.; Beaunoir, A.-M.S.; Beaulieu, C. Forest Floor Bacterial Community Composition and Catabolic Profiles in Relation to Landscape Features in Québec's Southern Boreal Forest. *Microb. Ecol.* **2007**, *54*, 10–20. [CrossRef] [PubMed]

5. Ramirez, K.S.; Craine, J.M.; Fierer, N. Consistent effects of nitrogen amendments on soil microbial communities and processes across biomes. *Glob. Change Biol.* **2012**, *18*, 1918–1927. [CrossRef]

6. Prescott, C.E.; Grayston, S.J. Tree species influence on microbial communities in litter and soil: Current knowledge and research needs. *For. Ecol. Manag.* **2013**, *309*, 19–27. [CrossRef]

7. Demenois, J.; Rey, F.; Ibanez, T.; Stokes, A.; Carriconde, F. Linkages between root traits, soil fungi and aggregate stability in tropical plant communities along a successional vegetation gradient. *Plant Soil* **2018**, *424*, 319–334. [CrossRef]

8. Behera, S.K.; Mishra, A.K.; Sahu, N.; Kumar, A.; Singh, N.; Kumar, A.; Bajpai, O.; Chaudhary, L.B.; Khare, P.B.; Tuli, R. The study of microclimate in response to different plant community association in tropical moist deciduous forest from northern India. *Biodivers. Conserv.* **2012**, *21*, 1159–1176. [CrossRef]

9. Delgado-Baquerizo, M.; Garcia-Palacios, P.; Milla, R.; Gallardo, A.; Maestre, F.T. Soil characteristics determine soil carbon and nitrogen availability during leaf litter decomposition regardless of litter quality. *Soil Biol. Biochem.* **2015**, *81*, 134–142. [CrossRef]

10. Wang, G.; Liu, G.; Xu, M. Above- and belowground dynamics of plant community succession following abandonment of farmland on the Loess Plateau, China. *Plant Soil* **2009**, *316*, 227–239. [CrossRef]

11. Yan, H.; Gu, X.; Shen, H. Microbial decomposition of forest litter: A review. *Chin. J. Ecol.* **2010**, *29*, 1827–1835.

12. Shin, N.-R.; Whon, T.W.; Bae, J.-W. Proteobacteria: Microbial signature of dysbiosis in gut microbiota. *Trends Biotechnol.* **2015**, *33*, 496–503. [CrossRef] [PubMed]

13. Barka, E.A.; Vatsa, P.; Sanchez, L.; Gaveau-Vaillant, N.; Jacquard, C.; Klenk, H.-P.; Clément, C.; Ouhdouch, Y.; van Wezel, G.P. Taxonomy, physiology, and natural products of Actinobacteria. *Microbiol. Mol. Biol. Rev.* **2016**, *80*, 1–43. [CrossRef]

14. Kielak, A.M.; Barreto, C.C.; Kowalchuk, G.A.; Van Veen, J.A.; Kuramae, E.E. The Ecology of Acidobacteria: Moving beyond Genes and Genomes. *Front. Microbiol.* **2016**, *7*, 744. [CrossRef] [PubMed]

15. Brockett, B.F.; Prescott, C.E.; Grayston, S.J. Soil moisture is the major factor influencing microbial community structure and enzyme activities across seven biogeoclimatic zones in western Canada. *Soil Biol. Biochem.* **2012**, *44*, 9–20. [CrossRef]

16. Finegan, B. Forest succession. *Nature* **1984**, *312*, 109–114. [CrossRef]

17. Walker, L.R.; Walker, J.; Hobbs, R.J. *Linking Restoration and Ecological Succession*; Springer: Berlin, Germamy, 2007.

18. Powers, J.S.; Marín-Spiotta, E. Ecosystem Processes and Biogeochemical Cycles in Secondary Tropical Forest Succession. *Annu. Rev. Ecol. Evol. Syst.* **2017**, *48*, 497–519. [CrossRef]

19. Allen, C.D.; Macalady, A.K.; Chenchouni, H.; Bachelet, D.; McDowell, N.; Vennetier, M.; Kitzberger, T.; Rigling, A.; Breshears, D.D.; Hogg, E.H.; et al. A global overview of drought and heat-induced tree mortality reveals emerging climate change risks for forests. *For. Ecol. Manag.* **2010**, *259*, 660–684. [CrossRef]

20. Camarero, J.J.; Gutiérrez, E. Pace and Pattern of Recent Treeline Dynamics: Response of Ecotones to Climatic Variability in the Spanish Pyrenees. *Clim. Chang.* **2004**, *63*, 181–200. [CrossRef]

21. Fernández-Alonso, M.J.; Yuste, J.C.; Kitzler, B.; Ortiz, C.; Rubio, A. Changes in litter chemistry associated with global change-driven forest succession resulted in time-decoupled responses of soil carbon and nitrogen cycles. *Soil Biol. Biochem.* **2018**, *120*, 200–211. [CrossRef]

22. Wang, Z.; Zhao, L.; Bai, Y.; Li, F.; Hou, J.; Li, X.; Jiang, Y.; Deng, Y.; Zheng, B.; Yang, W. Changes in plant debris and carbon stocks across a subalpine forest successional series. *For. Ecosyst.* **2021**, *8*, 40. [CrossRef]

23. Knapp, B.A.; Rief, A.; Seeber, J. Microbial communities on litter of managed and abandoned alpine pastureland. *Biol. Fertil. Soils* **2011**, *47*, 845–851. [CrossRef]

24. He, L.; Ivanov, V.Y.; Bohrer, G.; Thomsen, J.E.; Vogel, C.S.; Moghaddam, M. Temporal dynamics of soil moisture in a northern temperate mixed successional forest after a prescribed intermediate disturbance. *Agric. For. Meteorol.* **2013**, *180*, 22–33. [CrossRef]

25. Liu, J.; Jia, X.; Yan, W.; Zhong, Y.; Shangguan, Z. Changes in soil microbial community structure during long-term secondary succession. *Land Degrad. Dev.* **2020**, *31*, 1151–1166. [CrossRef]

26. Bell, C.W.; Acosta-Martinez, V.; McIntyre, N.E.; Cox, S.; Tissue, D.T.; Zak, J.C. Linking Microbial Community Structure and Function to Seasonal Differences in Soil Moisture and Temperature in a Chihuahuan Desert Grassland. *Microb. Ecol.* **2009**, *58*, 827–842. [CrossRef] [PubMed]

27. Landesman, W.J.; Nelson, D.M.; Fitzpatrick, M.C. Soil properties and tree species drive ß-diversity of soil bacterial communities. *Soil Biol. Biochem.* **2014**, *76*, 201–209. [CrossRef]

28. Zhang, J.; Elser, J.J. Carbon:Nitrogen:Phosphorus Stoichiometry in Fungi: A Meta-Analysis. *Front. Microbiol.* **2017**, *8*, 1281. [CrossRef]

29. Ouyang, S.; Xiang, W.; Gou, M.; Lei, P.; Chen, L.; Deng, X.; Zhao, Z. Variations in soil carbon, nitrogen, phosphorus and stoichiometry along forest succession in southern China. *Biogeosciences Discuss.* **2017**, 1–27. [CrossRef]

30. Shang, R.; Li, S.; Huang, X.; Liu, W.; Lang, X.; Su, J. Effects of Soil Properties and Plant Diversity on Soil Microbial Community Composition and Diversity during Secondary Succession. *Forests* **2021**, *12*, 805. [CrossRef]

31. Zhang, C.; Liu, G.; Xue, S.; Wang, G. Soil bacterial community dynamics reflect changes in plant community and soil properties during the secondary succession of abandoned farmland in the Loess Plateau. *Soil Biol. Biochem.* **2016**, *97*, 40–49. [CrossRef]

32. Chai, Y.; Cao, Y.; Yue, M.; Tian, T.; Yin, Q.; Dang, H.; Quan, J.; Zhang, R.; Wang, M. Soil Abiotic Properties and Plant Functional Traits Mediate Associations Between Soil Microbial and Plant Communities During a Secondary Forest Succession on the Loess Plateau. *Front. Microbiol.* **2019**, *10*, 895. [CrossRef] [PubMed]

33. Banning, N.C.; Gleeson, D.; Grigg, A.H.; Grant, C.D.; Andersen, G.; Brodie, E.L.; Murphy, D.V. Soil Microbial Community Successional Patterns during Forest Ecosystem Restoration. *Appl. Environ. Microbiol.* **2011**, *77*, 6158–6164. [CrossRef] [PubMed]

34. Yang, W.; Wu, F. *The Ecosystem Processes and Management of the Subalpine Coniferous Forest in the Upper Reaches of Yangtze River*; Science Press: Beijing, China, 2021.

35. Zhang, Y.; Duan, B.; Xian, J.; Korpelainen, H.; Li, C. Links between plant diversity, carbon stocks and environmental factors along a successional gradient in a subalpine coniferous forest in Southwest China. *For. Ecol. Manag.* **2011**, *262*, 361–369. [CrossRef]

36. Yang, B.; Pang, X.; Hu, B.; Bao, W.; Tian, G. Does thinning-induced gap size result in altered soil microbial community in pine plantation in eastern Tibetan Plateau? *Ecol. Evol.* **2017**, *7*, 2986–2993. [CrossRef]

37. Zhao, P.; Bao, J.; Wang, X.; Liu, Y.; Li, C.; Chai, B. Deterministic processes dominate soil microbial community assembly in subalpine coniferous forests on the Loess Plateau. *PeerJ* **2019**, *7*, e6746. [CrossRef]

38. Taylor, A.H.; Zisheng, Q.; Jie, L. Structure and dynamics of subalpine forests in the Wang Lang Natural Reserve, Sichuan, China. *Vegetatio* **1996**, *124*, 25–38. [CrossRef]

39. Chen, X.; Wang, X.; Li, J.; Kang, D. Species diversity of primary and secondary forests in Wanglang Nature Reserve. *Glob. Ecol. Conserv.* **2020**, *22*, e01022. [CrossRef]

40. Kang, D.; Lv, J.; Li, S.; Chen, X.; Wang, X.; Li, J. Community change characteristics of Abies faxoniana: A case study in the Wanglang Nature Reserve. *Ecol. Indic.* **2019**, *107*, 105594. [CrossRef]

41. Bao, S. *Agricultural and Chemistry Analysis of Soil*; China Agriculture Press: Beijing, China, 2000.

42. Li, S.; Huang, X.; Shen, J.; Xu, F.; Su, J. Effects of plant diversity and soil properties on soil fungal community structure with secondary succession in the Pinus yunnanensis forest. *Geoderma* **2020**, *379*, 114646. [CrossRef]

43. Carini, P.; Marsden, P.J.; Leff, J.W.; Morgan, E.E.; Strickland, M.S.; Fierer, N. Relic DNA is abundant in soil and obscures estimates of soil microbial diversity. *Nat. Microbiol.* **2016**, *2*, 16242. [CrossRef]

44. Yang, T.; Adams, J.M.; Shi, Y.; Sun, H.; Cheng, L.; Zhang, Y.; Chu, H. Fungal community assemblages in a high elevation desert environment: Absence of dispersal limitation and edaphic effects in surface soil. *Soil Biol. Biochem.* **2017**, *115*, 393–402. [CrossRef]

45. Zhang, J.; Kobert, K.; Flouri, T.; Stamatakis, A. PEAR: A fast and accurate Illumina Paired-End reAd mergeR. *Bioinformatics* **2014**, *30*, 614–620. [CrossRef] [PubMed]

46. Caporaso, J.G.; Kuczynski, J.; Stombaugh, J.; Bittinger, K.; Bushman, F.D.; Costello, E.K.; Fierer, N.; Peña, A.G.; Goodrich, J.K.; Gordon, J.I.; et al. QIIME allows analysis of high-throughput community sequencing data. *Nat. Methods* **2010**, *7*, 335–336. [CrossRef] [PubMed]

47. Bokulich, N.A.; Subramanian, S.; Faith, J.J.; Gevers, D.; Gordon, J.I.; Knight, R.; Mills, D.A.; Caporaso, J.G. Quality-filtering vastly improves diversity estimates from Illumina amplicon sequencing. *Nat. Methods* **2013**, *10*, 57–59. [CrossRef] [PubMed]

48. Edgar, R.C.; Haas, B.J.; Clemente, J.C.; Quince, C.; Knight, R. UCHIME improves sensitivity and speed of chimera detection. *Bioinformatics* **2011**, *27*, 2194–2200. [CrossRef]

49. Edgar, R.C. UPARSE: Highly accurate OTU sequences from microbial amplicon reads. *Nat. Methods* **2013**, *10*, 996–998. [CrossRef]

50. Wang, Q.; Garrity, G.M.; Tiedje, J.M.; Cole, J.R. Naive Bayesian classifier for rapid assignment of rRNA sequences into the new bacterial taxonomy. *Appl. Environ. Microbiol.* **2007**, *73*, 5261–5267. [CrossRef]

51. Kõljalg, U.; Nilsson, R.H.; Abarenkov, K.; Tedersoo, L.; Taylor, A.F.; Bahram, M.; Bates, S.T.; Bruns, T.D.; Bengtsson-Palme, J.; Callaghan, T.M.; et al. Towards a unified paradigm for sequence-based identification of fungi. *Mol. Ecol.* **2013**, *21*, 5271–5277. [CrossRef]

52. Altschul, S.F.; Gish, W.; Miller, W.; Myers, E.W.; Lipman, D.J. Basic local alignment search tool. *J. Mol. Biol.* **1990**, *215*, 403–410. [CrossRef]

53. Whittaker, R.H. Evolution and measurement of species diversity. *Taxon* **1972**, *21*, 213–251. [CrossRef]

54. Segata, N.; Izard, J.; Waldron, L.; Gevers, D.; Miropolsky, L.; Garrett, W.S.; Huttenhower, C. Metagenomic biomarker discovery and explanation. *Genome Biol.* **2011**, *12*, R60. [CrossRef] [PubMed]

55. DeSantis, T.Z.; Hugenholtz, P.; Larsen, N.; Rojas, M.; Brodie, E.L.; Keller, K.; Huber, T.; Dalevi, D.; Hu, P.; Andersen, G.L. Greengenes, a Chimera-Checked 16S rRNA Gene Database and Workbench Compatible with ARB. *Appl. Environ. Microbiol.* **2006**, *72*, 5069–5072. [CrossRef] [PubMed]

56. Oksanen, J.; Blanchet, F.G.; Kindt, R.; Legendre, P.; Minchin, P.R.; O'hara, R.; Simpson, G.L.; Solymos, P.; Stevens, M.H.H.; Wagner, H. Package 'vegan'. In *Community Ecology Package*; Version 2.4.5; R Core Team: Vienna, Austria, 2017.

57. R Development Core Team. *R: A Language and Environment for Statistical Computing. R Foundation for Statistical Computing*; R Core Team: Vienna, Austria, 2011. Available online: https://www.R-project.org/ (accessed on 26 December 2021).

58. Zhong, Z.; Zhang, X.; Wang, X.; Fu, S.; Wu, S.; Lu, X.; Ren, C.; Han, X.; Yang, G. Soil bacteria and fungi respond differently to plant diversity and plant family composition during the secondary succession of abandoned farmland on the Loess Plateau, China. *Plant Soil* **2020**, *448*, 183–200. [CrossRef]

59. Wang, G.; Liu, Y.; Cui, M.; Zhou, Z.; Zhang, Q.; Li, Y.; Ha, W.; Pang, D.; Luo, J.; Zhou, J. Effects of secondary succession on soil fungal and bacterial compositions and diversities in a karst area. *Plant Soil* **2021**, 1–12. [CrossRef]

60. Cui, Y.; Fang, L.; Guo, X.; Wang, X.; Wang, Y.; Li, P.; Zhang, Y.; Zhang, X. Responses of soil microbial communities to nutrient limitation in the desert-grassland ecological transition zone. *Sci. Total Environ.* **2018**, *642*, 45–55. [CrossRef]
61. Zhao, F.Z.; Bai, L.; Wang, J.Y.; Deng, J.; Ren, C.J.; Han, X.H.; Yang, G.H.; Wang, J. Change in soil bacterial community during secondary succession depend on plant and soil characteristics. *Catena* **2019**, *173*, 246–252. [CrossRef]
62. Liu, Y.; Zhu, G.; Hai, X.; Li, J.; Shangguan, Z.; Peng, C.; Deng, L. Long-term forest succession improves plant diversity and soil quality but not significantly increase soil microbial diversity: Evidence from the Loess Plateau. *Ecol. Eng.* **2020**, *142*, 105631. [CrossRef]
63. Kolton, M.; Harel, Y.M.; Pasternak, Z.; Graber, E.R.; Elad, Y.; Cytryn, E. Impact of Biochar Application to Soil on the Root-Associated Bacterial Community Structure of Fully Developed Greenhouse Pepper Plants. *Appl. Environ. Microbiol.* **2011**, *77*, 4924–4930. [CrossRef]
64. Kim, T.G.; Jeong, S.-Y.; Cho, K.-S. Functional rigidity of a methane biofilter during the temporal microbial succession. *Appl. Microbiol. Biotechnol.* **2014**, *98*, 3275–3286. [CrossRef]
65. Fierer, N.; Bradford, M.A.; Jackson, R.B. Toward an ecological classification of soil bacteria. *Ecology* **2007**, *88*, 1354–1364. [CrossRef]
66. Jangid, K.; Whitman, W.B.; Condron, L.M.; Turner, B.L.; Williams, M.A. Soil bacterial community succession during long-term ecosystem development. *Mol. Ecol.* **2013**, *22*, 3415–3424. [CrossRef] [PubMed]
67. Goldfarb, K.C.; Karaoz, U.; Hanson, C.A.; Santee, C.A.; Bradford, M.A.; Treseder, K.K.; Wallenstein, M.D.; Brodie, E.L. Differential Growth Responses of Soil Bacterial Taxa to Carbon Substrates of Varying Chemical Recalcitrance. *Front. Microbiol.* **2011**, *2*, 94. [CrossRef] [PubMed]
68. Shen, C.; Xiong, J.; Zhang, H.; Feng, Y.; Lin, X.; Li, X.; Liang, W.; Chu, H. Soil pH drives the spatial distribution of bacterial communities along elevation on Changbai Mountain. *Soil Biol. Biochem.* **2013**, *57*, 204–211. [CrossRef]
69. Delgado-Baquerizo, M.; Oliverio, A.M.; Brewer, T.E.; Benavent-González, A.; Eldridge, D.J.; Bardgett, R.D.; Maestre, F.T.; Singh, B.K.; Fierer, N. A global atlas of the dominant bacteria found in soil. *Science* **2018**, *359*, 320–325. [CrossRef]
70. Fuerst, J.A.; Sagulenko, E. Beyond the bacterium: Planctomycetes challenge our concepts of microbial structure and function. *Nat. Rev. Microbiol.* **2011**, *9*, 403–413. [CrossRef]
71. Brunner, I.; Plötze, M.; Rieder, S.; Zumsteg, A.; Furrer, G.; Frey, B. Pioneering fungi from the Damma glacier forefield in the Swiss Alps can promote granite weathering. *Geobiology* **2011**, *9*, 266–279. [CrossRef]
72. Xu, H.; Du, H.; Zeng, F.; Song, T.; Peng, W. Diminished rhizosphere and bulk soil microbial abundance and diversity across succession stages in Karst area, southwest China. *Appl. Soil Ecol.* **2021**, *158*, 103799. [CrossRef]
73. López-Lozano, N.E.; Heidelberg, K.B.; Nelson, W.C.; García-Oliva, F.; Eguiarte, L.E.; Souza, V. Microbial secondary succession in soil microcosms of a desert oasis in the Cuatro Cienegas Basin, Mexico. *PeerJ* **2013**, *1*, e47. [CrossRef]
74. Montagna, M.; Berruti, A.; Bianciotto, V.; Cremonesi, P.; Giannico, R.; Gusmeroli, F.; Lumini, E.; Pierce, S.; Pizzi, F.; Turri, F.; et al. Differential biodiversity responses between kingdoms (plants, fungi, bacteria and metazoa) along an Alpine succession gradient. *Mol. Ecol.* **2018**, *27*, 3671–3685. [CrossRef]
75. Qiang, W.; He, L.; Zhang, Y.; Liu, B.; Liu, Y.; Liu, Q.; Pang, X. Aboveground vegetation and soil physicochemical properties jointly drive the shift of soil microbial community during subalpine secondary succession in southwest China. *Catena* **2021**, *202*, 105251. [CrossRef]
76. Delgado-Baquerizo, M.; Bardgett, R.D.; Vitousek, P.M.; Maestre, F.T.; Williams, M.A.; Eldridge, D.J.; Lambers, H.; Neuhauser, S.; Gallardo, A.; García-Velázquez, L.; et al. Changes in belowground biodiversity during ecosystem development. *Proc. Natl. Acad. Sci. USA* **2019**, *116*, 6891–6896. [CrossRef] [PubMed]
77. Hendershot, J.N.; Read, Q.D.; Henning, J.A.; Sanders, N.J.; Classen, A.T. Consistently inconsistent drivers of microbial diversity and abundance at macroecological scales. *Ecology* **2017**, *98*, 1757–1763. [CrossRef] [PubMed]
78. Chen, W.; Wang, J.; Meng, Z.; Xu, R.; Chen, J.; Zhang, Y.; Hu, T. Fertility-related interplay between fungal guilds underlies plant richness–productivity relationships in natural grasslands. *New Phytol.* **2020**, *226*, 1129–1143. [CrossRef] [PubMed]
79. Nakayama, M.; Imamura, S.; Taniguchi, T.; Tateno, R. Does conversion from natural forest to plantation affect fungal and bacterial biodiversity, community structure, and co-occurrence networks in the organic horizon and mineral soil? *For. Ecol. Manag.* **2019**, *446*, 238–250. [CrossRef]
80. Cui, J.; Wang, J.; Xu, J.; Xu, C.; Xu, X. Changes in soil bacterial communities in an evergreen broad-leaved forest in east China following 4 years of nitrogen addition. *J. Soils Sediments* **2017**, *17*, 2156–2164. [CrossRef]
81. Colman, D.R.; Poudel, S.; Hamilton, T.L.; Havig, J.R.; Selensky, M.J.; Shock, E.L.; Boyd, E.S. Geobiological feedbacks and the evolution of thermoacidophiles. *ISME J.* **2018**, *12*, 225–236. [CrossRef] [PubMed]
82. Power, J.F.; Carere, C.R.; Lee, C.K.; Wakerley, G.L.; Evans, D.W.; Button, M.; White, D.; Climo, M.D.; Hinze, A.M.; Morgan, X.C.; et al. Microbial biogeography of 925 geothermal springs in New Zealand. *Nat. Commun.* **2018**, *9*, 2876. [CrossRef]
83. Zhang, W.; Parker, K.M.; Luo, Y.T.; Wan, S.; Wallace, L.L.; Hu, S. Soil microbial responses to experimental warming and clipping in a tallgrass prairie. *Glob. Chang. Biol.* **2005**, *11*, 266–277. [CrossRef]
84. Frey, S.; Drijber, R.; Smith, H.; Melillo, J. Microbial biomass, functional capacity, and community structure after 12 years of soil warming. *Soil Biol. Biochem.* **2008**, *40*, 2904–2907. [CrossRef]
85. Brown, J.H.; Gillooly, J.F.; Allen, A.P.; Savage, V.M.; West, G.B. Toward a metabolic theory of ecology. *Ecology* **2004**, *85*, 1771–1789. [CrossRef]

86. Gillooly, J.F.; Brown, J.H.; West, G.B.; Savage, V.M.; Charnov, E.L. Effects of Size and Temperature on Metabolic Rate. *Science* **2001**, *293*, 2248–2251. [CrossRef] [PubMed]
87. Suseela, V.; Conant, R.T.; Wallenstein, M.D.; Dukes, J.S. Effects of soil moisture on the temperature sensitivity of heterotrophic respiration vary seasonally in an old-field climate change experiment. *Glob. Chang. Biol.* **2012**, *18*, 336–348. [CrossRef]
88. BÁRCENAS-MORENO, G.; GÓMEZ-BRANDÓN, M.; Rousk, J.; Bååth, E. Adaptation of soil microbial communities to temperature: Comparison of fungi and bacteria in a laboratory experiment. *Glob. Chang. Biol.* **2009**, *15*, 2950–2957. [CrossRef]
89. Kaspar, T.C.; Bland, W.L. Soil temperature and root growth. *Soil Sci.* **1992**, *154*, 290–299. [CrossRef]
90. Li, Q.; Song, A.; Yang, H.; Müller, W.E.G. Impact of Rocky Desertification Control on Soil Bacterial Community in Karst Graben Basin, Southwestern China. *Front. Microbiol.* **2021**, *12*, 448. [CrossRef]
91. Schlatter, D.C.; Bakker, M.G.; Bradeen, J.M.; Kinkel, L.L. Plant community richness and microbial interactions structure bacterial communities in soil. *Ecology* **2015**, *96*, 134–142. [CrossRef]
92. Drenovsky, R.E.; Vo, D.; Graham, K.J.; Scow, K.M. Soil Water Content and Organic Carbon Availability Are Major Determinants of Soil Microbial Community Composition. *Microb. Ecol.* **2004**, *48*, 424–430. [CrossRef]
93. Sun, R.; Dsouza, M.; Gilbert, J.A.; Guo, X.; Wang, D.; Guo, Z.; Ni, Y.; Chu, H. Fungal community composition in soils subjected to long-term chemical fertilization is most influenced by the type of organic matter. *Environ. Microbiol.* **2016**, *18*, 5137–5150. [CrossRef]
94. Sariyildiz, T.; Anderson, J. Interactions between litter quality, decomposition and soil fertility: A laboratory study. *Soil Biol. Biochem.* **2003**, *35*, 391–399. [CrossRef]
95. Grayston, S.; Campbell, C.; Bardgett, R.; Mawdsley, J.; Clegg, C.; Ritz, K.; Griffiths, B.; Rodwell, J.; Edwards, S.; Davies, W.; et al. Assessing shifts in microbial community structure across a range of grasslands of differing management intensity using CLPP, PLFA and community DNA techniques. *Appl. Soil Ecol.* **2004**, *25*, 63–84. [CrossRef]
96. Franklin, R.B.; Mills, A.L. Importance of spatially structured environmental heterogeneity in controlling microbial community composition at small spatial scales in an agricultural field. *Soil Biol. Biochem.* **2009**, *41*, 1833–1840. [CrossRef]

Article

Variability in Soil Macronutrient Stocks across a Chronosequence of Masson Pine Plantations

Jie He [1,2], Quanhou Dai [1,2,*], Fengwei Xu [1,2], Youjin Yan [1,2] and Xudong Peng [1,2]

[1]   College of Forestry, Guizhou University, Guiyang 550025, China; hjieyh@163.com (J.H.);
     xfw21@sina.com (F.X.); gs.yanyj16@gzu.edu.cn (Y.Y.); bjpxd@126.com (X.P.)
[2]   Institute of Soil Erosion and Ecological Restoration, Guizhou University, South Jiaxiu Road,
     Guiyang 550025, China
*   Correspondence: qhdairiver@163.com

Abstract: Plantations play a vital role in the global nutrient cycle because they have large stocks of soil macronutrients. However, the impacts of plantations on soil macronutrient stocks combined with stand age and soil physicochemical properties have not been well quantified. We compared soil macronutrient stocks at soil depths of 0−20 and 20−40 cm across a 7-, 14-, 25-, and 30-year chronosequence of Masson pine (*Pinus massoniana* Lamb.) plantations. The results showed that the nitrogen (N), phosphorus (P), and potassium (K) stocks first increased and then decreased with stand age. The highest N and P stocks were observed in the 14-year-old plantation, and the 25-year-old plantation displayed the highest K stock. The C, N, and P stocks declined with increasing soil depth across all sites, whereas the reverse trend was found in the K stock. Carbon stocks were highest for all plantations, followed by the K, N, and P stocks. Plantation soils exhibited a higher C:P ratio and a lower P:K ratio at various soil depths. The dominant controlling factors for the soil macronutrient stocks varied significantly at different stand ages and soil depths according to statistical analysis. For the total soil system, the C stock was affected by the available nutrients, organic matter, and stoichiometry; the available nutrients and organic matter were the determinant factors of the N and P stocks. Aggregate stability could be the primary parameter affecting the K stock. Organic matter explained most of the variation in soil macronutrient stocks, followed by the P:K ratio and available K. Collectively, our results suggest that the response of soil macronutrient stocks to stand age and soil depth will be dependent on different soil physicochemical properties, and P and K may be important limiting factors in Masson pine plantation ecosystems.

Keywords: soil macronutrient stocks; Masson pine plantations; controlling factors; chronosequence

Citation: He, J.; Dai, Q.; Xu, F.; Yan, Y.; Peng, X. Variability in Soil Macronutrient Stocks across a Chronosequence of Masson Pine Plantations. *Forests* **2022**, *13*, 17. https://doi.org/10.3390/f13010017

Academic Editors: Fuzhong Wu, Zhenfeng Xu and Wanqin Yang

Received: 16 November 2021
Accepted: 20 December 2021
Published: 23 December 2021

Publisher's Note: MDPI stays neutral with regard to jurisdictional claims in published maps and institutional affiliations.

1. Introduction

Soil organic carbon (SOC) is the main part of terrestrial carbon reservoirs and an important part of soil fertility. As of 2004, the size of soil carbon pool was 3.3 times that of atmospheric pool and 4.5 times that of biotic pool [1]. Carbon sequestration implies transferring atmospheric carbon dioxide into long-lived pools and achieving the purpose of mitigating CO_2 emissions. The role of organic carbon is self-evident, but macronutrients such as nitrogen (N), phosphorus (P), and potassium (K) are equally as important as carbon (C) [2]. Metabolic processes in soils are often nutrient limited by N and P [3,4]. Sequestering soil carbon relies on the availability of stabilizing elements such as N and P, which are essential for stabilizing organic C pool [5]. Changing the N and P availabilities in soils may alter the magnitude of the imbalance between SOC decomposition and formation processes [6]. It is estimated that 80 million tons (Mt) of N, 20 Mt of P, and 15 Mt of K are required to sequester 1 Gt of C in world soils [1].

Soil organic carbon, nitrogen, phosphorus, potassium, and their stoichiometry, as critical regulatory indicators, have been shown to modulate forest ecosystem functions [7,8].

Nitrogen cycle processes are closely coupled with other macronutrients, and the content and availability of N directly influence plant growth and the net primary productivity of terrestrial ecosystems [9,10]. Phosphorus can only be released slowly by weathering of parent material, internal nutrient recycling, and reallocation within the soil profile in a natural environment, and it may have become the limiting nutrient in almost all terrestrial ecosystems [11]. Potassium is another essential nutrient required for plants [8]. The application of K fertilizer could increase water use efficiency, promote carbon sequestration, and reduce runoff and soil loss under plant cover [12]. Furthermore, the stoichiometry of macronutrients is a useful indicator of the intensity of nutrient flux from deadwood to the soil [13].

An age series of Masson plantations may alter community structure and soil properties following reforestation, which in turn influences soil macronutrient stocks and stoichiometric traits [14]. There is a complex response of soil macronutrient stocks and stoichiometry to vegetation restoration. The soil C stock and C:N:P stoichiometry increased, but the N stocks first increased and then decreased with forest restoration age, while the P stocks showed little variation in a study by Xu et al. [15]. However, the P stock may become a limiting factor for C sequestration as forests age [16]. The N, P, and K stocks in the 20 cm soil layer did not differ significantly between years [17]. Similarly, many studies have indicated that soil nutrient variability is likely to be small with forest stand age [18,19]. Some researchers have reported that the soil P availability declined continuously with the development of larch plantations [20]. Therefore, there still exists some controversy regarding the effects of stand age on soil macronutrient stocks.

Plantations can provide both economic and environmental benefits, and ecosystems play important roles in maintaining regional ecological protection functions, including curbing erosion, accelerating the geochemical cycling of elements, and ultimately increasing productivity. Due to the previous extensive destruction of forests and the recognition of important ecological service functions, many provinces have established a pattern of rapid afforestation of progressively larger regions over the past 10–15 years [21]. As a native tree species, Masson pine (*Pinus massoniana* Lamb.) has a wide distribution range and large forest area in central and southern China and is one of the main plantation varieties; it is always referred to as a pioneer tree species due to its rapid growth and strong adaptability [22,23]. Accurate estimation of C, N, P, and K stocks in Masson plantation ecosystems is crucial because they are the most important macronutrient pools in Earth systems, and there still exists some controversy regarding the effects of stand age on soil macronutrient stocks. In addition, plantations of different stand ages significantly influence soil physicochemical properties by changing the interior environment, litter inputs, root exudates, and so on [24]. An improved understanding of how soil physicochemical properties will affect C, N, P, and K stocks at various stand ages and soil depths may enable more accurate prediction of soil macronutrient stocks in plantation ecosystems.

Many of the C stock estimates of terrestrial ecosystems have been reported by researchers [25–28], and all of which are essential for understanding C-cycling patterns and their influences on terrestrial ecosystems. However, more studies on the response of N, P, and K stocks to plantation stand age are necessary, particularly those that attach less importance to K stock research. Furthermore, the response of soil organic carbon to soil properties has been observed by many authors [29–31], but few studies have shown the substantial drivers of soil properties on N, P, and K [32], and knowledge of the potential mechanisms remains limited. To estimate how soil macronutrient stocks change, four sample stands of Masson pine were compared across a 7-, 14-, 25-, and 30-year chronosequence. In addition, we investigated the effects of soil physicochemical properties on the storage of macronutrients. Therefore, the objectives of this study were to (1) estimate the soil macronutrient stocks and their changes with stand age and (2) determine the influence of soil physicochemical properties on the C, N, P, and K stocks at various stand ages and soil depths.

2. Materials and Methods

2.1. Experimental Design

This study was carried out in 4 sample stands of Masson pine (*P. massoniana* Lamb.), located in Dushan County in Guizhou Province within state-owned forest farms, covering an area of more than 18,860 km^2 and extending at least 50 km from north to south (107°27′–107°30′ E, 25°41′–25°41′ N) (Figure 1). The mean annual precipitation in this area is approximately 1346 mm, the mean annual temperature is 15 °C, and the altitude is 830–1479 m. The soil in the region is classified as yellow soil. Masson pine mostly forms uniform stands with a small admixture of other tree species, usually fir. The understorey vegetation and some basic information of the sites are shown in Table 1.

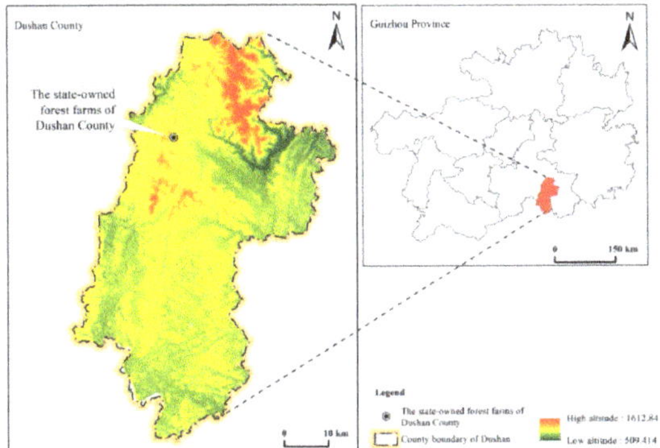

Figure 1. Location of the state-owned forest farms in Dushan County, Guizhou Province.

Table 1. Basic characteristics of the four Masson pine plantations.

Stand Age (a)	Slope Direction	Slope (°)	Elevation (m)	Soil Type	Bulk Density (g/cm^{-3})	Carbon (g/kg)	Nitrogen (g/kg)	Phosphorus (g/kg)	Canopy Density	Average Diameter at Breast Height (cm)	Average Tree Height (m)	Main Species
7	NW 35°	28	1072~1076	yellow soil	1.28	20.46	0.93	0.17	0.5	3.25	2.78	*Dicranopteris linearis, Imperata cylindrica* (L.) Beauv., *Miscanthus floridulus* (Lab.) Warb. ex Schum. et Laut., *Nandina domestica*
14	NE 13°	35	1063~1066	yellow soil	1.12	23.19	1.54	0.24	0.7	11.03	10.67	
25	NE 10°	25	1075~1080	yellow soil	1.12	20.23	1.25	0.19	0.8	14.07	12.6	
30	NE 45°	30	1067~1068	yellow soil	1.08	22.80	1.00	0.16	0.9	25.10	15.43	

We established three 30 × 30 m standard quadrats in each plantation, covering four stand age classes (7-, 14-, 25-, and 30-year-old secondary forest stands) in the forest farms. Three points were randomly selected in each Masson pine plantation quadrat, and soil samples were collected from 0−20 and 20−40 cm soil layers after removal of plant residues, gravel, or other debris. We collected a total of five random samples from each plot as per the S-shaped sampling method and then placed them into an aluminum specimen box to ensure that the main structure was maintained during transport to the laboratory. Three ring-knife samples from each soil layer were collected at the same time.

2.2. Soil Analyses and Calculations

Soil samples were divided into two groups related to research indicators. One part of the soil samples was broken into blocks with a diameter approximately 10 mm according to the natural structure, and litter stones and roots were removed. When air-dried, these samples were used to determine soil aggregation characteristics. The other soil samples were air-dried and sieved to 2 mm for chemical analyses. The wet-sieving method was applied to determine the composition of water-stable aggregates in different Masson pine plantations. Aggregated soils successively passed through a column of sieves with 5, 3, 2, 1, 0.5, and 0.25 mm diameters to quantify the losses of sediment of different sizes.

Soil bulk density (SBD), saturated hydraulic conductivity (Ks), and soil porosity were measured using the ring-knife method [33]. The method of Walkley and Black [34] was employed to measure the soil organic carbon and organic matter contents. Soil pH in water 1:2.5 (soil:water) was measured after shaking for 5 min [35]. The ammonium acetate saturation (AMAS) method was used to study the cation-exchange capacity (CEC) [36]. Soil total nitrogen (N) and the available N (AN) were measured by Kjeldahl digestion and alkaline hydrolysis diffusion method, respectively; phosphorus (P) and the available P (AP) were measured using molybdenum blue colorimetric analysis; potassium (K) and the available K (AK) were determined using a flame photometer [14].

The soil aggregate stability was characterized by mean weight diameter (MWD), fractal dimension (FD), geometric mean diameter (GMD), and proportion of >0.25 mm water-stable aggregates (WSA > 0.25 mm). Stability parameters of aggregates were calculated using the following formulae [37]:

$$\text{MWD} = \sum_{i=1}^{n} x_i \times w_i \tag{1}$$

$$\text{GMD} = \exp\left[\frac{\sum_{i=1}^{n}(w_i \times \ln x_i)}{\sum_{i=1}^{n} w_i}\right] \tag{2}$$

$$R_{0.25} = \frac{M_{t>0.25}}{M_t} \times 100\% \tag{3}$$

$$\frac{M_{(r<x_i)}}{M_t} = \left(\frac{x_i}{d_{max}}\right)^{3-D} \tag{4}$$

where x_i is the mean diameter (mm) of the soil aggregate size fractions, w_i is the proportion of all soil in the ith size fraction (%), M_t is the total mass of aggregates (g), $M_{t>0.25}$ is the mass of aggregates larger than 0.25 mm (g), $M_{(r<xi)}$ is the mass of aggregates smaller than x_i (g), and d_{max} is the maximum diameter of the soil aggregate size fractions (mm).

The calculation formula of soil macronutrient density (Mg C·ha^{-1}) in a soil layer is as follows [38]:

$$\text{SMD}_i = \sum_{i=1}^{n} C_i \times D_i \times T_i \times (1 - G_i) \times 10^{-1} \tag{5}$$

$$\text{SMS}_i = \text{SMD}_i \times S \tag{6}$$

where SMD_i is the soil macronutrient density in the i layer of soil (Mg C·ha^{-1}), C_i is the soil macronutrient concentration in the i layer of soil (g·kg^{-1}), T_i, D_i, and G_i are the soil thickness (cm), soil bulk density (g·cm^{-3}), and volume percentage of gravel that is larger than 2 mm in soil, respectively. 10^{-1} is the unit conversion factor. SMS_i is the soil macronutrient stock in the i layer of soil (Mg C) and S is the soil acreage of the calculation grid.

2.3. Statistical Analysis

Multivariate statistical analyses were performed using CANOCO 5.0 (Microcomputer Power, Ithaca, NY, USA) and Statistical Product and Service Solutions 22.0. The Shapiro–Wilk test was used to verify the statistical distribution, and nonnormally distributed values were log10 transformed. One-way analysis of variance (ANOVA) was used to analyze the statistically significant differences and the variance between treatments. Stepwise multiple linear regression and redundancy analysis (RDA) were conducted to determine the strength of possible relationships between soil macronutrient stocks and soil physicochemical properties at different stand ages and soil depths, respectively.

3. Results

3.1. Soil Macronutrient Stocks and Allocation

There is no significant difference in C stocks of the 0−20 cm soil layer, but the N, P, and K stocks in the 0−20 and 20−40 cm soil layers increased first and then decreased with forest stand age (Figure 2). The 14- and 25-year-old plantations displayed the highest P stock and K stock, respectively, whereas the lowest values of N and P stocks were all observed in the 30-year-old plantation. The K stocks were higher in the 20−40 cm soil layer than in the 0−20 cm soil layer except in the 14-year-old plantation; the C, N, and P stocks declined with increasing soil depth across all sites. The sequence of soil macronutrient stocks was as follows: C stock > K stock > N stock > P stock, except in the 20−40 cm layer of the 25-year-old plantation.

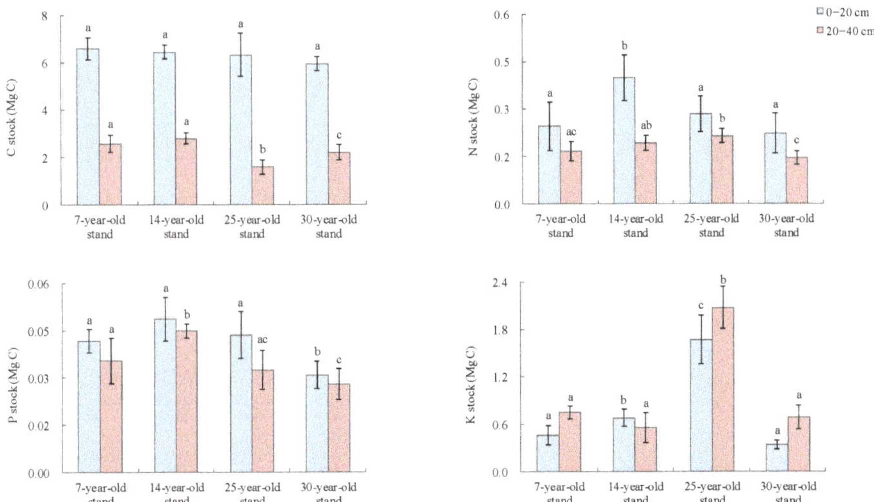

Figure 2. Soil macronutrient stocks across a chronosequence of Masson pine plantations. Different lowercase letters indicate significant differences in plantations with different stand ages ($p < 0.05$).

3.2. Variations of Soil Macronutrient Stoichiometry

Under all plantation types, C:N, C:P, N:P, C:K, N:K, and P:K ratios were higher in the 0−20 cm soil layers than in the 20−40 cm, except for P:K in the 14-year-old plantation and N:P in the 25-year-old plantation (Figure 3). The ratios of C:N and C:P first decreased and then increased, with a maximum level in the 30-year-old plantation. The ratios of C:K in the 0−20 and 20−40 cm soil layers were highest in the 30- and 14-year-old plantations, respectively. The variation in the N:K ratio was similar to that in the C:K ratio, but the N:K ratio was much smaller, ranging from 0.10 to 0.67 for all depths. The lowest C:K, N:K, and P:K ratios were all observed in the 25-year-old plantation. The stoichiometric traits of the soil were ranked as C:P > C:N > C:K > N:P > N:K > P:K in the 0−20 cm soil layer of the 7-, 14-, and 30-year-old plantations, whereas in the 20−40 cm soil layer, the sequence was C:P > C:N > N:P > C:K > N:K > P:K except for the 14-year-old plantation (Figure 3).

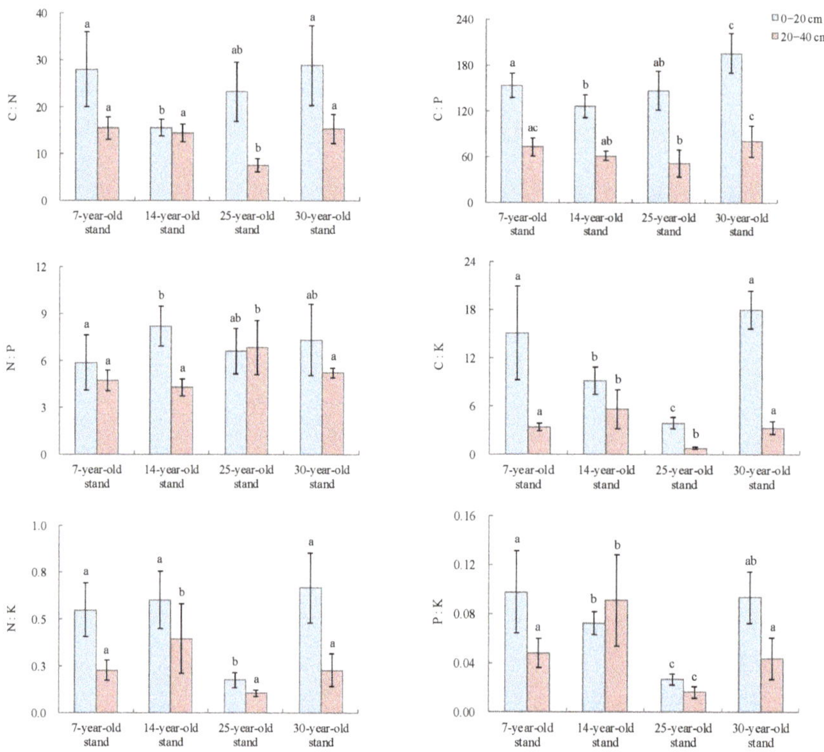

Figure 3. The macronutrient stoichiometry of soil at different depths in Masson pine plantations. Different lowercase letters indicate significant differences in plantations with different stand ages (*p* < 0.05).

3.3. Relationships between Macronutrient Stocks and Soil Physicochemical Properties

3.3.1. Controls of Macronutrient Stocks at Various Plantations

Stepwise multiple regression was performed to determine soil physicochemical properties affecting the variability of soil macronutrient stocks within each Masson pine plantation (Table 2). For the 7-year-old plantation, 97.5% of the variation in the C stock could be explained by the AP, MWD, and organic matter together. The proportion of >0.25 mm was the most dominant factor for the C stock in the 14-, 25-, and 30-year-old plantations. Soil bulk density showed a positive correlation with N and P stocks in the 14-year-old plantation. The available N contributed 42.8% of the N stock variation in the 7-year-old plantation, and the available K explained most of the P stock variation in the 7- and 30-year-old plantations. The available P and organic matter were the most dominant controls for the N stock in the 25- and 30-year-old plantations, and the available P showed a negative relationship with the N stock. Most of the variation in the K stock in different plantations could be explained by soil aggregate stability.

Table 2. Models of soil macronutrient stocks at different stand ages.

Plantations	Equation	Adjusted R^2	p Value
7-year-old plantation	$C_{stock} = -0.861 + 3.281AP + 2.282MWD - 0.051OM$	0.975	0.000
	$N_{stock} = 0.091 + 0.001AN$	0.428	0.012
	$P_{stock} = 0.02 + 0.00044AK$	0.338	0.028
	$K_{stock} = -0.288 - 0.049WSA > 0.25\ mm + 0.783pH + 0.552AP - 0.207D - 0.005OM$	0.980	0.000
14-year-old plantation	$C_{stock} = -2.110 + 0.161WSA > 0.25\ mm$	0.975	0.000
	$N_{stock} = -1.924 + 0.012WSA > 0.25\ mm + 0.319SBD + 0.288pH$	0.930	0.000
	$P_{stock} = -0.007 + 0.015Ks + 0.035SBD + 0.000441CEC$	0.804	0.001
	$K_{stock} = 0.401 + 0.491GMD$	0.268	0.049
25-year-old plantation	$C_{stock} = 12.499 - 0.119WSA > 0.25\ mm$	0.934	0.000
	$N_{stock} = 0.191 + 0.004OM - 0.083AP$	0.792	0.000
	$P_{stock} = 0.029 + 0.000258OM$	0.528	0.004
	$K_{stock} = 4.891 - 1.183D$	0.424	0.013
30-year-old plantation	$C_{stock} = -6.606 + 0.053WSA > 0.25\ mm + 0.214CEC + 1.128pH$	0.990	0.000
	$N_{stock} = 0.122 + 0.004OM - 0.078AP$	0.830	0.000
	$P_{stock} = 0.021 + 0.000296AK$	0.460	0.009
	$K_{stock} = 0.767 - 0.150MWD$	0.740	0.000

Independent variables considered include the available N, P, and K; soil aggregate stability (MWD, GMD, WSA > 0.25 mm and D); and organic matter (OM), pH, CEC, SBD, and Ks.

3.3.2. Controls of Macronutrient Stocks at Various Soil Depths

Redundancy analysis was used to explore the correlations between macronutrient stocks and soil physicochemical properties at various soil depths (Figure 4). The first ordination axis explained 87.2% of the variation, while the second axis accounted for only 5.8% in the 0−20 cm soil layer. The available K and P:K ratio explained most of the variation in soil macronutrient stocks. Negative correlations were observed between the C, N, P, and K stocks and Ks and MWD. The C:N ratio, C:P ratio, and available K were the most important factors associated with the N, P, and K stocks, respectively. Axes 1 and 2 captured 88.8% and 7.7% of the total variance in the 20−40 cm soil layer, respectively, and the variance could be explained mainly by the geometric mean diameter and P:K ratio. The stability of aggregates was strongly associated with C, N, and K stocks, while the P stock showed a positive correlation with the C:K, N:K, and P:K ratios.

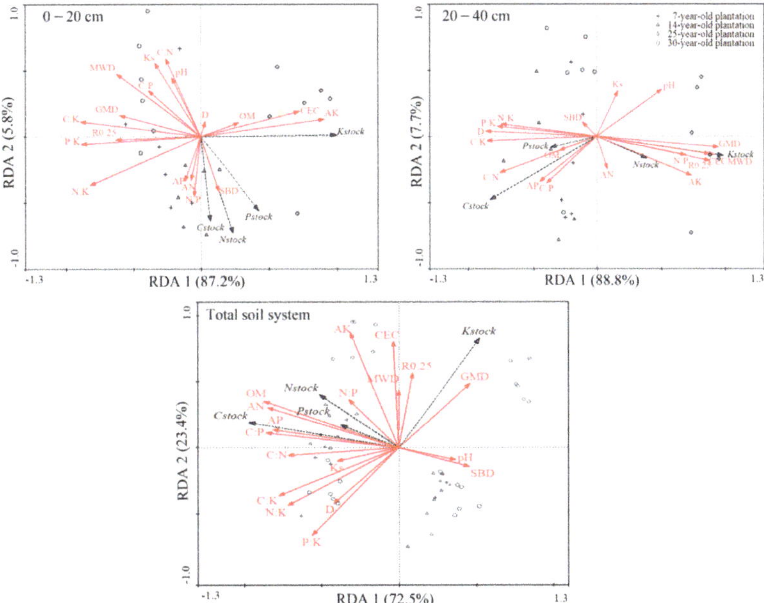

Figure 4. Ordination plots of redundancy analysis (RDA) of macronutrient stocks and soil physicochemical properties at various soil depths.

For the total soil system, axis 1 explained 72.5% of the variability in macronutrient stocks, primarily related to organic matter, the C:P ratio, the available N, and the available P. Axis 2 described 23.4% of the variation, mainly associated with the available K, CEC, and the P:K ratio. The stocks of C, N, and P were positively correlated with the available N, P, K, and organic matter and negatively related to pH. Moreover, there was a strong positive relationship between the C stock and the C, N, P, and K stoichiometry. The K stock was positively correlated with CEC, MWD, GMD, and WSA > 0.25 mm but negatively correlated with the C:K, N:K, and P:K ratios and D (Figure 4).

4. Discussion

4.1. Changes in Soil Macronutrient Stocks

Significant spatial and temporal variability in soil macronutrient stocks was observed in this study. Deep root distribution, rhizosphere effects, and higher organic inputs may increase SOC [39]. Although the SOC concentration increased gradually with stand age, there was no statistically significant difference in C stocks of the 0−20 cm soil layer between different stands. As revealed by Wellock et al. [40], there was a significant decline in the C density of the surface soil after 27 years of afforestation. In our study, the stocks of N, P, and K increased first and then decreased with stand age in both the 0−20 and 20−40 cm soil layers. These results are different from other studies, which have shown that not only the N stock but also the C stock in mineral soil increased with increasing stand age, while the soil P stock exhibited a trend of increasing first and then decreasing [41], or that soil P stocks tended to increase but N stocks slightly decreased with stand age [42].

The increasing trend of macronutrient stocks in the 14-year-old plantation may be illustrated by the increase in litter quantity and decomposition, which returns nutrients to the soil [43]. This was the main reason why the C, N, and P stocks declined with increasing soil depth across all sites. The decreasing trend in macronutrient stocks could be because of the nutrient requirements of plantations during the vigorous growing stage [44]. Carbon, nitrogen, and phosphorus are essential elements for soil organisms. With increasing stand age, the microenvironment becomes more hospitable for soil organisms. A large number of soil organisms and strong biological activity can increase macronutrient stocks by accelerating the decomposition of litter and can decrease macronutrient stocks by immobilizing elements for growth or transformation [45,46]. The pH plays an important role in nitrogen biogeochemical cycling through autotrophic nitrification [47], and the mechanisms of P sorption are also affected in different ways by pH [48]. Lower pH is beneficial for phosphate in binding to Fe and Al precipitates [43], and organic acids could form strong bonds with metal ions through metal chelation, which might affect the P stock [49]. Therefore, bacterial biomass and mineral concentration should be considered together with pH in the study of the circulation of soil phosphorus [50]. This is the reason for the highest P stock in the 14-year-old plantation.

The C stocks were highest in each soil layer for all plantations, followed by the K, N, and P stocks; therefore, K plays an important role in terrestrial ecosystem. Soil acidification also promotes the release of mineral K, and potassium deficiency occurs when the available K is absorbed by plants; moreover, cation exchange capacity is another important controlling factor for soil K availability [51]. Increased pH and cation exchange capacity facilitated the accumulation of K in the 25-year-old plantation. The lower K stock was observed in the 30-year-old plantation, probably because a large amount of organic matter and root exudates increased the release of mineral K in the soil [52]. The other reason was that more K-solubilizing bacteria facilitate the conversion of potassium from mineral K into available K [8]. Additionally, the K stock was higher in the 20−40 cm soil layer than in the 0−20 cm soil layer, which could be because of the leaching process. The soil structure improved as the vegetation grew, which was conducive to leaching K from the soil surface into the deeper soil layers.

4.2. Factors Controlling Soil Macronutrient Stocks

The spatial patterns of soil macronutrients were significantly different at various soil depths across a chronosequence of Masson pine plantations. Soil macronutrients and their stoichiometric ratios are important parameters of soil quality. Climate, topography, vegetation, and soil properties interact to influence the characteristics of soil nutrients, especially climate and soil properties [9,10]. The results of the present study showed that soil macronutrient stocks were mainly affected by soil physicochemical properties under similar climatic conditions, but the dominant controlling factors of soil macronutrient stocks varied at various soil depths and stand ages.

Effects of soil physicochemical properties on the variability in soil macronutrient stocks within each Masson pine plantation were revealed by stepwise multiple regression. Soil aggregates promote the stabilization of organic matter by regulating the availability of oxygen and water; hence, the formation and stability of aggregates are crucial for the SOC stock [53,54]. In our research, the proportion of >0.25 mm was the key factor for the C stock in the 14-, 25-, and 30-year-old plantations. The stocks of C and K in the 7-year-old plantation were controlled by the available P. Similar results were reported by Zhong et al. [55], who reported that lower availability of P could result in a decrease in leaf photosynthetic capacity and nutrient concentrations, which adversely affects the organic carbon input to soils from litter. Ks and SBD showed a positive correlation with N and P stocks in the 14-year-old plantation. The SBD is an important parameter for the calculation of N and P stocks, and a higher Ks indicates better hydrothermal conditions, which is beneficial for promoting microorganisms to biologically sequester N and P [56].

The available N, P, and K were the key factors controlling macronutrient stocks in the 7-year-old plantation. In addition, the available P and organic matter were the most dominant controls for the N stock in the 25- and 30-year-old plantations. The implication of these results is that the available nutrients were not only the principal factor affecting the soil macronutrient stocks in the 7-year-old plantation but also the key factor driving the distribution of N stock within the 25- and 30-year-old plantations. Soil acidification is prone to the leaching of elements and thus these soils tend to be nutritionally poor [57]. This might be a good explanation for the positive correlations between C stock and pH in the 30-year-old plantation. Moreover, soil aggregate stability could be the primary parameters affecting K stock across a chronosequence of Masson pine plantations.

The method of redundancy analysis was used to analyze the correlations between macronutrient stocks and soil physicochemical properties at various soil depths. Macronutrient stocks in the soil surface layer could be affected by many factors, and the variance could be explained mainly by the available K and P:K ratio in our research. The present research showed that the C:P ratio was significantly higher than the other ratios within each plantation, followed by the C:N ratio, whereas the N:K and P:K ratios were at the lowest level regardless of soil depth. Litter input is an important means for vegetation to regulate soil nutrients [58] through the priming effect and nutrient input [59]. Therefore, the observed accumulation of C, N, and P stocks in the 0−20 cm soil layer was higher than that in the 20−40 cm soil layer, which exacerbated the K limitation in this soil layer. Phosphorus was also an important limiting factor in our research. Therefore, K availability to plants and the balance of P and K were more essential to macronutrient stocks in the surface soil with abundant nutrients. However, Ks had an adverse effect on macronutrient stocks. Root growth and root exudates can improve soil structure in the 20−40 cm soil layer, and in addition to the P:K ratio, the stability of aggregates is strongly associated with C, N, and K stocks in this soil layer. Studies have also reported that soil aggregate stability has positive effects on C and N sequestration [60], but research on K is rare. For the total soil system, we found that organic matter explained most of the variation in soil macronutrient stocks, followed by the P:K ratio and available K. Specifically, the C stock was affected by the available nutrients, organic matter, and stoichiometry; and the N and P stocks were influenced by the available nutrients and organic matter. In addition, the C, N, and P stocks may decrease with soil acidification. The K stock was positively correlated with aggregate stability, which requires further study. Therefore, the response of soil macronutrient stocks to stand age and soil depth relies upon different soil physicochemical properties, and P and K may be the critical limitations of Masson pine plantation ecosystems in our study area.

Author Contributions: Conceptualization, Q.D.; data curation, J.H.; investigation, F.X.; methodology, F.X.; writing—original draft, J.H.; writing—review and editing, Q.D., Y.Y. and X.P. All authors have read and agreed to the published version of the manuscript.

Funding: This research was funded by China Postdoctoral Science Foundation (2020M673296), Natural Science Foundation of China (42167044, 42007054), the High-level Innovative Talents in Guizhou Province of Guizhou Province (Qian Ke He Platform Talents (2018)5641), the Science and Technology Projects of Guizhou Province (Qian Ke He Platform Talents (2017)5788), the first-class discipline Construction Project of Guizhou Province (GNYL (2017)007), and the Cultivation project of Guizhou University (Cultivation (2019) No.10 of Guizhou University).

Data Availability Statement: The data presented in this study are available on request from the corresponding author.

Conflicts of Interest: All authors declared that they have no conflict of interest to this work. We declare that we do not have any commercial or associative interest that represents a conflict of interest in connection with the work submitted.

References

1. Lal, R. Soil Carbon Sequestration Impacts on Global Climate Change and Food Security. *Science* **2004**, *304*, 1623–1627. [CrossRef] [PubMed]
2. Węgiel, A.; Bielinis, E.; Polowy, K. Macronutrient stocks in Scots pine stands of different densities. *Forests* **2018**, *9*, 593. [CrossRef]
3. Shaver, G.R.; Chapin, F.S. Long-Term Responses to Factorial, NPK Fertilizer Treatment by Alaskan Wet and Moist Tundra Sedge Species. *Ecography* **1995**, *18*, 259–275. [CrossRef]
4. Street, L.E.; Mielke, N.; Woodin, S.J. Phosphorus Availability Determines the Response of Tundra Ecosystem Carbon Stocks to Nitrogen Enrichment. *Ecosystems* **2018**, *21*, 1155–1167. [CrossRef]
5. Kirkby, C.A.; Kirkegaard, J.A.; Richardson, A.E.; Wade, L.J.; Blanchard, C.; Batten, G. Stable soil organic matter: A comparison of C:N:P:S ratios in Australian and other world soils. *Geoderma* **2011**, *163*, 197–208. [CrossRef]
6. Bradford, M.A.; Fierer, N.; Reynolds, J.F. Soil Carbon Stocks in Experimental Mesocosms Are Dependent on the Rate of Labile Carbon, Nitrogen and Phosphorus Inputs to Soils. *Funct. Ecol.* **2008**, *22*, 964–974. [CrossRef]
7. Chen, L.L.; Wang, K.X.; Baoyin, T. Effects of grazing and mowing on vertical distribution of soil nutrients and their stoichiometry (C:N:P) in a semi-arid grassland of North China. *Catena* **2021**, *206*, 105507. [CrossRef]
8. Pramanik, P.; Goswami, A.J.; Ghosh, S.; Kalita, C. An indigenous strain of potassium-solubilizing bacteria Bacillus pseudomycoides enhanced potassium uptake in tea plants by increasing potassium availability in the mica waste-treated soil of North-east India. *J. Appl. Microbiol.* **2018**, *126*, 215–222. [CrossRef]
9. Tian, L.; Zhao, L.; Wu, X.; Fang, H.; Zhao, Y.; Hu, G.; Yue, G.; Sheng, Y.; Wu, J.; Chen, J.; et al. Soil moisture and texture primarily control the soil nutrient stoichiometry across the Tibetan grassland. *Sci. Total Environ.* **2018**, *622–623*, 192–202. [CrossRef]
10. Xu, L.; He, N.P.; Yu, G.R. Nitrogen storage in China's terrestrial ecosystems. *Sci. Total Environ.* **2020**, *709*, 136201. [CrossRef]
11. Gao, X.L.; Li, X.G.; Zhao, L.; Kuzyakov, Y. Shrubs magnify soil phosphorus depletion in Tibetan meadows: Conclusions from C:N:P stoichiometry and deep soil profiles. *Sci. Total Environ.* **2021**, *785*, 147320. [CrossRef]
12. Birendra, N.G.; Om Pal, S.K.; Ranjan, B.; Kuldeep, S.D.; Prasant, K.M. Effect of potassium on soil conservation and productivity of maize/cowpea based crop rotations in the north-west Indian Himalayas. *J. Mt. Sci.* **2016**, *13*, 754–762. [CrossRef]
13. Piaszczyk, W.; Błońska, E.; Lasota, J.; Lukac, M. A comparison of C:N:P stoichiometry in soil and deadwood at an advanced decomposition stage. *Catena* **2019**, *179*, 1–5. [CrossRef]
14. Huang, R.; Lan, T.; Song, X.; Li, J.; Ling, J.; Deng, O.P.; Wang, C.Q.; Gao, X.S.; Li, Q.Q.; Tang, X.Y.; et al. Soil labile organic carbon impacts C:N:P stoichiometry in urban park green spaces depending on vegetation types and time after planting. *Appl. Soil Ecol.* **2021**, *163*, 103926. [CrossRef]
15. Xu, H.W.; Qu, Q.; Li, P.; Guo, Z.Q.; Wulan, E.; Xue, S. Stocks and Stoichiometry of Soil Organic Carbon, Total Nitrogen, and Total Phosphorus after Vegetation Restoration in the Loess Hilly Region, China. *Forests* **2019**, *10*, 27. [CrossRef]
16. Guo, Y.; Abdalla, M.; Espenberg, M.; Hastings, A.; Hallett, P.; Smith, P. A systematic analysis and review of the impacts of afforestation on soil quality indicators as modified by climate zone, forest type and age. *Sci. Total Environ.* **2021**, *757*, 143824. [CrossRef] [PubMed]
17. McMahon, D.E.; Vergütz, L.; Valadares, S.V.; Silva, I.R.d.; Jackson, R.B. Soil nutrient stocks are maintained over multiple rotations in Brazilian Eucalyptus plantations. *For. Ecol. Manag.* **2019**, *448*, 364–375. [CrossRef]
18. Yang, K.; Zhu, J.J.; Xu, S.; Zheng, X. Conversion from temperate secondary forests into plantations (*Larix* spp.): Impact on belowground carbon and nutrient pools in northeastern China. *Land Degrad. Dev.* **2018**, *29*, 4129–4139. [CrossRef]

19. Lucas-Borja, M.E.; Hedo, J.; Cerdá, A.; Candel-Pérez, D.; Viñegla, B. Unravelling the importance of forest age stand and forest structure driving microbiological soil properties, enzymatic activities and soil nutrients content in Mediterranean Spanish black pine(*Pinus nigra* Ar. ssp. salzmannii) Forest. *Sci. Total Environ.* **2016**, *562*, 145–154. [CrossRef]

20. Zeng, F.P.; Chen, X.; Huang, B.; Chi, G.Y. Distribution Changes of Phosphorus in Soil–Plant Systems of Larch Plantations across the Chronosequence. *Forests* **2018**, *9*, 563. [CrossRef]

21. Wang, J.; Feng, L.; Palmer, P.I.; Liu, Y.; Fang, S.X.; Bösch, H.; O'Dell, C.W.; Tang, X.P.; Yang, D.X.; Liu, L.X.; et al. Large Chinese land carbon sink estimated from atmospheric carbon dioxide data. *Nature* **2020**, *586*, 720–723. [CrossRef] [PubMed]

22. Zhang, X.; Zhao, Z.Y.; Chen, T.; Zhao, T.Y.; Song, L.H.; Mei, L. Fertilization and clear-cutting effect on greenhouse gases emission of pinewood nematode damaged Masson pine plantation. *Ecosyst. Health Sustain.* **2021**, *7*, 1868271. [CrossRef]

23. Luo, S.P.; He, B.H.; Zeng, Q.P.; Li, N.J.; Yang, L. Effects of seasonal variation on soil microbial community structure and enzyme activity in a Masson pine forest in Southwest China. *J. Mt. Sci.* **2020**, *17*, 1398–1409. [CrossRef]

24. Wang, H.X.; Wu, C.Y.; Chen, D.S.; Liu, H.Y.; Sun, X.M.; Zhang, S.G. Changes in soil carbon and nutrients and related extracellular enzymes in successive rotations of Japanese larch plantations. *Catena* **2021**, *204*, 105386. [CrossRef]

25. Tamang, M.; Chettri, R.; Vineeta, V.; Shukla, G.; Bhat, J.A.; Kumar, A.; Kumar, M.; Suryawanshi, A.; CabralPinto, M.; Chakravarty, S. Stand Structure, Biomass and Carbon Storage in Gmelina arborea Plantation at Agricultural Landscape in Foothills of Eastern Himalayas. *Land* **2021**, *10*, 387. [CrossRef]

26. Gaurav, M.; Avishek, S.; Krishna, G.; Jyoti, N.A.; Rattan, L.; Rosa, F. Changes in soil carbon stocks under plantation systems and natural forests in Northeast India. *Ecol. Model.* **2021**, *446*, 109500. [CrossRef]

27. Ma, Q.L.; Wang, X.Y.; Chen, F.; Wei, L.Y.; Zhang, D.K.; Jin, H.J. Carbon Sequestration of Sand-fixing Plantation of Haloxylon ammodendron in Shiyang River Basin: Storage, Rate and Potential. *Glob. Ecol. Conserv.* **2021**, *28*, e01607. [CrossRef]

28. Yao, X.; Yu, K.Y.; Deng, Y.B.; Zeng, Q.; Lai, Z.J.; Liu, J. Spatial distribution of soil organic carbon stocks in Masson pine (*Pinus massoniana*) forests in subtropical China. *Catena* **2019**, *178*, 189–198. [CrossRef]

29. Luo, S.; Gao, Q.; Wang, S.; Tian, L.; Zhou, Q.; Li, X.; Tian, C. Long-term fertilization and residue return affect soil stoichiometry characteristics and labile soil organic matter fractions. *Pedosphere* **2020**, *30*, 703–713. [CrossRef]

30. Li, L.; Liu, Y.; Xiao, T.H.; Hou, F.J. Different responses of soil C:N:P stoichiometry to stocking rate and nitrogen addition level in an alpine meadow on the Qinghai-Tibetan Plateau. *Appl. Soil Ecol.* **2021**, *165*, 103961. [CrossRef]

31. Wang, M.; Wang, S.Z.; Cao, Y.W.; Jiang, M.; Wang, G.D.; Dong, Y.M. The effects of hummock-hollow microtopography on soil organic carbon stocks and soil labile organic carbon fractions in a sedge peatland in Changbai Mountain, China. *Catena* **2021**, *201*, 105204. [CrossRef]

32. Wang, L.J.; Wang, P.; Sheng, M.Y.; Tian, J. Ecological stoichiometry and environmental influencing factors of soil nutrients in the karst rocky desertification ecosystem, southwest China. *Glob. Ecol. Conserv.* **2018**, *16*, e00449. [CrossRef]

33. Kang, B.; Liu, S.R.; Cai, D.X.; Lu, L.H.; He, R.M.; Gao, Y.X.; Di, W.Z. Soil physical and chemical characteristics under different vegetation restoration patterns in China south subtropical area. *Chin. J. Appl. Ecol.* **2010**, *21*, 2479–2486.

34. Walkley, A.; Black, I.A. An examination of the Degtjareff method for determining soil organic matter, and a proposed modification of the chromic acid titration method method. *Soil Sci.* **1934**, *37*, 29–38. [CrossRef]

35. Martinez, P.; Buurman, P.; do Nascimento, D.L.; Almquist, V.; Vidal-Torrado, P. Substantial changes in podzol morphology after tree-roots modify soil porosity and hydrology in a tropical coastal rainforest. *Plant Soil* **2021**, *463*, 77–95. [CrossRef]

36. Bao, S.D.; Jiang, R.F.; Yang, C.G. *Soil and Agricultural Chemistry Analysis*, 3rd ed.; China Agricultural Press: Beijing, China, 2008.

37. Bai, Y.X.; Zhou, Y.C.; He, H.Z. Effects of rehabilitation through afforestation on soil aggregate stability and aggregate-associated carbon after forest fires in subtropical China. *Geoderma* **2020**, *376*, 114548. [CrossRef]

38. Huang, X.F.; Zhou, Y.C.; Zhang, Z.M. Carbon Sequestration Anticipation Response to land use change in a mountainous karst basin in China. *J. Environ. Manag.* **2018**, *228*, 40–46. [CrossRef]

39. Alidoust, E.; Afyuni, M.; Hajabbasi, M.A.; Mosaddeghi, M.R. Soil carbon sequestration potential as affected by soil physical and climatic factors under different land uses in a semiarid region. *Catena* **2018**, *171*, 62–71. [CrossRef]

40. Wellock, M.L.; Rashad, R.; Laperle, C.M.; Matthias, P.; Gerard, K. Changes in ecosystem carbon stocks in a grassland ash (*Fraxinus excelsior*) afforestation chronosequence in Ireland. *J. Plant Ecol.* **2014**, *7*, 429–438. [CrossRef]

41. Cao, J.X.; Pan, H.; Chen, Z.; Shang, H. Dynamics in Stoichiometric Traits and Carbon, Nitrogen, and Phosphorus Pools across Three Different-Aged Picea asperata Mast. Plantations on the Eastern Tibet Plateau. *Forests* **2020**, *11*, 1346. [CrossRef]

42. Wu, H.L.; Xiang, W.H.; Yang, S.O.; Xiao, W.F.; Li, S.G.; Chen, L.; Lei, P.F.; Deng, X.W.; Zeng, Y.L.; Zeng, L.X.; et al. Tree growth rate and soil nutrient status determine the shift in nutrient-use strategy of Chinese fir plantations along a chronosequence. *For. Ecol. Manag.* **2020**, *460*, 117896. [CrossRef]

43. Wu, H.L.; Xiang, W.H.; Chen, L.; Yang, S.O.; Xiao, W.F.; Li, S.G.; Forrester, D.I.; Lei, P.F.; Zeng, Y.L.; Deng, X.W.; et al. Soil Phosphorus Bioavailability and Recycling Increased with Stand Age in Chinese Fir Plantations. *Ecosystems* **2020**, *23*, 973–988. [CrossRef]

44. Terrer, C.; Phillips, R.P.; Hungate, B.A.; Rosende, J.; Ridge, J.P.; Craig, M.E.; van Groenigen, K.J.; Keenan, T.F.; Sulman, B.N.; Stocker, B.D.; et al. A trade-off between plant and soil carbon storage under elevated CO_2. *Nature* **2021**, *591*, 599–603. [CrossRef]

45. Ye, X.M.; Zhang, Y.; Chen, F.S.; Wang, G.G.; Fang, X.M.; Lin, X.F.; Wan, S.Z.; He, P. The effects of simulated deposited nitrogen on nutrient dynamics in decomposing litters across a wide quality spectrum using a 15 N tracing technique. *Plant Soil* **2019**, *442*, 141–156. [CrossRef]

46. Marañón-Jiménez, S.; Peñuelas, J.; Richter, A.; Sigurdsson, B.D.; Fuchslueger, L.; Leblans, N.I.W.; Janssens, I.A. Coupled carbon and nitrogen losses in response to seven years of chronic warming in subarctic soils. *Soil Biol. Biochem.* **2019**, *134*, 152–161. [CrossRef]
47. Zhang, Y.; Zhao, W.; Zhang, J.B.; Cai, Z.C. N_2O production pathways relate to land use type in acidic soils in subtropical China. *J. Soils Sediments* **2017**, *17*, 306–314. [CrossRef]
48. Penn, C.J.; Camberato, J.J. A Critical Review on Soil Chemical Processes that Control How Soil pH Affects Phosphorus Availability to Plants. *Agriculture* **2019**, *9*, 120. [CrossRef]
49. Osmolovskaya, N.; Dung, V.V.; Kuchaeva, L. The role of organic acids in heavy metal tolerance in plants. *Biol. Commun.* **2018**, *63*, 9–16. [CrossRef]
50. Dinesh, A.; Tianyi, J.; Taiki, K.; Takamitsu, K.; Kenzo, K.; Araki, K.S.; Motoki, K. Relationship among Phosphorus Circulation Activity, Bacterial Biomass, pH, and Mineral Concentration in Agricultural Soil. *Microorganisms* **2017**, *5*, 79. [CrossRef]
51. Liu, K.L.; Han, T.F.; Huang, J.; Asad, S.; Li, D.M.; Yu, X.H.; Huang, Q.H.; Ye, H.C.; Hu, H.W.; Zhang, H.M. Links between potassium of soil aggregates and pH levels in acidic soils under long-term fertilization regimes. *Soil Tillage Res.* **2020**, *197*, 104480. [CrossRef]
52. Zörb, C.; Senbayram, M.; Peiter, E. Potassium in agriculture—Status and perspectives. *J. Plant Physiol.* **2014**, *171*, 656–669. [CrossRef] [PubMed]
53. Six, J.; Paustian, K. Aggregate-associated soil organic matter as an ecosystem property and a measurement tool. *Soil Biol. Biochem.* **2014**, *68*, A4–A9. [CrossRef]
54. Yang, S.Y.; Jansen, B.; Absalah, S.; van Hall, R.L.; Kalbitz, K.; Cammeraat, E.L.H. Lithology- and climate-controlled soil aggregate-size distribution and organic carbon stability in the Peruvian Andes. *Soil* **2020**, *6*, 1–15. [CrossRef]
55. Zhong, Z.K.; Zhang, X.Y.; Wang, X.; Dai, Y.Y.; Chen, Z.X.; Han, X.H.; Yang, G.H.; Ren, C.J.; Wang, X.J. C:N:P stoichiometries explain soil organic carbon accumulation during afforestation. *Nutr. Cycl. Agroecosyst.* **2020**, *117*, 243–259. [CrossRef]
56. Han, W.Y.; Xu, J.M.; Wei, K.; Shi, R.Z.; Ma, L.F. Soil carbon sequestration, plant nutrients and biological activities affected by organic farming system in tea (*Camellia sinensis* (L.) O. Kuntze) fields. *Soil Sci. Plant Nutr.* **2013**, *59*, 727–739. [CrossRef]
57. Oulehle, F.; Wright, R.F.; Svoboda, M.; Bače, R.; Matějka, K.; Kaňa, J.; Hruška, J.; Couture, R.-M.; Kopáček, J. Effects of Bark Beetle Disturbance on Soil Nutrient Retention and Lake Chemistry in Glacial Catchment. *Ecosystems* **2019**, *22*, 725–741. [CrossRef]
58. Bing, H.J.; Wu, Y.H.; Zhou, J.; Sun, H.Y.; Luo, J.; Wang, J.P.; Yu, D. Stoichiometric variation of carbon, nitrogen, and phosphorus in soils and its implication for nutrient limitation in alpine ecosystem of Eastern Tibetan Plateau. *J. Soils Sediments* **2016**, *16*, 405–416. [CrossRef]
59. Chao, L.; Liu, Y.Y.; Freschet, G.T.; Zhang, W.D.; Yu, X.; Zheng, W.H.; Guan, X.; Yang, Q.P.; Chen, L.C.; Dijkstra, F.A.; et al. Litter carbon and nutrient chemistry control the magnitude of soil priming effect. *Funct. Ecol.* **2019**, *33*, 876–888. [CrossRef]
60. Zhu, G.Y.; ShangGuan, Z.P.; Deng, L. Variations in soil aggregate stability due to land use changes from agricultural land on the Loess Plateau, China. *Catena* **2021**, *200*, 105181. [CrossRef]

Article

Budget of Plant Litter and Litter Carbon in the Subalpine Forest Streams

Jianfeng Hou, Fei Li, Zhihui Wang, Xuqing Li and Wanqin Yang *

School of Life Science, Taizhou University, Taizhou 318000, China; mehoujianfeng@163.com (J.H.);
flissrs@yeah.net (F.L.); 17816890031@163.com (Z.W.); xuqing.li@outlook.com (X.L.)
* Correspondence: scyangwq@163.com; Tel.: +86-4518-219-1992

Abstract: Investigations on the budget of plant litter and litter carbon in forest streams can provide a key scientific basis for understanding the biogeochemical linkages of terrestrial–aquatic ecosystems and managing forest catchments. To understand the biogeochemical linkages among mountain forests, riparian vegetation, and aquatic ecosystems, the changes in litter input and output from the subalpine streams with stream characteristics and critical periods were investigated in an ecologically important subalpine coniferous forest catchment in the upper reaches of the Yangtze River. The annual litter input to the stream was 20.14 g m^{-2} and ranged from 2.47 to 103.13 g m^{-2} for 15 streams during the one-year investigation. Simultaneously, the litter carbon input to the stream was 8.61 mg m^{-2} and ranged from 0.11 to 40.57 mg m^{-2}. Meanwhile, the annual litter output varied from 0.02 to 22.30 g m^{-2}, and the average value was 0.56 g m^{-2}. Correspondingly, the litter carbon output varied from 0.01 to 1.51 mg m^{-2}, and the average value was 0.16 mg m^{-2}. Furthermore, the average ratio of litter carbon input to output was 270.01. The maximum and minimum values were observed in the late growing season and the snowmelt season, respectively. Additionally, seasonal variations in temperature, together with the stream length, dominated the input of litter and litter carbon to the stream, while the precipitation, temperature, water level, and sediment depth largely determined their output. Briefly, the seasonal dynamics of litter and litter carbon were dominated by stream characteristics and precipitation as well as temperature patterns.

Keywords: litter input and output; litter carbon budget; mountain forest stream; riparian zone

Citation: Hou, J.; Li, F.; Wang, Z.; Li, X.; Yang, W. Budget of Plant Litter and Litter Carbon in the Subalpine Forest Streams. *Forests* **2021**, *12*, 1764. https://doi.org/10.3390/f12121764

Academic Editor: Marcello Vitale

Received: 23 November 2021
Accepted: 12 December 2021
Published: 13 December 2021

Publisher's Note: MDPI stays neutral with regard to jurisdictional claims in published maps and institutional affiliations.

1. Introduction

In the geosphere, streams cover less than 3% of the forest catchment area, but function as the bridges linking terrestrial–aquatic biogeochemical cycles [1]. In particular, plant litter from neighboring forests and riparian zones is the major source and carrier of carbon in forest streams and plays crucial roles not only in maintaining stream ecosystem productivity, but also in maintaining the structure and function of the butted aquatic ecosystem [2,3]. Additionally, litter decomposition in the forest stream ecosystem can contribute significantly to the global carbon cycle [4–6]. Therefore, understanding the budget of litter and litter carbon in forest streams can provide a key scientific basis for managing forest catchments and predicting the global carbon cycle.

The forest stream might act as a sink of plant litter and bioelements in the forest catchment. Theoretically, litter input to the stream is hierarchically regulated by three interactive factors: climate, forest type, and stream characteristics [7]. First, the climate has been considered the primary factor influencing litter production [8]. Generally, average litter production decreases gradually from tropical zones to boreal alpine zones along the climate gradient [9,10]. Compared to cold temperate zones, evergreen broadleaved forests in tropical regions often have larger amounts of litter production due to the higher temperature and moisture [11,12], implying that more litter can enter the stream. Second, the dynamics of litter input vary greatly with forest types, as different tree species have different phenological phases [7,13], which in turn determine the quantity and dynamics of litter input to the forest stream [13,14]. For instance, on a local scale, evergreen and deciduous forests usually show higher litter production than dark coniferous forests in the subalpine forest region [15,16]. Plant species composition in the riparian zone differs greatly from that in the mountain forest, and the litter production of shrub and herb species in the riparian zone is lower than that in the mountain forest [7,14], implying that the litter input to the stream in the riparian zone might be lower than that directly in the forest. Third, litter input is also modulated by the stream length and width [17], and longer and wider streams can receive more litter along the stream [18,19]. Although litter input to the stream has been systematically investigated in northern America [20–22], litter input to streams has not been fully investigated around the world, limiting our understanding of the biogeochemical linkages of mountain forest and riparian zones with streams and rivers.

The litter and litter carbon output from streams are known as the major carbon sources of butted rivers. That is to say, the forest stream also acts as the source of litter and litter carbon [23,24]. In theory, the litter output from forest streams is usually regulated by the stream litter quality and quantity, stream biological community, stream characteristics, and climate [8,25]. To begin with, the magnitude of litter input to the stream determines the size of the litter source of a butted river [4,5]. Meanwhile, the scouring action of stream water on litter can directly accelerate litter fragmentation [26], which may lead to the output and confluence characteristics of litter varying with the seasons [27–29]. For example, streams with lower flow rates and slower velocities always accompany faster litter decomposition and litter deposition, leading to smaller amounts of output, and vice versa [30]. In addition, the length and width of the streams, together with their microtopography, might significantly influence litter output [29]. Finally, forest stream characteristics, such as discharge and velocity, are always regulated by seasonal precipitation (rainfall, snowfall, and snowmelt), theoretically modulating the output of litter and litter carbon from the streams. In particular, the plant rhythm with seasonal changes accompanies the seasonal dynamics of precipitation and temperature, which play important roles in controlling the output of litter from a forest stream [7,31]. Therefore, investigations into the litter output from forest streams could facilitate a better understanding of the biogeochemical linkages of mountain forests and riparian zones with aquatic ecosystems.

Carbon is the basic component in both terrestrial and aquatic ecosystems. The dynamic pattern of litter carbon in forest stream ecosystems can reveal terrestrial–aquatic carbon biogeochemical linkages [30,32]. Past investigations have found that the carbon fractions derived from upstream and neighboring ecosystems are the two major sources of dissolved carbon (DC) in forest stream ecosystems [33,34]. Most investigations of carbon biogeochemical linkages between mountain forests and butted aquatic ecosystems have employed the small-scale runoff field method [35]. However, this method has difficulty revealing the roles of forest streams in terrestrial–aquatic carbon biogeochemical linkages, especially in geographically fragile mountainous regions [36]. First, due to geological fragmentation and serious soil percolation [37], surface runoff is rarely observed in most rainfall and snowfall events, while percolating water becomes an important biogeochemical link between mountain forest ecosystems and aquatic ecosystems in fragile mountainous regions. Second, forest streams can be directly involved in the biological carbon cycle rather than indirectly involved through surface runoff, since litter from forest and riparian vegetation is a major source of carbon input to the butted aquatic ecosystem [38]. Third, the riparian zone is an important domain in the forest stream ecosystem. The decomposition of allochthonous organic materials (e.g., foliar litter) in riparian zones is often a critical factor affecting the continued availability of carbon resources in these ecosystems [39]. Hence, the systematic investigation of litter carbon dynamics in streams and riparian zones will provide baseline data for further understanding of the biogeochemical linkages of terrestrial–aquatic ecosystems.

As the second largest forest region in China, the subalpine forest region in the eastern Qinghai-Tibet Plateau is the most important freshwater conservation area and headwater region of the Yangtze River, and plays paramount and irreplaceable roles in holding water, conserving soil, and maintaining the safety of water resources and downstream aquatic ecosystems [40,41]. These forest stream ecosystems are typically cold ecosystems that experience considerable seasonal freezing and thawing events, and seasonal changes are associated with distinct changes in environmental conditions [42,43]. Therefore, a deep investigation of the budget of litter and litter carbon in these forest stream ecosystems is key to revealing the carbon biogeochemical linkages between subalpine forests and aquatic ecosystems.

Although the concentration and storage of carbon in woody and nonwoody debris have been investigated in subalpine forest streams [38,44], little information is available on estimating the budget of litter and litter carbon in these subalpine forest streams. Therefore, we hypothesized that (1) the seasonal input and output of litter and litter carbon might have different dynamic patterns; (2) the maximum and minimum values of the indices mentioned above would appear in the litterfall and snowmelt periods, respectively; and (3) the litter and litter carbon input to the streams would be higher than the outputs from the streams in forest stream ecosystems.

2. Material and Methods

2.1. Site Description

This study was conducted at the Long-Term Research Station of Alpine Forest Ecosystem in the Bipenggou Valley (31°14′~31°19′ N, 102°53′~102°57′ E, 2458~4619 m above sea level (masl)), Li County, Southwest China, which is located in the alpine gorge area with frequent geological breaks, clear seasonal snow cover (the maximum snow depth was about 35 cm), and frequent freeze/thaw cycles [7] (Figure 1). The mean annual precipitation is approximately 850 mm, and the annual mean air temperature is approximately 3 °C, with maximum and minimum temperatures of 23 °C (July) and −18 °C (January), respectively. The frozen season lasts from November to April, and thaw begins in late April. This subalpine forest is dominated by Minjiang fir (*Abies faxoniana* Rehder & E.H.Wilson), larch (*Larix mastersiana* Rehder & E.H.Wilson), and cypress (*Sabina saltuaria* Rehder & E.H.Wilson), and is interspersed with shrubs of azaleas (*Rhododendron* spp.), willow (*Salix* spp.), and barberry (*Berberis sargentiana* C.K.Schneid). The herbaceous plants consist mainly of ferns (*Cystopteris montana* (Lam.) Bernh. ex Desv) [7]. The concentrations of carbon (C), nitrogen (N), and phosphorus (P) in the surface soil (5 cm depth) was 126.0, 5.8, and 1.2 g kg^{-1}, respectively.

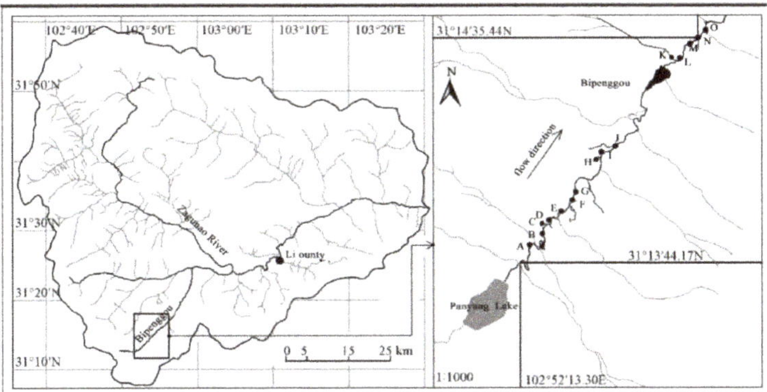

Figure 1. The investigated streams in the Bipenggou Valley, located in the upper reaches of the Yangtze River. The letters A–O indicate the 15 sampling streams [7].

2.2. Experimental Design

Based on previous investigations [7,38], the input and output of litter and litter carbon in 15 representative permanent streams in the forest catchment were investigated. Three sampling sites for collecting litter were set up in the upper, middle, and end of every stream, and the streams' characters (length, width, sediment depth, and flow velocity) were also measured at every sampling time. The length was measured from the source to the estuary using a flexible rule along the banks of the streams. Meanwhile, the width was measured horizontally from one bank to another at every sampling site, the depth was measured vertically using a rule, and the flow velocity was measured using a flowmeter (Martin Marten Z30, Current-meter, Barcelona, Spain) every 30 min. Every index mentioned above was measured three times, then the mean was taken at every sampling time for every sampling site. These streams are in a typical subalpine forest catchment at elevations of 3600~3700 m, and the total area of the investigated forest catchment was 4.3 km^2. At every site, a button thermometer (iButton DS1923-F5, Maxim/Dallas Semiconductor, Sunnyvale, CA, USA) was set to record the temperature every 2 h, and the precipitation was measured using a rainfall monitor (ZXCAWS600, Zxweather, Beijing, China) for real-time monitoring.

2.3. Monitoring the Input and Output of Litter and Litter Carbon

In order to collect litter, according to the stream length, a quadratic litter collector (0.8 m × 0.8 m) was randomly installed at the source, middle, and end of the stream (when the stream width < 0.8 m, one litter collector was positioned; when the stream width > 0.8 m and <1.6 m, two litter collectors were positioned; the stream widths are shown in Table 1), and each was installed 0.5 m above the water or ground surface (Figure 2). To avoid litter decay in the litter collectors caused by rainfall, the litter samples were collected every 15 days, but the litter was collected only once in the cold winter since litterfall in winter was rare. All of the collected litter samples were put into precleaned polyethylene bags and transported to the lab. The samples were dried to a constant weight and stored at 65 °C for less than one week until analysis.

Table 1. Basic characteristics of 15 representative subalpine forest streams in the investigated subalpine forest catchment.

Stream	Elevation (m)	Length (m)	Width (m)	Water Level (cm)	Flow Velocity (m^3/s)	Main Plants
A	3668	220	0.63 ± 0.16	8.57 ± 3.20	0.11 ± 0.11	*A. faxoniana, Cyperus* spp., *S saltuaria*
B	3667	66	0.69 ± 0.14	5.15 ± 1.60	0.07 ± 0.07	*A. faxoniana, Cyperus* spp., *S. saltuaria*
C	3658	13	0.63 ± 0.22	7.16 ± 3.83	0.01 ± 0.02	*A. faxoniana, Cyperus* spp., *S. saltuaria*
D	3658	92.4	0.86 ± 0.19	4.81 ± 1.08	0.06 ± 0.07	*A. faxoniana, Cyperus* spp., *S. saltuaria*
E	3657	47	0.34 ± 0.30	3.73 ± 3.43	0.01 ± 0.01	*A. faxoniana, Cyperus* spp., *S. saltuaria*
F	3640	65	1.11 ± 0.30	6.90 ± 1.44	0.11 ± 0.12	*S. saltuaria, R. lapponicum*
G	3640	186	1.02 ± 0.33	8.96 ± 1.91	0.06 ± 0.06	*S. saltuaria, Carex* spp., *R. weginzowii*
H	3634	108	0.82 ± 0.28	6.73 ± 4.38	0.04 ± 0.07	*S. saltuaria, Carex* spp., *S. rufopilosa*
I	3634	256	1.02 ± 0.22	7.19 ± 1.80	0.13 ± 0.12	*S. saltuaria, R. weginzowii, Carex* spp.
J	3634	18	1.29 ± 1.00	3.85 ± 3.05	0.04 ± 0.07	*S. saltuaria, R. weginzowii, Carex* spp.
K	3611	36	1.00 ± 0.24	7.93 ± 2.32	0.10 ± 0.08	*R. lapponicum, S. saltuaria*
L	3611	11	0.93 ± 0.58	3.70 ± 1.89	0.03 ± 0.05	*R. lapponicum, S. saltuaria*
M	3610	12	0.85 ± 0.23	6.26 ± 1.42	0.03 ± 0.08	*R. lapponicum, S. saltuaria*
N	3607	28	0.84 ± 0.32	11.72 ± 4.88	0.15 ± 0.14	*S. mastersiana, Cyperus* spp., *S. rufopilosa*
O	3607	17	0.65 ± 0.41	4.34 ± 3.84	0.02 ± 0.04	*S. mastersiana, S. rufopilosa, Cyperus* spp.

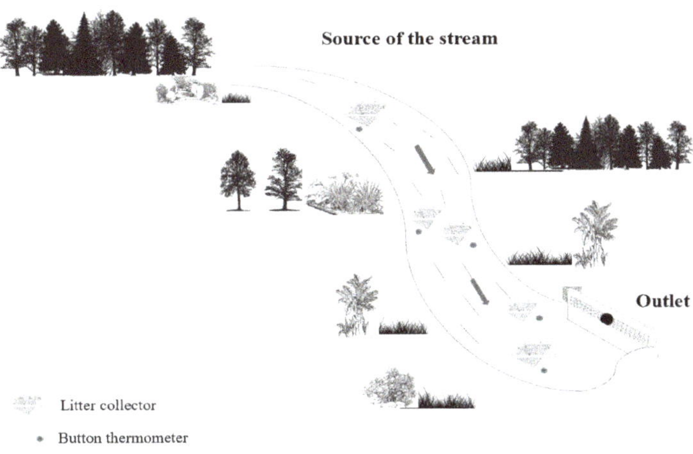

Figure 2. Litter and litter carbon input and output monitoring system in the investigated streams of the Bipenggou Valley, located in the upper reaches of the Yangtze River.

To make sure the amount of litter collected was accurate, a litter collector (0.8 m × 0.8 m) was installed at the start of every stream (the width of all stream sources were less than 0.8 m). Two litter interception dams with different mesh specifications were set at the outlet of each stream, and the total amount of litter intercepted by the two interception dams was the output amount of litter (Figure 2). The interception dam with 3 cm × 3 cm apertures was in front, which mainly collected litter with a larger diameter, and the 100-mesh interception dam was installed behind, which was mainly used to collect litter of a smaller diameter. The cycles of measuring the litter output were consistent with the litter input measurement. During the sampling time, all litter collected from the interceptor was quickly brought back to the laboratory, dried to a constant weight at 65 °C, then weighed and recorded as the litter output.

We divided one year into five different periods, i.e., the snowmelt season (SMS: April to May), early growing season (EGS: May to June), growing season (GS: July to August), later growing season (LGS: September to October), and seasonal snow cover (SSC: November to April the following year) based on phenological changes, seasonal precipitation, and temperature dynamics [40]. Litter was collected nine times in the growing seasons and four times in the non-growing seasons. Specifically, litter in the collectors was collected during the LGS at approximately 15-day intervals. The quantity of inputted and outputted litter and litter carbon at each period was calculated as the cumulant values of this stream, and the 15 streams were treated as 15 repeats during data analysis.

The temperatures in the in-stream and riparian zones varied considerably with the ecosystem type during the two years of the experiment [7]. The temperature fluctuated sharply with critical periods, with the average daily temperatures ranging from 0 °C to 10.7 °C. However, the in-stream and riparian zone temperatures were almost always above 0 °C throughout the investigation period. Similarly, the water varied substantially between the stream and riparian zones for a comparable period, and varied significantly within the stream or riparian zone among different periods (Table 1).

2.4. Analytical Methods and Calculations

The concentration of litter carbon was determined using the potassium dichromate oxidation–external heating method [45]. Following this, 0.01 g of the dried litter that had been sieved through a 0.15-mm sifter was placed in the bottom of a 100-mL Erlenmeyer flask. The required amount of H_2SO_4 (5 mL) and 10 mL of potassium dichromate solution were added. After attaching a reflux condenser, the mixture was boiled at 220–230 °C for 15 min on an electric stove. After cooling and rinsing the condenser with water, 3 to 5 drops of N-phenylanthranilic acid were added. The titration was performed with a 0.2 M solution of ferrous sulfate salt at room temperature. With the addition of one drop, the color shifted from violet to bright green:

$$C_S = (C_C \times M_L)/S_S$$

where C_S is the litter carbon stock; C_C is the carbon concentration, g kg^{-1}; M_L is the litter stock, g; and S_S is the area of the stream, m^2.

Linear mixed effect models were used to analyze the relationships of the input and output of litter and litter carbon with climate and stream characteristics in the subalpine forest stream among different sampling periods. First, we tested the normality of residuals, homoscedasticity of errors, and independence of errors to determine whether our data met the assumptions of the analyses. Second, the sampling period was treated as a fixed effect, and then we conducted a repeated measures analysis of variance (ANOVA) to examine the variability of the different variables (litter input, litter carbon input, litter output, and litter carbon output) at different critical periods. Third, to better illustrate the correlations of the input and output of litter and litter carbon with the explanatory variables, these variables were treated as fixed factors and the stream was included as a random factor. We used linear and quadratic models to fit the relationships of four indices of litter with the changes in the various explanatory variables. The relationships between the ratios of input to output of litter and litter carbon vs. stream characteristics were also tested by the linear mixed effect models. All analyses were conducted in R using the LME4 package [46].

3. Results

3.1. Litter and Litter Carbon Input to the Stream

The litter input to the stream showed two peaks (Figure 3A). The maximum peak (101.40 g m^{-2}) was observed in the LGS and then decreased gradually, reaching the minimum value (18.91 g m^{-2}) in the SMS. The second peak (59.44 g m^{-2}) was found in the GS (Figure 3A), and these values were all averages per period. Meanwhile, the litter input to the stream ranged from 2.47 to 103.13 g m^{-2}, and the annual value was 20.14 g m^{-2} for the 15 investigated streams during this one-year investigation.

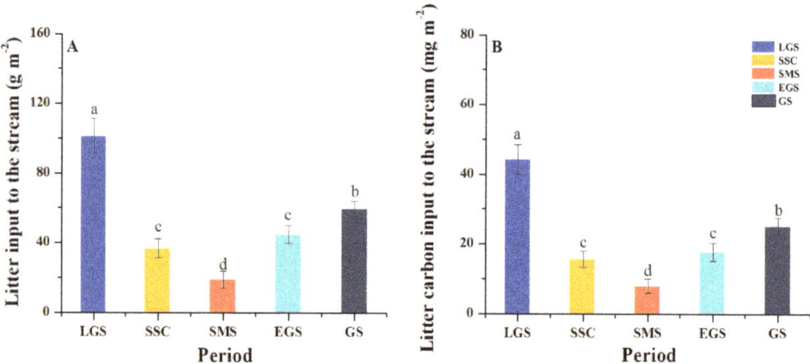

Figure 3. Dynamics of litter (**A**) and litter carbon (**B**) input to the subalpine forest streams in the upper reaches of the Yangtze River. LGS, SSC, SMS, EGS and GS indicate the sampling periods, i.e., later growing season (LGS: September to October), seasonal snow cover (SSC: November to April next year), snowmelt season (SMS: April to May), early growing season (EGS: May to June), and growing season (GS: July to August). The vertical coordinate is the mean of litter input accumulation of 15 streams during this period. Different lowercase letters indicate the significant difference among different periods ($p < 0.05$), while the same letter indicates no significant difference among each other.

Like litter input, the dynamic pattern of litter carbon in the streams increased gradually from September to November (LGS), reached its maximum value (44.39 mg m^{-2}), and then decreased gradually from April to June (SMS) to reach its minimum value (7.98 mg m^{-2}) in SMS (Figure 3B). The second peak (25.12 mg m^{-2}) of litter carbon input to the stream was observed in the GS (Figure 3B), and these values were all average of per period. For all streams, the litter carbon input to the streams ranged from 0.11 to 40.57 mg m^{-2}, and the annual value was 8.61 mg m^{-2} for all streams in this one-year investigation.

3.2. Litter and Litter Carbon Output from the Stream

The dynamic pattern of litter output from the stream also showed a similar pattern to that of the input. The maximum value (1.69 g m^{-2}) appeared in the LGS, and then the value decreased gradually, reaching the minimum value in the SMS (0.46 g m^{-2}). The second peak (1.72 g m^{-2}) was observed in the EGS (Figure 4A), and these values were all average of per period. The litter output from the all streams ranged from 0.02 to 22.30 g m^{-2}, and the annual average value was 0.56 g m^{-2} during this one-year investigation.

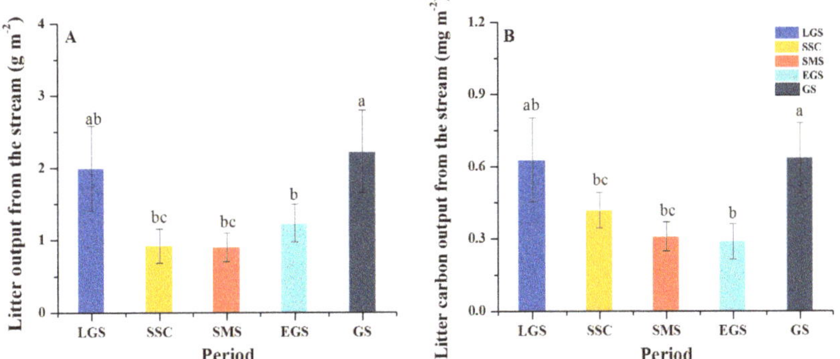

Figure 4. Dynamics of litter (**A**) and litter carbon (**B**) output from the subalpine forest streams in the upper reaches of the Yangtze River from 11 July 2015, to 2 August 2016. LGS, SSC, SMS, EGS and GS indicate the sampling periods, i.e., later growing season (LGS: September to October), seasonal snow cover (SSC: November to April next year), snowmelt season (SMS: April to May), early growing season (EGS: May to June), and growing season (GS: July to August). The vertical coordinate is the mean of litter input accumulation of 15 streams during this period. Different lowercase letters indicate the significant difference among different periods ($p < 0.05$), while the same letter indicates no significant difference among each other.

Consistent with the pattern of litter input to the stream, the output of litter carbon from the streams also had two peaks (Figure 4B), which also appeared in the LGS (0.63 mg m^{-2}) and EGS (0.64 mg m^{-2}), and the lowest point was observed in the SMS (0.31 mg m^{-2}) (Figure 4B), and these values were all average of per period. Additionally, the litter carbon output from all streams ranged from 0.01 to 4.53 g m^{-2}, and the annual value was 0.45 g m^{-2} during this one-year investigation.

3.3. The Ratios of the Input to Output of Litter and Litter Carbon

During the investigated period, the average ratios of input to output of litter and litter carbon were 188.17 and 270.01 at the forest catchment level, respectively. The ratios of litter input to output ranged from 23.85 to 619.67 (Figure 5A), and the ratios of the input to output of litter carbon ranged from 16.80 to 1147.91 (Figure 5B). Meanwhile, the ratios of both litter input to output and litter carbon input to output showed similar dynamic patterns with the two peaks appearing in the LGS (272.74, 329.70) and EGS (190.46, 304.13), and the lowest point observed in the SMS (37.99, 67.11) for litter and litter carbon, respectively. Additionally, the ratios of input to output for both litter and litter carbon varied greatly with the investigated streams, and the averaged ratios of input to output ranged from 16.42 to 577.91 for litter (Figure 6A), and from 16.80 to 1147.91 for litter carbon, respectively (Figure 6B).

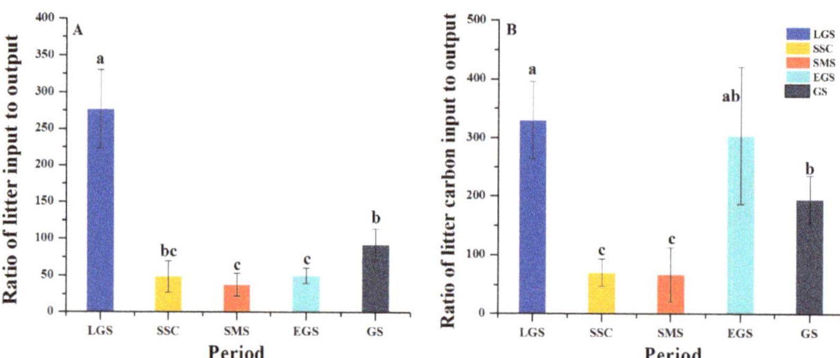

Figure 5. The ratios of input to output of litter (**A**) and litter carbon (**B**) in the subalpine forest streams in the upper reaches of the Yangtze River from 11 July 2015, to 2 August 2016. The value of each dot is the ratio of the average for the investigated streams during the sampling time. Different lowercase letters indicate the significant difference among different periods ($p < 0.05$), while the same letter indicates no significant difference among each other.

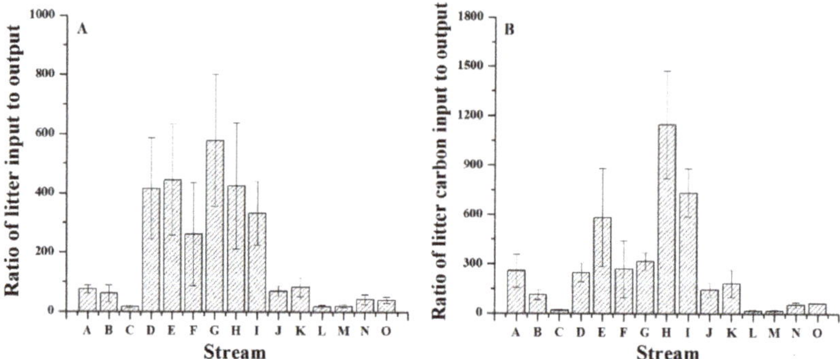

Figure 6. The ratios of input to output of litter (**A**) and litter carbon (**B**) in the subalpine forest streams in the upper reaches of the Yangtze River from 11 July 2015, to 2 August 2016. Each bar is the average of 13 sampling times for each forest stream. A–O are the sampled streams in the study.

3.4. Relationships of Litter and Litter Carbon Input/Output with Relative Variables

Litter input to the streams was significantly and positively correlated with the temperature and length (Table 2). However, litter input was slightly and negatively correlated with the stream width, but slightly and positively related to the precipitation, water level, sediment depth, and flow velocity of the forest streams (Table 2). Correspondingly, litter carbon input to the streams was also showed the similar relationships with relative variables (Table 2).

Table 2. Relationships of litter and litter carbon input with stream characteristics in the subalpine forest catchment.

	Factors	d.f.	*p*	r^2
Litter input	Precipitation	73	0.97	<0.01
	Temperature	73	<0.01 **	0.32
	Sediment depth	73	0.27	<0.01
	Water level	73	0.50	0.04
	Flow velocity	73	0.27	<0.01
	Width	73	−0.29	<0.01
	Length	73	<0.01 **	0.11
Litter carbon input	Precipitation	73	0.63	0.02
	Temperature	73	<0.01 **	0.30
	Sediment depth	73	0.21	0.04
	Water level	73	0.22	0.01
	Flow velocity	73	0.21	0.01
	Width	73	−0.43	<0.01
	Length	73	0.01 *	0.06

* $p < 0.05$, ** $p < 0.01$. d.f.: degree of freedom.

Similarly, litter output from the streams was significantly and positively correlated with the precipitation, temperature, litter input, and flow velocity, and slightly and positively with the water level, but significantly and negatively with sediment depth, and slightly and positively with stream width and length. Correspondingly, litter carbon output from the streams was also the similar relationships with the variables mentioned above (Table 3).

Table 3. Relationships of litter and litter carbon output with stream characteristics in the subalpine forest streams.

	Factors	d.f.	*p*	r^2
Litter output	Precipitation	73	0.002 **	0.15
	Temperature	73	0.02 *	0.08
	Sediment depth	73	−0.04 *	0.05
	Water level	73	0.24	0.02
	Flow velocity	73	0.02 *	0.07
	Width	73	−0.68	<0.01
	Length	73	−0.07	0.04
	Litter input	73	0.002 **	0.18
Litter carbon output	Precipitation	73	0.04 *	0.01
	Temperature	73	0.003 **	0.08
	Sediment depth	73	−0.03 *	0.04
	Water level	73	0.33	<0.01
	Flow velocity	73	0.04 *	0.06
	Width	73	−0.41	<0.01
	Length	73	−0.26	<0.01
	Litter carbon input	73	<0.001 **	0.22

* $p < 0.05$, ** $p < 0.01$. d.f.: degree of freedom.

4. Discussion

Our results demonstrated that both the input and output of litter and litter carbon in the subalpine forest streams showed two peaks with critical periods, which was not consistent with our first hypothesis that the seasonal input and output of litter and litter carbon might have different dynamic patterns. Climate factors (temperature and precipitation) together with stream characteristics (sediment depth, length, and flow velocity) drive the seasonal sink-source pattern of litter and litter carbon in the subalpine forest streams. Meanwhile, the results showed that the maximum and minimum values of all the indices mentioned above appeared in the later growing season and snowmelt periods, respectively, which was consistent with our second hypothesis that the maximum and minimum values of the indices mentioned above would appear in the litterfall and snowmelt periods, respectively. In particular, the litter and litter carbon input to the streams were higher than the output from the streams in the forest-stream ecosystem, which was consistent with our third hypothesis that the litter and litter carbon input to the streams would be higher than the outputs from the streams in forest-stream ecosystems, implying that subalpine forest streams play an important role in carbon sink in the subalpine forest region.

4.1. Dynamics of Litter and Carbon Input in the Subalpine Forest Streams

Worldwide, the annual litter input to forest streams varies from 3 to 1000 g m^{-2} y^{-1}, and the highest and lowest values are observed in a coniferous forest in North Carolina and in a shrub/grass-covered riparian zone of Deep Cr, Idaho, respectively (Table 4) [20–22,47–50]. In this study, the annual litter input to the subalpine forest streams was 20.14 g m^{-2} and ranged from 2.47 to 103.13 g m^{-2} for our one-year investigation. Meanwhile, litter as the carrier of carbon biogeochemical linkages between forest and stream ecosystems [7], our results also showed the average litter carbon input to forest streams was 8.61 mg m^{-2} and ranged from 0.11 to 40.57 mg m^{-2}. Annual litter and its carbon input to the forest streams are dominated by the litter production of mountain forest and riparian vegetation through which the streams flow. That is, the higher the litter production in neighboring forest and riparian vegetation, the higher the litter and more litter carbon input to the butted streams. Meanwhile, the streams are linked more closely with the forest ecosystems, and more litter can be inputted to the streams since longer and wider streams can receive more litter. For instance, Wenger et al. (1999) have found that the interception and conversion efficiencies of riparian zones for litter depend on the width of the riparian zones [51]. Previous investigations have also found that subalpine deciduous forests have higher litter production than those in the subalpine shrub and grass vegetation [15]. The subalpine forest region in this study consists mainly of dark coniferous tree species and deciduous coniferous and broadleaved tree species, and litter production varies greatly with forest types [52]. Higher litter production is observed in the Minjiang fir-dominated primary coniferous forest and *Larix masteriana*-dominated deciduous coniferous forest, and the lower value is observed in the rhododendron shrub forest [15]. Consequently, annual litter and litter carbon input to the subalpine forest streams also varied greatly with the investigated streams, resulting from the investigated streams having different length and width, different forest types and riparian vegetation types along the streams (Table 1). In this study, both litter and litter carbon input were significantly and positively correlated with the stream length (Table 2), indicating that the longer the stream, the more litter input to the streams, which was consistent with the litter input. However, the stream width correlated slightly and negatively with litter input owing to the wider streams flowing through the riparian zones where shrub and herb species produce less plant litter. These results suggest that the litter input to the streams from mountain forests plays more important roles in maintaining the structure and function of the subalpine forest stream ecosystem, since plant litter is a pivotal component of stream food webs and ecosystem functioning [3].

Table 4. Litter Input to Forest Streams Reported Worldwide.

Site	Tree Spices	Input (g m^{-2} y^{-1})	References
Upper Reaches of Yangtze River, China	Coniferous forest	262	This study
Ogeechee River, Georgia, USA	Coniferous forest	537	Triska et al. (1982)
Satellite Br, North Carolina, USA	Mixed deciduous forest	629	Benfield (1997)
Devil's Club Cr, Oregon, USA	Coniferous forest	736	Benfield (1997)
Deep Cr, Idaho, USA	Shrub/grass cover	3	Benfield (1997)
Mack Cr, Oregon, USA	Coniferous forest	730	Benfield (1997)
Sycamore Cr, Arizona, USA	Shrub cover	20	Benfield (1997)
North Carolina, USA	Deciduous forest	1000	Webster et al. (1997)
Coast Range of Oregon, USA	Deciduous forest	613	Hart (2006)
Domtar Inc. White River Forest, Canada	Mixed wood forest	167	Muto (2008)
Brazil	Atlantic forest	131	Tonin et al. (2017)
Brazil	Amazon forest	165	Tonin et al. (2017)
Brazil	Cerrado savanna,	213	Tonin et al. (2017)
Wit River, Western Cape, South Africa	typical mountain fynbos	69	Railoun (2018)
Du Toit's River, Western Cape, South Africa	typical mountain fynbos	68	Railoun (2018)

Litter and litter carbon input to the forest streams also showed two peaks with critical periods, and the maximum and minimum values were observed in the late growing season and seasonal snowmelt season, respectively. A reasonable explanation is that seasonal climate change controls the plant rhythm and in turn dominates the seasonal dynamics of litterfall in mountain forests and riparian vegetation. Two peaks of litterfall in the subalpine forest have been observed in the early and late growing seasons [15,53]. The litterfall peak that occurs in the early growing season contributes to the abnormal litter due to windstorms harming young leaves, while the maximum litterfall peak in the late growing season contributes mainly to the plant rhythm due to plant leaf senescence and falling in autumn [52]. Consequently, higher litter input to the forest streams was observed in the later growing season and early growing season. Correspondingly, lower litterfall was accompanied by lower litter input in the snowmelt season. Meanwhile, stream characteristics are often regulated by seasonal precipitation and then influence the seasonal dynamic pattern of litter input. Heavier rainfall can increase the velocity and width of stream flow, leading to more forest floor litter input to the forest stream. Johnson and Jones (2001) have revealed that the stream water level is a limiting factor for the interception and transformation of elements, and more litter can be inputted to the forest stream when the stream water level rises [54]. Briefly, the litter input to forest streams is affected by seasonal climate change, stream characteristics and plant growth rhythms.

4.2. Seasonal Dynamics of Litter and Carbon Output from Subalpine Forest Streams

The annual litter output from the investigated subalpine forest streams varied from 0.02 to 22.30 g m^{-2}, and the average value was 0.56 g m^{-2} for the investigated streams during this one-year investigation. Correspondingly, the annual litter carbon output varied from 0.01 to 1.51 mg m^{-2}, and the average value was 0.16 mg m^{-2}. In theory, the litter and litter carbon output from the streams depend mainly on three pathways. Firstly, the stocks of litter in the stream ecosystem determine the output from the streams. In other words, the higher the reserves of litter in the streams, the higher the litter and litter outputs from the streams, implying that the litter and litter input to the streams manipulate the litter outputs from the streams [3,15,30]. Secondly, the decomposition and deposition of plant litter in the stream largely regulate the output of litter. For instance, Yue et al. (2016) have found that plant litter incubated in the subalpine forest streams decomposes completely after two years [7]. Our sediment data at the bottom of the streams (unpublished) showed that more than 95% of the litter was locked in the stream floor as sediments: the more litter that is stored in the form of sediment, the less the output of litter from the forest streams, leading to a lower carbon output. The significantly negative relationships found between litter and litter carbon output and sediment depth in this study were consistent with this (Table 3). Thirdly, litter loss in deeper streams was generally significantly higher for a specific litter species [7,55]. Given these cases, stream characteristics, particularly the sediment depth and flow velocity, also play key roles in regulating litter carbon output [30]. For example, Bilby et al. (2003) also found that as stream width increased from 5 m to 15 m, the water velocity also increased; as the litter carbon captured by falling wood dropped by 80%, more carbon was outputted from the streams in the study of the Coastal Pacific Northwest of the United States [56]. The positive relationship between litter output and water level and flow velocity in this study was consistent with their results. Overall, forest streams are key bonds of biological matter linkage between mountain forest ecosystems and downstream.

In aquatic habitats, contrary to our expectation, the litter and its carbon output from the forest streams also presented similar seasonal patterns to those of the input. The results can be explained by the following reasons. On the one hand, the factors influencing litter carbon input to streams also modulate litter carbon output from the streams. For example, the peak and nadir values of litter carbon output are also observed in the later growing season and snow melting season, respectively. On the other hand, stream characteristics also play key roles in manipulating the output of litter and litter carbon from the streams, and these characteristics are always regulated by seasonal precipitation and plant rhythm. In this study, the highest values of litter carbon output from the forest streams were observed in the later growing season, owing to the higher flow velocity and litterfall peak in this period, while the lower flow rate, together with less litter production in the snow-covered period, might contribute more to the occurrence of the lowest values during the snowmelt season. Seasonal precipitation, together with seasonal melting of snow cover, dominates the periodic fluctuations in the depth, velocity, and width of forest streams and determines the seasonal dynamics of litter and litter carbon output from the streams. The forest streams act as the source of litter and litter carbon in the butted river, and the output dynamics are complexly regulated by the sink of litter and litter carbon in the streams and the stream characteristics, as affected by seasonal precipitation.

4.3. Litter Carbon Budget in the Subalpine Forest Stream

The average ratio of litter input to output in the subalpine forest catchment was 188.17, and the average ratio of input to output of litter carbon was 270.01, indicating that forest streams act as the sink of litter carbon in the subalpine forest region. Owing to this huge difference between the input and output, the variables that had a significant effect on litter and litter caron input could explain the significant variability of these ratios in the sampled streams to a great extent. Possible reasons for this wide gap between litter carbon input and output include the following. First, although much litter was inputted to the forest streams, more than 95% of the litter input was deposited at the bottom of the streams in the form of sediment, as mentioned above, resulting in very little carbon being exported from the streams in the form of litter. Second, the effects of abrasion caused by sediment transport in the streams could be another driver of the higher litter loss, which was also regulated by seasonal precipitation (rainfall and snowfall). This process could accelerate litter decomposition rates in aquatic ecosystems [55], leading to less litter carbon being output in the form of litter. Generally, the decomposition and deposition of plant litter in the stream largely regulate the gap between the input and output of litter carbon. For all of the reasons above, forest streams are indispensable carbon sinks in the subalpine forest stream ecosystem. However, the processes and mechanisms of the stream litter budget need to be fully investigated in the future.

Although the carbon output from the streams was smaller than the carbon input, the dissolved carbon output from the streams was a noteworthy part of the forest ecosystem [36,57]. The litter carbon output from the forest streams in this study was much less than that in the investigation of east Finland (0.62–0.94 g m^{-2} a^{-1}) [58], which might be mainly due to the streams in our research having a shorter length and less anti-interference ability, or since much of the carbon stored in the litter was retained in the more complex ecosystems of rivers and lakes rather than in small streams, resulting in a smaller amount of carbon in the forest catchments [59]. Aquatic litter carbon (stream and riparian zone) could connect forest ecosystems with river ecosystems [16,60]. Therefore, investigations of the litter carbon budget in the forest streams can provide baseline data for further studies on the interconnections between individual aquatic and terrestrial ecosystems within a forest stream ecosystem.

5. Conclusions

Both the input and the output of litter and litter carbon in the subalpine forest streams showed similar patterns with critical periods, as affected by seasonal changes in temperature and precipitation together with stream characteristics (length, flow velocity, sediment depth, etc.). The seasonal temperature pattern and the stream length dominated the input of litter and litter carbon to the stream. Meanwhile, the seasonal temperature and precipitation patterns, sediment depth, and flow velocity determined the output of litter and litter carbon from the stream. Additionally, the litter and litter carbon input to the streams were higher than the output from the streams in the forest stream ecosystem, indicating that the subalpine forest streams play a crucial role in linking the carbon biogeochemical cycle between subalpine forests and downstream rivers. In-depth investigations on the carbon biogeochemical linkages in forest–riparian–river meta-ecosystems should be conducted, particularly concerning the linkages between the carbon biogeochemical cycle in the meta-ecosystems and forest characteristics, topographical features, and catchment shapes under different climate change scenarios.

Author Contributions: Conceptualization, W.Y. and J.H.; methodology, J.H. and W.Y.; software, J.H. and F.L.; validation, J.H. and W.Y.; formal analysis, J.H.; investigation, J.H., F.L., Z.W., X.L.; resources, W.Y.; data curation, W.Y. and J.H.; writing—original draft preparation, J.H.; writing—review and editing, W.Y. and J.H.; visualization, J.H.; supervision, W.Y.; project administration, W.Y.; funding acquisition, W.Y. and F.L. All authors have read and agreed to the published version of the manuscript.

Funding: This study was supported by the National Natural Science Foundation of China (32071554, 32001139).

Institutional Review Board Statement: Not applicable.

Informed Consent Statement: Not applicable.

Data Availability Statement: All data generated or analyzed during this study are included in this published article. No new data were created or analyzed in this study. Data sharing is not applicable to this article. Data in Table 4 are available through Triska et al. 1982, Benfield 1997, Webster et al. 1997, Hart 2006, Muto 2008, Tonin et al. 2017, and Railoun 2018.

Acknowledgments: The authors of this study would like to thank all people involved in the initial sampling assignments.

Conflicts of Interest: The authors declare no conflict of interest.

References

1. Sutherland, A.B.; Meyer, J.L.; Gardiner, E.P. Effects of land cover on sediment regime and fish assemblage structure in four southern Appalachian streams. *Freshw. Biol.* **2002**, *47*, 1791–1805. [CrossRef]
2. Gessner, M.O.; Swan, C.M.; Dang, C.K.; McKie, B.G.; Bardgett, R.D.; Wall, D.H.; Hattenschwiler, S. Diversity meets decomposition. *Trends Ecol. Evol.* **2010**, *25*, 372–380. [CrossRef] [PubMed]
3. Wallace, J.B.; Eggert, S.L.; Meyer, J.L.; Webster, J.R. Multiple trophic levels of a forest stream linked to terrestrial litter inputs. *Science* **1997**, *277*, 102–104. [CrossRef]
4. Tranvik, L.J.; Downing, J.A.; Cotner, J.B.; Loiselle, S.A.; Striegl, R.G.; Ballatore, T.J.; Ballatore, T.J.; Dillon, P.; Finlay, K.; Fortino, K.; et al. Lakes and reservoirs as regulators of carbon cycling and climate. *Limnol. Oceanograohy* **2009**, *54*, 2298–2314. [CrossRef]
5. Battin, T.J.; Kaplan, L.A.; Findlay, S.; Hopkinson, C.S.; Marti, E.; Packman, A.I.; Newbold, J.D.; Sabater, F. Biophysical controls on organic carbon fluxes in fluvial networks. *Nat. Geosci.* **2008**, *1*, 95–100. [CrossRef]
6. Loreau, M.; Mouquet, N.; Holt, R.D. Meta-ecosystems: A theoretical framework for a spatial ecosystem ecology. *Ecol. Lett.* **2003**, *6*, 673–679. [CrossRef]
7. Yue, K.; Yang, W.Q.; Peng, C.H.; Peng, Y.; Zhang, Z.; Huang, C.P.; Tan, Y.; Wu, F.Z. Foliar litter decomposition in an alpine forest meta-ecosystem on the eastern Tibetan Plateau. *Sci. Total Environ.* **2016**, *566*, 279–287. [CrossRef]
8. Coûteaux, M.M.; Bottner, P.; Berg, B. Litter decomposition, climate, and liter quality. *Trends Ecol. Evol.* **1995**, *10*, 63–66. [CrossRef]

9. Abril, M.; Muñoz, I.; Menéndez, M. Heterogeneity in leaf litter decomposition in a temporary Mediterranean stream during flow fragmentation. *Sci. Total Environ.* **2016**, *553*, 330–339. [CrossRef] [PubMed]

10. Ferreira, V.; Canhoto, C. Future increase in temperature may stimulate litter decomposition in temperate mountain streams: Evidence from a stream manipulation experiment. *Freshw. Biol.* **2015**, *60*, 881–892. [CrossRef]

11. Berg, B.; Meentemeyer, V. Litterfall in some European coniferous forests as dependent on climate: A synthesis. *Can. J. For. Res.* **2001**, *31*, 292–301. [CrossRef]

12. Louzada, J.N.C.; Schoereder, J.H.; Marco, J.P. Litter decomposition in semidecidous forest and Eucalyptus spp. crop in Brazil: A comparison. *For. Ecol. Manag.* **1997**, *94*, 31–36. [CrossRef]

13. Bachega, L.R.; Bouillet, J.P.; Piccolo, M.C.; Saint-André, L.; Bouvet, J.M.; Nouvellon, Y.; Goncalves, J.L.M.; Robin, A.; Laclau, J.P. Decomposition of Eucalyptus grandis and Acacia mangium leaves and fine roots in tropical conditions did not meet the home field advantage hypothesis. *For. Ecol. Manag.* **2016**, *359*, 33–43. [CrossRef]

14. Cornwell, W.K.; Cornelissen, J.H.C.; Amatangelo, K.; Dorrepaal, E.; Eviner, V.T.; Godoy, O.; Hobbie, S.E.; Hoorens, B.; Kurokawa, H.; Pérez-Harguindeguy, N.; et al. Plant species traits are the predominant control on litter decomposition rates within biomes worldwide. *Ecol. Lett.* **2008**, *11*, 1065–1071. [CrossRef] [PubMed]

15. Wei, X.; Yang, Y.; Shen, Y.; Chen, Z.; Dong, Y.; Wu, F.; Zhang, L. Litter production and its dynamic pattern in a dark coniferous forest in the alpine gorge region. *Chin. J. Appl. Environ. Biol.* **2017**, *23*, 745–752.

16. Guo, J.; Yu, L.H.; Fang, X.; Xiang, W.H.; Deng, X.W.; Lu, X. Litter production and turnover in four types of subtropical forests in China. *Acta Ecol. Sin.* **2015**, *35*, 4668–4677.

17. Rueda-Delgado, G.; Wantzen, K.M.; Tolosa, M.B. Leaf-litter decomposition in an Amazonian floodplain stream: Effects of seasonal hydrological changes. *J. N. Am. Benthol. Soc.* **2006**, *25*, 233–249. [CrossRef]

18. Taylor, B.R.; Chauvet, E.E. Relative influence of shredders and fungi on leaf litter decomposition along a river altitudinal gradient. *Hydrobiologia* **2014**, *721*, 239–250. [CrossRef]

19. Tiegs, S.D.; Akinwole, P.O.; Gessner, M.O. Litter decomposition across multiple spatial scales in stream networks. *Oecologia* **2009**, *161*, 343–351. [CrossRef]

20. Benfield, E.F. Comparison of Litterfall Input to Streams. *Freshw. Sci.* **1997**, *16*, 104–108. [CrossRef]

21. Webster, J.R. Stream organic matter budgets-introduction. *J. N. Am. Benthol. Soc.* **1997**, *16*, 2–13. [CrossRef]

22. Triska, F.J.; Sedell, J.R.; Gregory, S.V. Coniferous forest streams. In *Analysis of Coniferous Forest Ecosystems in the Western United States*; Edmonds, R.L., Ed.; Hutchinson Ross: Stroudsburg, PA, USA, 1982; pp. 292–332.

23. Soininen, J.; Bartels, P.; Heino, J.; Miska, L.; Hillebrand, H. Toward more integrated ecosystem research in aquatic and terrestrial environments. *BioScience* **2015**, *65*, 174–182. [CrossRef]

24. Wang, Y.Q.; Wang, Y.J. Evolution of study on the forest stream water quality. *Res. Soil Water Conserv.* **2004**, *10*, 242–246.

25. Bünemann, E.; Bossio, D.; Smithson, P.; Frossard, E.; Oberson, A. Microbial community composition and substrate use in a highly weathered soil as affected by crop rotation and P fertilization. *Soil Biol. Biochem.* **2004**, *36*, 889–901. [CrossRef]

26. Perakis, S.S.; Hedin, L.O. Nitrogen loss from unpolluted South American forests mainly via dissolved organic compounds. *Nature* **2002**, *415*, 416–419. [CrossRef] [PubMed]

27. Zhang, M.H.; Cheng, X.L.; Geng, Q.H.; Shi, Z.; Luo, Y.Q.; Xu, X. Leaf litter traits predominantly control litter decomposition in streams worldwide. *Glob. Ecol. Biogeogr.* **2019**, *28*, 1469–1486. [CrossRef]

28. Baldy, V.; Gobert, V.; Guerold, F.; Chauvet, E.; Lambrigot, D.; Charcosset, J.Y. Leaf litter breakdown budgets in streams of various trophic status: Effects of dissolved inorganic nutrients on microorganisms and invertebrates. *Freshw. Biol.* **2007**, *52*, 1322–1335. [CrossRef]

29. Leroy, C.J.; Whitham, T.G.; Keim, P.M.; Jane, C. Plant genes link forests and streams. *Ecology* **2006**, *87*, 255–261. [CrossRef] [PubMed]

30. Zhang, Y.; Yang, W.Q.; Tan, B.; Liang, Z.Y.; Wu, F.Z. Storage and content of organic carbon in alpine forest stream in the upper reaches of the Yangtze River. *Fresenius Environ. Bull.* **2017**, *26*, 3402–3408.

31. Yuan, Z.Q.; Li, B.H.; Bai, X.J.; Lin, F.; Shi, S.; Ye, J.; Wang, X.G.; Hao, Z.Q. Composition and seasonal dynamics of litterfalls in a broad-leaved Korean pine (*Pinus koraiensis*) mixed forest in Changbai Mountains, Northeast China. *Chin. J. Appl. Ecol.* **2010**, *21*, 2171–2178.

32. Skalny, A.V. Bioelementology as an interdisciplinary integrative approach in life sciences: Terminology, classification, perspectives. *J. Trace Elem. Med. Biol.* **2011**, *25*, S3–S10. [CrossRef] [PubMed]

33. Finlay, J.C. Controls of stream water dissolved inorganic carbon dynamics in a forested watershed. *Biogeochemistry* **2003**, *62*, 231–252. [CrossRef]

34. Easthouse, K.B.; Mulder, J.; Christophersen, N.; Seip, H.M. Dissolved organic carbon fractions in soil and stream water during variable hydrological conditions at Birkenes, Southern Norway. *Water Resourc. Res.* **1992**, *28*, 1585–1596. [CrossRef]

35. Gomes, T.F.; Broek, M.V.; Govers, G.; Silva, R.W.C.; Moraes, J.M.; Camargo, P.B.; Mazzi, E.A.; Martinelli, L.A. Runoff, soil loss, and sources of particulate organic carbon delivered to streams by sugarcane and riparian areas: An isotopic approach. *Catena* **2019**, *181*, 1–9. [CrossRef]

36. Zhang, X.Y.; Yu, G.R.; Zheng, Z.M.; Wang, Q.F. Carbon Emission and Spatial Pattern of Soil Respiration of Terrestrial Ecosystems in China: Based on Geostatistic Estimation of Flux Measurement. *Prog. Geogr.* **2012**, *1*, 97–108.

37. Duan, L.; Zheng, W.; Li, M. Geologic hazards on the western Sichuan plateau and their controls. *Sediment. Geol. Tethyan Geol.* **2005**, *25*, 95–99.

38. Zhang, H.; Yang, W.Q.; Wang, M.; Liao, S. Carbon, nitrogen and phosphorus storage of woody debris in headwater streams in an alpine forest in upper reaches of the Mingjiang River. *Acta Ecol. Sin.* **2016**, *36*, 1967–1974.

39. Alvarez, S.; Guerrero, M. Enzymatic activities associated with decomposition of particulate organic matter in two shallow ponds. *Soil Biol. Biochem.* **2000**, *32*, 1941–1951. [CrossRef]

40. Yang, W.Q.; Wu, F.Z. *Ecosystem Processes and Management of Subalpine Coniferous Forest in the Upper Reaches of the Yangtze River*; Science Press: Beijing, China, 2021.

41. Sun, H.L. *Ecological and Environmental Problems in the Upper Reaches of the Yangtze River*; China Environmental Press: Beijing, China, 2008.

42. Wu, F.Z.; Peng, C.H.; Zhu, J.; Zhang, J.; Tan, B.; Yang, W.Q. Impact of freezing and thawing dynamics on foliar litter carbon release in alpine/subalpine forests along an altitudinal gradient in the eastern Tibetan Plateau. *Biogeosciences* **2014**, *11*, 6471–6481.

43. Wu, F.Z.; Yang, W.Q.; Zhang, J.; Deng, R. Litter decomposition in two subalpine forests during the freeze–thaw season. *Acta Oecol.* **2010**, *36*, 135–140. [CrossRef]

44. Zhang, C.; Yang, W.Q.; Zhang, H.; Zhang, B.; Yue, K.; Peng, Y.; Wu, F.Z. Standing biomass and carbon-storage of non-woody debris and their distribution in the alpine forest streams of Western Sichuan in the upper reaches of Minjiang River. *Ecol. Environ. Sci.* **2014**, *23*, 1509–1514.

45. Lu, R.K. *Soil Agrochemical Analysis Methods*; China Agricultural Science and Technology Press: Beijing, China, 1999.

46. Bates, D.; Maechler, M.; Bolker, B.; Walker, S. Fitting linear mixed-effects models using lme4. *J. Stat. Softw.* **2015**, *67*. [CrossRef]

47. Hart, J.J.; Welch, R.M.; Norvell, W.; Kochian, L.V. Characterization of cadmium uptake, translocation and storage in near-isogenic lines of durum wheat that differ in grain cadmium concentration. *New Phytol.* **2006**, *172*, 261–271. [CrossRef] [PubMed]

48. Muto, E.A. *The Characteristics and Fate of Leaf Litter Inputs to Boreal Shield Streams in Relation to Riparian Stand Structure*; The University of Guelph: Guelph, ON, Canada, 2008.

49. Tonin, A.M.; Goncalves, J.F.; Bambi, P.; Couceiro, S.R.M.; Feitoza, L.A.M.; Fontana, L.E.; Hamada, N.; Hepp, L.U.; Lezan-Kowalczuk, V.G.; Leite, G.F.M.; et al. Plant litter dynamics in the forest-stream interface: Precipitation is a major control across tropical biomes. *Sci. Rep.* **2006**, *7*, 10799. [CrossRef] [PubMed]

50. Railoun, M.Z. *Impacts of the Invasive Tree Acacia Mearnsii on Riparian and Instream Aquatic Environments in the Cape Floristic Region, South Africa*; Stellenbosch University: Stellenbosch, South Africa, 2018.

51. Wenger, S.A. *A Review of the Scientific Literature in Riparian Buffer Width, Extent and Vegetation*; University of Georgia Press: Athens, GA, USA, 1999; pp. 6–51.

52. Yang, W.Q.; Wang, K.Y.; Kellomaki, S.; Gong, H.D. Litter dynamics of three subalpine forests in Western Sichuan. *Pedosphere* **2005**, *15*, 653–659.

53. Wei, X.Y.; Yang, Y.L.; Shen, Y.; Chen, Z.H.; Dong, Y.L.; Wu, F.Z.; Zhang, L. Effects of litterfall on the accumulation of extracted soil humic substances in subalpine forests. *Front. Plant Sci.* **2020**, *11*, 254. [CrossRef] [PubMed]

54. Johnson, S.L.; Jones, J.A. Stream temperature responses to forest harvest and debris flows in western Cascades, Oregon. *Can. J. Fish. Aquat. Sci.* **2000**, *57*, 30–39. [CrossRef]

55. Graça, M.A.; Ferreira, V.; Canhoto, C.; Encalada, A.C.; Guerrero-Bolaño, F.; Wantzen, K.M.; Boyero, L. A conceptual model of litter breakdown in low order streams. *Int. Rev. Hydrobiol.* **2015**, *100*, 1–12. [CrossRef]

56. Bilby, R.E. Role of organic debris dams in regulating the export of dissolved organic and particulate matter from a forested watershed. *Ecology* **1981**, *62*, 1234–1243. [CrossRef]

57. Drewitt, G.B.; Black, T.A.; Nesic, Z.; Humpgreys, E.R.; Morgenstern, K. Measuring Forest floor CO_2 fluxes in a Douglas-fir forest. *Agric. For. Meteorol.* **2002**, *110*, 299–317. [CrossRef]

58. Rantakari, M.; Mattsson, T.; Kortelainen, P.; Mattsson, T.; Piirainen, S. Organic and inorganic carbon concentrations and fluxes from managed and unmanaged boreal first-order catchments. *Sci. Total Environ.* **2010**, *408*, 1649–1658. [CrossRef] [PubMed]

59. Cole, J.J.; Prairie, Y.T.; Caraco, N.F.; McDowell, W.H.; Tranvik, L.J.; Striegl, R.G.; Duarte, C.M.; Kortelaien, P.; Downing, J.A.; Middelburg, J.J.; et al. Plumbing the global carbon cycle: Integrating inland waters into the terrestrial carbon budget. *Ecosystems* **2007**, *10*, 172–185. [CrossRef]

60. Wallace, T.A.; Ganf, G.G.; Brookes, J.D. A comparison of phosphorus and DOC leachates from different types of leaf litter in an urban environment. *Freshw. Biol.* **2008**, *53*, 1902–1913. [CrossRef]

MDPI
St. Alban-Anlage 66
4052 Basel
Switzerland
Tel. +41 61 683 77 34
Fax +41 61 302 89 18
www.mdpi.com

Forests Editorial Office
E-mail: forests@mdpi.com
www.mdpi.com/journal/forests